Intersubband Transitions in Quantum Wells

Physics and Device Applications I

SEMICONDUCTORS
AND SEMIMETALS
Volume 62

Semiconductors and Semimetals

A Treatise

Edited by R. K. Willardson
CONSULTING PHYSICIST
SPOKANE, WASHINGTON

Eicke R. Weber
DEPARTMENT OF MATERIALS SCIENCE
AND MINERAL ENGINEERING
UNIVERSITY OF CALIFORNIA
AT BERKELEY

Intersubband Transitions in Quantum Wells

Physics and Device Applications I

SEMICONDUCTORS
AND SEMIMETALS

Volume 62

Volume Editors

H. C. LIU

INSTITUTE FOR MICROSTRUCTURAL SCIENCES
NATIONAL RESEARCH COUNCIL
OTTOWA, ONTARIO, CANADA

FEDERICO CAPASSO

BELL LABORATORIES, LUCENT TECHNOLOGIES
MURRAY HILL, NEW JERSEY

ACADEMIC PRESS
San Diego   San Francisco   New York   Boston
London   Sydney   Tokyo

This book is printed on acid-free paper.

COPYRIGHT © 2000 BY ACADEMIC PRESS

ALL RIGHTS RESERVED.

NO PART OF THIS PUBLICATION MAY BE REPRODUCED OR TRANSMITTED IN ANY FORM OR BY ANY MEANS, ELECTRONIC OR MECHANICAL, INCLUDING PHOTOCOPY, RECORDING, OR ANY INFORMATION STORAGE AND RETRIEVAL SYSTEM, WITHOUT PERMISSION IN WRITING FROM THE PUBLISHER.

Requests for permission to make copies of any part of the work should be mailed to: Permissions Department, Harcourt, Inc., 6277 Sea Harbor Drive, Orlando, Florida, 32887-6777

The appearance of the code at the bottom of the first page of a chapter in this book indicates the Publisher's consent that copies of the chapter may be made for personal or internal use of specific clients. This consent is given on the condition, however, that the copier pay the stated per-copy fee through the Copyright Clearance Center, Inc. (222 Rosewood Drive, Danvers, Massachusetts 01923), for copying beyond that permitted by Sections 107 or 108 of the U.S. Copyright Law. This consent does not extend to other kinds of copying, such as copying for general distribution, for advertising or promotional purposes, for creating new collective works, or for resale. Copy fees for pre-1999 chapters are as shown on the title pages; if no fee code appears on the title page, the copy fee is the same as for current chapters. 0080-8784/00 $30.00

ACADEMIC PRESS *A Harcourt Science & Technology Company*
525 B Street, Suite 1900, San Diego, CA 92101-4495, USA
http://www.academicpress.com

ACADEMIC PRESS
24-28 Oval Road, London NW1 7DX, UK
http://www.hbuk.co.uk/ap/

International Standard Book Number: 0-12-752171-2
International Standard Serial Number: 0080-8784

PRINTED IN THE UNITED STATES OF AMERICA
99 00 01 02 03 04 EB 9 8 7 6 5 4 3 2 1

Contents

PREFACE xi
LIST OF CONTRIBUTORS xiii

Chapter 1   The Basic Physics of Intersubband Transitions   1

Manfred Helm

I. INTRODUCTION 1
II. THE INTERSUBBAND ABSORPTION COEFFICIENT 5
III. THE SYMMETRIC QUANTUM WELL 14
IV. EXPERIMENTAL GEOMETRIES AND THEIR ELECTROMAGNETICS 19
V. ASYMMETRIC QUANTUM WELLS 26
VI. MULTIQUANTUM WELLS AND SUPERLATTICES 32
VII. NONPARABOLICITY AND MANY-BODY EFFECTS 39
   1. *Nonparabolicity* 40
   2. *Self-Consistent Coulomb Potential* 42
   3. *Many-Body Effects on the Energy* (*Exchange and Correlation*) 45
   4. *Collective Effects on the Absorption* (*Depolarization and Exciton Shift*) 47
VIII. MECHANISMS FOR IN-PLANE ABSORPTION 54
   1. *Indirect-Gap Semiconductors* 54
   2. *Spatial Variation of the Effective Mass* 56
   3. *Coupling to the Valence Band* 57
IX. INTERSUBBAND ABSORPTION IN THE VALENCE BAND 59
X. LINE BROADENING AND RELAXATION 73
XI. OTHER PHENOMENA RELATED TO INTERSUBBAND TRANSITIONS 80
   1. *Magnetic-Field Effects* 80
   2. *Parabolic Quantum Wells* 84
   3. *Impurities* 86
   4. *Photon Drag Effect* 88
XII. CONCLUDING REMARKS AND OUTLOOK 90
   REFERENCES 91

Chapter 2 Quantum Interference Effects in Intersubband Transitions . 101

Jerome Faist, Carlo Sirtori, Federico Capasso, Loren N. Pfeiffer, Ken W. West, Deborah L. Sivco, and Alfred Y. Cho

    I. INTRODUCTION 101
    II. EXPERIMENTAL SETUP 104
    III. BOUND STATE ABOVE A QUANTUM WELL THROUGH ELECTRON WAVE INTERFERENCE 104
    IV. QUANTUM INTERFERENCE IN A COUPLED-WELL SYSTEM 108
    V. FANO INTERFERENCE IN INTERSUBBAND ABSORPTION 112
    VI. CONTROL OF QUANTUM INTERFERENCE BY TUNNELING TO A CONTINUUM . . . 116
        1. *Absorption Experiments* 118
        2. *Emission Experiments* 122
    REFERENCES 127

Chapter 3 Quantum Well Infrared Photodetector Physics and Novel Devices 129

H. C. Liu

    I. INTRODUCTION 129
        1. *Background* 129
        2. *Simple Description of Intersubband Transition* 132
        3. *Wavelengths Covered by GaAs-Based Structures* 135
    II. DETECTOR PHYSICS 137
        1. *Dark Current* 137
        2. *Photocurrent* 150
        3. *Detector Performance* 160
        4. *Design of an Optimized Detector* 167
    III. NOVEL STRUCTURES AND DEVICES 169
        1. *Multicolor and Multispectral Detectors* 169
        2. *High-Frequency Detectors* 176
        3. *Integration with LEDs* 183
    IV. CONCLUSIONS 188
        1. *Concluding Remarks* 188
        2. *List of Symbols* 190
    REFERENCES 193

Chapter 4 Quantum Well Infrared Photodetector (QWIP) Focal Plane Arrays 197

S. D. Gunapala and S. V. Bandara

    I. INTRODUCTION 198
    II. COMPARISON OF VARIOUS TYPES OF QWIPs 201
        1. *n-Doped Bound-to-Bound QWIPs* 201
        2. *n-Doped Bound-to-Continuum QWIPs* 202
        3. *n-Doped Bound-to-Quasibound QWIPs* 203
        4. *n-Doped Broadband QWIPs* 205
        5. *n-Doped Bound-to-Bound Miniband QWIPs* 206

6. n-Doped Bound-to-Continuum Miniband QWIPs 207
7. n-Doped Bound-to-Miniband QWIPs 208
8. n-Doped Asymmetrical $GaAs-Al_xGa_{1-x}As$ QWIPs 209
9. p-Doped QWIPs 210
10. Single-Quantum-Well Infrared Photodetectors 211
11. Indirect Bandgap QWIPs 212
12. n-Doped $In_{0.53}Ga_{0.47}As-In_{0.52}Al_{0.48}As$ QWIPs 213
13. n-Doped $In_{0.53}Ga_{0.47}As-InP$ QWIPs 214
14. InGaAsP Quaternary QWIPs 215
15. n-Doped $GaAs-Ga_{0.5}In_{0.5}P$ QWIPs 216
16. n-Doped $GaAs-Al_{0.5}In_{0.5}P$ QWIPs 217
17. n-Doped $In_{0.15}Ga_{0.85}As-GaAs$ QWIPs 217
18. p-Doped $In_{0.53}Ga_{0.47}As-InP$ QWIPs 218
III. FIGURES OF MERIT 220
1. Absorption Spectra 221
2. Dark Current 222
3. Responsivity 225
4. Dark Current Noise 228
5. Noise Gain and Photoconductive Gain 229
6. Quantum Efficiency 232
7. Detectivity 234
IV. LIGHT COUPLING 237
1. Random Reflectors 238
2. Two-Dimensional Periodic Gratings 239
3. Corrugated Structure 240
4. Microlenses 242
V. IMAGING FOCAL PLANE ARRAYS 246
1. Effect of Nonuniformity 246
2. 128 × 128 VLWIR Imaging Camera 249
3. 256 × 256 LWIR Imaging Camera 251
4. 640 × 486 LWIR Imaging Camera 254
5. Dualband (MWIR and LWIR) Detectors 260
6. Dualband (LWIR and VLWIR) Detectors 263
7. High-Performance QWIPs for Low-Background Applications 264
8. Broadband QWIPs for Thermal Infrared Imaging Spectrometers 267
VI. APPLICATIONS 270
1. Fire Fighting 270
2. Volcanology 271
3. Medicine 271
4. Defense 273
5. Astronomy 274
VII. SUMMARY 275
REFERENCES 278

INDEX 283
CONTENTS OF VOLUMES IN fTHIS SERIES 291

Preface

Research on intersubband transitions in quantum wells has led to several practical devices, such as the QWIP (quantum well infrared photodetector) and the QCL (quantum cascade laser). These are two of the success stories in using quantum wells for practical device applications. Research activities in this area have been very intense over the past ten years, resulting in many new devices that are presently being developed for the market. We therefore feel that the time is right to collect a comprehensive review of the various topics related to intersubband transitions in quantum wells. We hope that this volume will provide a good reference for researchers in this and related fields and for those individuals — graduate students, scientists, and engineers — who are interested in learning about this subject.

The eight chapters in Volumes 62 and 66 of the Academic Press Semiconductors and Semimetals serial cover the following topics: Chapters 1 and 2 in Volume 62 discuss the basic physics and related phenomena of intersubband transitions. Chapters 3 and 4 in Volume 62 present the physics and applications of QWIP. Chapter 1 in Volume 66 reviews the development of QCL. Chapter 2 in Volume 66 studies nonlinear optical processes. Chapters 3 and 4 in Volume 66 introduce two related topics: photon-assisted tunneling and optically excited Bloch oscillation.

We thank all the contributors who have devoted their valuable time and energy in putting together a timely volume. We also thank Dr. Zvi Ruder of Academic Press for providing assistance and keeping us on schedule.

H. C. LIU
FEDERICO CAPASSO

List of Contributors

Numbers in parentheses indicate the pages on which the authors' contribution begins.

S. V. BANDARA (197), *Jet Propulsion Laboratory, California Institute of Technology, Pasadena, California*

FEDERICO CAPASSO (101), *Bell Laboratories, Lucent Technologies, Murray Hill, New Jersey*

ALFRED Y. CHO (101), *Bell Laboratories, Lucent Technologies, Murray Hill, New Jersey*

JEROME FAIST (101), *Institute of Physics, University of Neuchâtel, Neuchâtel, Switzerland*

S. D. GUNAPALA (197), *Jet Propulsion Laboratory, California Institute of Technology, Pasadena, California*

MANFRED HELM (1), *Institut für Halbleiter- und Festkörperphysik, Johannes-Kepler-Universität Linz, Linz, Austria*

H. C. LIU (129), *Institute for Microstructural Sciences, National Research Council, Ottawa, Ontario, Canada*

LOREN N. PFEIFFER (101), *Bell Laboratories, Lucent Technologies, Murray Hill, New Jersey*

CARLO SIRTORI (101), *Thomson-CSF, Laboratoire Central de Recherches, Orsay, France*

DEBORAH L. SIVCO (101), *Bell Laboratories, Lucent Technologies, Murray Hill, New Jersey*

KEN W. WEST (101), *Bell Laboratories, Lucent Technologies, Murray Hill, New Jersey*

CHAPTER 1

The Basic Physics of Intersubband Transitions

Manfred Helm

INSTITUT FÜR HALBLEITER- UND FESTKÖRPERPHYSIK
JOHANNES-KEPLER-UNIVERSITÄT LINZ
LINZ, AUSTRIA

I. INTRODUCTION	1
II. THE INTERSUBBAND ABSORPTION COEFFICIENT	5
III. THE SYMMETRIC QUANTUM WELL	14
IV. EXPERIMENTAL GEOMETRIES AND THEIR ELECTROMAGNETICS	19
V. ASYMMETRIC QUANTUM WELLS	26
VI. MULTIQUANTUM WELLS AND SUPERLATTICES	32
VII. NONPARABOLICITY AND MANY-BODY EFFECTS	39
1. *Nonparabolicity*	40
2. *Self-Consistent Coulomb Potential*	42
3. *Many-Body Effects on the Energy (Exchange and Correlation)*	45
4. *Collective Effects on the Absorption (Depolarization and Exciton Shift)*	47
VIII. MECHANISMS FOR IN-PLANE ABSORPTION	54
1. *Indirect-Gap Semiconductors*	54
2. *Spatial Variation of the Effective Mass*	56
3. *Coupling to the Valence Band*	57
IX. INTERSUBBAND ABSORPTION IN THE VALENCE BAND	59
X. LINE BROADENING AND RELAXATION	73
XI. OTHER PHENOMENA RELATED TO INTERSUBBAND TRANSITIONS	80
1. *Magnetic-Field Effects*	80
2. *Parabolic Quantum Wells*	84
3. *Impurities*	86
4. *Photon Drag Effect*	88
XII. CONCLUDING REMARKS AND OUTLOOK	90
REFERENCES	91

I. Introduction

The term intersubband transitions has been used to describe optical transitions between quasi-two-dimensional electronic states in semiconductors ("subbands"), which are formed due to the confinement of the electron wave function in one dimension. The formation of such low-dimensional electronic systems has been one of the major topics of semiconductor physics for the past two and a half decades (Ando et al., 1982), and in this context, the term "band-structure engineering" was coined. This development has been mainly triggered by epitaxial crystal growth techniques such as molecular beam epitaxy, which provides atomic-layer control of layer thicknesses.

In these terms, the conceptually simplest band-structure engineered system that can be fabricated is a quantum well, which consists of a thin semiconductor layer (of the order of 100 Å) embedded in a semiconductor with a larger bandgap (see Fig. 1). Depending on the relative band offsets of the two semiconductor materials, both electrons and holes can be confined in one direction in the conduction band and the valence band, respectively, and one obtains allowed energy levels that are quantized along the growth direction. These energy levels can be tuned by the quantum well depth and thickness. Whereas, of course, optical transitions can take place between valence and conduction band states, in this chapter the term "intersubband transitions" is used solely for transitions between quantized levels within the conduction (or valence) band (Fig. 1 schematically shows both interband and intersubband transitions).

The first experimental evidence for quantized states in a semiconductor quantum well was presented by Dingle et al. in 1974 through optical bandgap spectroscopy in a GaAs/AlGaAs structure. Later Esaki and Sakaki

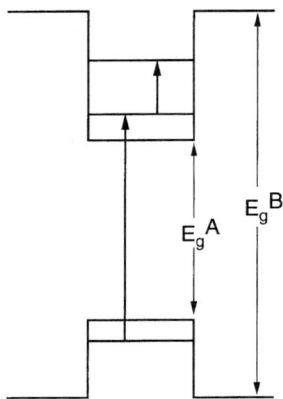

FIG. 1. Schematic of a quantum well made from a semiconductor with energy gap E_g^A embedded in a semiconductor with gap E_g^B. Subbands in the conduction and valence band are indicated as well as interband and intersubband transitions.

(1977) and Smith *et al.* (1983) showed the possibility of using intersubband transitions in quantum wells for infrared light detection. More than 10 years after the Dingle *et al.* report, intersubband absorption in a GaAs–AlGaAs quantum well was observed by West and Eglash (1985). Although this is often quoted as the beginning of intersubband spectroscopy, researchers put a lot of effort into the investigation of another quasi-two-dimensional electron system much earlier. Fowler *et al.* (1966) showed that the conducting channel in a Si-MOSFET (metal-oxide semiconductor field effect transistor) is effectively a two-dimensional electron gas that is formed at the interface between the Si and the oxide. Extensive investigation of this system followed, and Kamgar *et al.* (1974) reported the observation of intersubband absorption in a Si accumulation layer. Similar research was carried out on the GaAs–AlGaAs system, and in 1983, intersubband absorption was reported in an inversion layer at the interface of a GaAs–AlGaAs heterostructure (Schlesinger *et al.*, 1983).

These works remained relatively unknown to many researchers who later entered the field; one reason for this may be the fact that in these accumulation and inversion layers, the absorption wavelength was in the far infrared (FIR), at wavelengths longer than 40 μm, whereas in the work of West and Eglash (1985) the absorption occurred in the technologically more important range around 10 μm. Nevertheless the physical mechanism is, of course, the same in both types of structures.

After 1985, the number of works on intersubband absorption in quantum wells increased dramatically, most often motivated by the high technological potential of intersubband transitions for novel infrared detectors (Levine, 1993), emitters, and nonlinear optical elements. Research in these three areas has been so successful that each is covered in separate chapters of this volume. For example, focal-plane arrays of intersubband detectors (quantum well infrared photodetectors, QWIPs) with high detectivities have been fabricated (see Chapter 4) by a number of groups and intersubband lasers (quantum cascade lasers) (Faist *et al.*, 1994a; see also the corresponding chapter in this volume) were finally demonstrated in 1994.

Intersubband transitions have been observed in many different material systems. Apart from GaAs–AlGaAs the most important are strained InGaAs–AlGaAs structures, InGaAs–InAlAs lattice matched to InP, and InAs–AlGaSb structures. Much work has also been done with quantum wells based on group IV elements such as Si–SiGe. Both conduction and valence-band quantum wells have been employed. Besides the usual way of introducing *n*- or *p*-type doping into the quantum wells, a number of investigations were devoted to photoinduced intersubband absorption (Olszakier *et al.*, 1989; Yang *et al.*, 1990; Abramovich *et al.*, 1994; Julien and Boucaud, 1997). Here, electrons and holes are created in undoped material using above-bandgap radiation and the intersubband absorption is studied simultaneously.

The photon energy (or wavelength) range accessible with intersubband transitions is limited through the magnitude of the conduction- or valence-band offset of the heterojunction on the high-energy side, and by the typical intersubband linewidth on the low-energy side. Currently, the spectral range covered by intersubband transitions extends over more than two orders of magnitude, from 200 (Helm et al., 1991) to 2 µm (Chui et al., 1994). Even the technologically important wavelength of 1.55 µm has been reached (Smet et al., 1994). The development of intersubband physics research over the past decade can be followed in the proceedings of the Intersubband Physics Workshop, which has been held every two years since 1991. These were published as books (Rosencher et al., 1992; H. C. Liu et al., 1994; Li and Su, 1998) and in *Superlattices and Microstructures*, volume 19 (1996), respectively.

This chapter covers the more basic aspects of intersubband transitions with a main emphasis on linear absorption. Intersubband emission is covered in Chapter 5 by Faist et al. (see also the review by Perera et al., 1997). A more compact account of some of the material covered here is due to Loehr and Manasreh (1993). The organization of the chapter is as follows.

In Section II, we derive an expression for the intersubband absorption coefficient based on Fermi's golden rule for the induced transition rate in the framework of the one-band effective mass approximation. In Section III this is applied to the case of a symmetric single quantum well. Section IV discusses several experimental geometries that enable coupling of the electromagnetic wave to the intersubband transition, which in most cases requires a polarization component perpendicular to the quantum well (QW) layers. Section V discusses asymmetrically shaped QW potentials, such as those induced by a vertical electric field or by variation of alloy composition. Due to the symmetry breaking, more transitions become allowed than in the symmetric situation. In Section VI we consider multiple quantum wells and superlattices, especially with regard to the formation of extended minibands with a finite dispersion along the growth direction. Section VII is devoted to the discussion of effects going beyond the most simple single-particle, one-band model, which are nonparabolicity and the inclusion of Coulomb and many-body effects in the intersubband absorption. In some situations, radiation polarized parallel to the layers can also be absorbed. These are discussed in Section VIII. The most important one exhibits constant-energy ellipsoids that are tilted away from the QW confinement direction. Intersubband absorption can take place not only in the conduction band but also in the valence band of *p*-type doped QWs. Due to the complicated valence–

band structure, this requires a more sophisticated theoretical treatment on the basis of a multilevel $\mathbf{k} \cdot \mathbf{p}$ Luttinger–Kohn type Hamiltonian, which is presented in Section IX. In Section X, we discuss line broadening and intersubband relaxation. In a loosely connected manner, Section XI presents several other effects also related to intersubband transitions, including magnetic-field effects, parabolic quantum wells, impurities, and the photon drag effect.

II. The Intersubband Absorption Coefficient

In this section, we derive an expression for the intersubband absorption coefficient based on Fermi's golden rule for the induced transition rate. On a more sophisticated level, the absorption spectrum can be calculated in a full many-body framework using linear response theory. Here, however, we present the single-particle approach, to which several many-body corrections can be added afterward. (For more discussion and references on this issue, see Section VII). For the moment, we restrict ourselves to the case of a single electronic band; that is, we make use of the envelope-function approach in the effective-mass approximation.

The total wave function $\psi_i(\mathbf{r})$ can then be written as the product between the lattice-periodic Bloch function of band v at the center (or another extremum) of the Brillouin zone $u_v(\mathbf{r})$ and an envelope function $f_i(\mathbf{r})$, which is supposed to vary slowly over one lattice period.

$$\psi_i(\mathbf{r}) = f_i(\mathbf{r}) u_v(\mathbf{r}) \tag{1}$$

where i denotes the quantum numbers of the problem. This provides a reasonably good description for the conduction band of many semiconductors including GaAs. More accurate descriptions, which include several bands, are briefly discussed in Section VIII and more extensively in Section IX in connection with valence-band intersubband transitions. Equation (1) is simply a generalization of the usual Bloch ansatz for a bulk crystal, where the envelope function $f_i(\mathbf{r})$ reduces to a plane wave $e^{i\mathbf{k}\cdot\mathbf{r}}$. The envelope function $f_i(\mathbf{r})$ will depend on the shape of the quantum well potential or other external potentials such as electric and magnetic fields.

Under the assumption that the lattice-periodic function is the same in all constituent materials a Schrödinger equation only for the envelope function can be derived:

$$\frac{-\hbar^2}{2m^*} \nabla^2 f_i(\mathbf{r}) + V(\mathbf{r}) f_i(\mathbf{r}) = E_i f_i(\mathbf{r}) \tag{2}$$

When z is chosen the growth direction, the free motion in the x and y directions, can be separated:

$$f_{n\mathbf{k}_\perp}(\mathbf{r}) = \frac{1}{\sqrt{A}} e^{i\mathbf{k}_\perp \cdot \mathbf{r}} \varphi_n(z) \tag{3}$$

where $\mathbf{k}_\perp$ denotes the two-dimensional vector (k_x, k_y) and A is the sample area. Then Eq. (3) reduces to a one-dimensional Schrödinger equation of the textbook form

$$\frac{-\hbar^2}{2m^*} \frac{d^2 \varphi_n}{dz^2} + V(z)\varphi_n(z) = E_n \varphi_n(z) \tag{4}$$

This equation must be solved in each material layer $(A, B, \ldots)$ of the heterostructure, and the solutions have to be connected with the following matching conditions at each interface z_0:

$$\varphi^A(z_0) = \varphi^B(z_0) \quad \text{and} \quad \frac{1}{m^{*A}} \frac{d\varphi^A}{dz}(z_0) = \frac{1}{m^{*B}} \frac{d\varphi^B}{dz}(z_0) \tag{5}$$

The latter condition ensures conservation of the probability current, but as a consequence, the envelope function has a kink at each interface when the effective mass is discontinuous. In this case, the first term in Eq. (4) is better written in the form $-(\hbar^2/2)(d/dz)[1/m^*(z)](d/dz)\varphi_n(z)$.

The solution of Eqs. (2) and (4) leads to energy eigenvalues of the form

$$E_{n,\mathbf{k}_\perp} = E_n + \frac{\hbar^2 \mathbf{k}_\perp^2}{2m^*} \tag{6}$$

where the subband energies E_n depend on the shape of $V(z)$.

In simple cases, such as the finite, symmetric quantum well, Eq. (4) can be solved analytically, but for more complicated structures the solution must be obtained numerically. The most common method for an arbitrary one-dimensional potential is the well-documented, so-called transfer matrix method (Kane, 1969; Chuang, 1995). In this method, the potential is assumed to be piecewise constant, thus having plane-wave solutions (with either real or imaginary k-vector). The set of matching conditions in Eq. (5) is formulated in a 2×2 matrix at each interface. The solution is then achieved by multiplication of all matrices together with the requirement that the wave functions decay exponentially at each end of the structure.

Now let us turn to the calculation of the absorption coefficient. We start from Fermi's golden rule for the transition rate form a state i to a state f

induced by an external electromagnetic field

$$W_{if} = \frac{2\pi}{\hbar} |\langle \psi_i | H' | \psi_f \rangle|^2 \delta(E_f - E_i - \hbar\omega) \tag{7}$$

where H' is the interaction Hamiltonian, $H' = (e/2m^*)(\mathbf{A} \cdot \mathbf{p} + \mathbf{p} \cdot \mathbf{A})$ (the electron charge is $-e$). It can be shown (Bastard, 1988) that in a one-band effective-mass model, the interaction is correctly described when the effective mass m^* is used in this expression and not the free-electron mass. A linearly polarized, plane electromagnetic wave is described by

$$\mathbf{E} = E_0 \mathbf{e} \cos(\mathbf{q} \cdot \mathbf{r} - \omega t) \tag{8}$$

where $\mathbf{E}$ is the electric field, $\mathbf{e}$ is the polarization vector, and $\mathbf{q}$ is the propagation vector. The corresponding vector potential, which is related to the electric field by $\mathbf{E} = -\partial \mathbf{A}/\partial t$, can be written as

$$\mathbf{A} = \frac{iE_0 \mathbf{e}}{2\omega} e^{i(\mathbf{q}\cdot\mathbf{r} - \omega t)} + \text{c.c.} \tag{9}$$

At this point, we can employ the dipole approximation, which requires the wavelength of the radiation to be much larger than any characteristic dimension of electronic origin. This is the lattice period for interband transitions and the quantum well width for intersubband transitions; in both cases, the dipole approximation is very well fulfilled. (Note, however, that electric quadrupole transitions were discussed by Sa'ar, 1993.) Then $\mathbf{p}$ commutes with $\mathbf{A}$, which leads to $H' = (e/m^*)\mathbf{A} \cdot \mathbf{p}$, and we obtain

$$W_{if} = \frac{2\pi}{\hbar} \frac{e^2 E_0^2}{4m^{*2}\omega^2} |\langle i | \mathbf{e} \cdot \mathbf{p} | f \rangle|^2 \delta(E_f - E_i - \hbar\omega) \tag{10}$$

Before deriving the absorption coefficient from the transition rate, we take a closer look at the matrix element in Eq. (10). Due to the properties of the periodic Bloch functions and the slowly varying envelope functions (Eq. (1)), the complete matrix element can be split up in the following way (Bastard, 1988):

$$\langle i | \mathbf{e} \cdot \mathbf{p} | f \rangle = \mathbf{e} \cdot \langle u_v | \mathbf{p} | u_{v'} \rangle \langle f_n | f_{n'} \rangle + \mathbf{e} \cdot \langle u_v | u_{v'} \rangle \langle f_n | \mathbf{p} | f_{n'} \rangle \tag{11}$$

where v and v' and n and n' are the band and subband indices of the initial and final states, respectively. The first term describes interband transitions, which are accompanied by a change of the band index v (e.g., transitions from the valence band to the conduction band). It consists of a dipole matrix element of Bloch functions, which dictates the interband polarization selection rules, and an overlap integral of the envelope functions, which gives rise to selection rules concerning the electron and hole subband quantum numbers. If the initial and the final bands are the same, as in the case of intersubband transitions in the conduction band, this term vanishes. Then the second term, which describes transitions between subbands in the same band, becomes relevant. It consists of an overlap integral of Bloch functions (which vanishes for $v \neq v'$ and is unity for $v = v'$), and a dipole matrix element of the envelope functions. This last term is the only one relevant for intersubband transitions treated in the one-band model. In a multiband model, which is necessary to describe intersubband transitions in the valence band, however, both terms are important. We come back to this point in Section IX.

Thus let us evaluate the dipole matrix element of the envelope functions:

$$\langle f_{n\mathbf{k}_\perp}|\mathbf{e}\cdot\mathbf{p}|f_{n'\mathbf{k}'_\perp}\rangle = \frac{1}{A}\int d^3r e^{-i\mathbf{k}_\perp\cdot\mathbf{r}}\varphi_n^*(z)[e_x p_x + e_y p_y + e_z p_z]e^{i\mathbf{k}'_\perp\cdot\mathbf{r}}\varphi_{n'}(z) \quad (12)$$

Only the term proportional to e_z yields a contribution at finite frequency. The other terms, proportional to e_x and e_y vanish, except when initial and final states are identical ($n = n'$ and $\mathbf{k}_\perp = \mathbf{k}'_\perp$). The physical meaning of these terms is the free-carrier absorption in the two-dimensional electron gas, which is finite only at zero frequency when no scattering processes are included. This is due to the impossibility of conserving energy and momentum simultaneously during the absorption of a photon by an electron. Thus it is the matrix element

$$\langle n|p_z|n'\rangle = \int dz\varphi_n^*(z)p_z\varphi_{n'}(z) \quad (13)$$

which determines the intersubband absorption in a one-band model. Thus we obtain the result that the electric field of the radiation must have a z component (i.e., a component perpendicular to the semiconductor layers) to couple to the intersubband transition. This is the famous polarization selection rule for intersubband transitions, which, of course, has consequences for how practical experiments are performed. We discuss several frequently employed coupling schemes and sample geometries in Section IV.

Next we briefly note that within the dipole approximation, a different interaction Hamiltonian H' for the electron–photon coupling can be employed, namely, $H' = -e\mathbf{E}\cdot\mathbf{r}$. The r matrix elements are related to the p

matrix elements through

$$\mathbf{p}_{nn'} = im^*\omega_{nn'}\mathbf{r}_{nn'} \tag{14}$$

where $\omega_{nn'} = (E_n - E_{n'})/\hbar$. Note, however, that the use of $e\mathbf{E}\cdot\mathbf{r}$ can lead to wrong results in a system with nonnormalizable, unbounded wave functions, such as in the Kronig–Penney model for superlattices. In this case, the $\mathbf{A}\cdot\mathbf{p}$ interaction should be used.

As in all areas of optical spectroscopy, it is very useful to define the dimensionless oscillator strength $f_{nn'}$ by

$$f_{nn'} = \frac{2}{m^*\hbar\omega_{n'n}}|\langle n|p_z|n'\rangle|^2 = \frac{2m^*\omega_{n'n}}{\hbar}|\langle n|z|n'\rangle|^2 \tag{15}$$

since it facilitates the comparison of transition strengths in different physical systems and also obeys the sum rule

$$\sum_{n'} f_{nn'} = 1 \tag{16}$$

valid for all initial states n and the sum extending over all final states n'. Absorption processes are counted positive, emission processes with a negative sign, which is implicitly taken care of already in the definition in Eq. (15). Sometimes the free-electron mass m_0 is used in Eq. (15) instead of the effective mass m^* (West and Eglash, 1985). The such defined oscillator strenghs are then not of the order of unity for allowed transitions, but rather of the order m_0/m^*, which also has to be substituted for the right-hand side of Eq. (16). This is especially necessary for multiband (or nonparabolic) models, where interband and intersubband transitions cannot be separated in the sum rule (Khurgin, 1993; Sirtori et al., 1994). In a similar manner, the position dependence of the effective mass can be incorporated. It has been shown that m_0/m^* has to be replaced by $\langle n|m_0/m^*(z)|n\rangle$ in this case (Davé and Taylor, 1994); that is, an average of m^* over the ground state wave function must be performed. Of course, the absorption coefficient, if calculated consistently, remains unaffected by these definitions.

It is worth again summarizing the main issue related to the effective-mass approximation (with constant effective mass). As long as m^* is assumed constant and z-independent in all formulas known from elementary quantum mechanics, m_0 can be substituted for m^* and one can work with the envelope function alone. The incorporation of the position (Davé and Taylor, 1994) or energy (Sirtori et al., 1994) dependence (nonparabolicity) requires more sophistication.

Having achieved a very simple form of the matrix element, we can proceed to evaluate the absorption coefficient α. The absorption coefficient is usually defined through the ratio of the absorbed electromagnetic energy per unit time and volume, $\hbar\omega \cdot W_{if}/V$, and the intensity of the incident radiation, $I = (1/2)\varepsilon_0 c\eta E_0^2$, summed over all occupied initial and empty final states. Here η denotes the refractive index of the material, for the moment taken to be real and constant. This leads to a dimension of inverse length for α. In the case of quasi-two-dimensional layers, this concept must be modified, and we have several possible ways to do that. The most natural way is to define a dimensionless absorption coefficient, α_{2D}, by dividing through the area A instead of the volume. Here α_{2D} is simply a measure of the fraction of electromagnetic energy absorbed by a 2D layer. When we also allow for stimulated emission in addition to absorption, sum over all possible combinations of initial and final states, and take into account their occupation via the Fermi–Dirac distribution function, we can write the very general expression for the absorption coefficient

$$\alpha_{2D} = \frac{\hbar\omega}{IA} \sum_{n,n'} \sum_{\mathbf{k}_\perp} \frac{2\pi}{\hbar} |\langle n|(e/m^*)\mathbf{A}\cdot\mathbf{p}|n'\rangle|^2 [f(E_n(\mathbf{k}_\perp)) - f(E_{n'}(\mathbf{k}_\perp))]$$
$$\times \delta(E_{n'}(\mathbf{k}_\perp) - E_n(\mathbf{k}_\perp) - \hbar\omega) \qquad (17)$$

Now we express both I and A by the electric field amplitude E_0 (which then cancels) and change the summation into a two-dimensional integration in the usual way (including a factor of 2 for the spins).

$$\alpha_{2D} = \frac{\pi e^2}{\varepsilon_0 c\eta\omega m^{*2}} \sum_{n,n'} \frac{2}{(2\pi)^2} \int d^2k_\perp |\langle n|p_z|n'\rangle|^2 [f(E_n) - f(E_{n'})] \cdot \delta(E_{n'} - E_n - \hbar\omega)$$
$$(18)$$

Here we have already accounted for the polarization selection rule and suppressed the k-dependence of the energies E_n and $E_{n'}$. This formula can be significantly simplified, if we assume a parabolic in-plane dispersion. Then the two-dimensional integration over the Fermi–Dirac distributions (where E_F is the Fermi energy and k is Boltzmann's constant) can be performed analytically (which is possible only in 2D!) and we get the final result

$$\alpha_{2D} = \frac{e^2 kT}{2\varepsilon_0 c\eta\hbar} \sum_{n,n'} f_{nn'} \ln\left[\frac{1 + \exp((E_F - E_n)/kT)}{1 + \exp((E_F - E_{n'})/kT)}\right] \frac{\Gamma/\pi}{(E_{n'} - E_n - \hbar\omega)^2 + \Gamma^2}$$
$$(19)$$

We have also replaced the energy-conserving δ function by a normalized Lorentzian with half width at half maximum (HWHM) of Γ and the momentum matrix element by the oscillator strength using Eq. (15) and $\omega_{n'n} \approx \omega$, so that the prefactor in Eq. (19) becomes independent of ω. (Note that problems can arise in this connection for broad absorption lines such that $\hbar\omega_{n'n} \approx \Gamma$. This was thoroughly discussed by Cohen-Tannoudji et al., 1989).

A main feature of intersubband transitions is their δ-function-like density of states, which is a consequence of the same curvature of the initial and final subband (corrections due to nonparabolicity will be discussed in Section VII). Mathematically, this is reflected through the cancellation of the $\hbar^2 k_\perp^2 / 2m^*$ terms in the energy conserving δ function (Eq. (17)). This is why intersubband transitions largely behave like atomic transitions, although we have a two-dimensional rather than a zero-dimensional system.

At zero temperature, where the ln approaches $(E_F - E_1)/kT$ (assuming only one subband is occupied) a particularly simple result is obtained, which can be used to estimate the peak absorption strength of a particular structure with certain carrier density, oscillator strength and linewidth. We also leave out the sum over all transitions with the exception of $1 \to 2$, which is usually the most important one:

$$\alpha_{2D}(T=0) = \frac{n_s e^2 \hbar}{2\varepsilon_0 c \eta m^*} f_{12} \frac{\Gamma}{(E_2 - E_1 - \hbar\omega)^2 + \Gamma^2} \quad (20)$$

Here n_s is the areal electron concentration. At the resonance, $E_2 - E_1 = \hbar\omega$, this yields the useful formula for GaAs (with $m^* = 0.068\, m_0$ and $\eta = 3.4$)

$$\alpha_{2D} = 0.15\, n_s[10^{12}\,\text{cm}^{-2}] \frac{f_{12}}{\Gamma[\text{meV}]} \quad (21)$$

Note, however, that this is valid only as long as $\alpha_{2D} \ll 1$ (strictly when $\alpha_{2D} \cdot \sin^2\theta/\cos\theta \ll 1$, where θ is the angle between the growth axis and the light propagation direction; for more details, see later), which is, fortunately, fulfilled for nearly all realistic cases. From Eq. (21) we see that a 2D layer with $n_s = 10^{12}\,\text{cm}^{-2}$, $f_{12} = 1$ and $\Gamma = 1$ meV, which is about what can be achieved to date, leads to $\alpha_{2D} = 0.15$. If $\alpha_{2D} \gtrsim 1$ (which actually happens in cyclotron resonance on high-mobility 2D electron gases), the simple linear absorption theory breaks down and we must treat the 2D system as a highly conducting, quasi-metallic sheet with a frequency-dependent surface conductivity σ^{2D}, and we must determine the transmitted intensity using the correct electromagnetic boundary conditions. The microscopic properties are then included in σ^{2D}. For normal incidence this results to (Höpfel and Gornik, 1986).

$$T = \frac{4\eta_1\eta_2}{|\eta_1 + \eta_2 + (\sigma^{2D}/\varepsilon_0 c)|^2}$$

$$R = \frac{|\eta_1 - \eta_2 - (\sigma^{2D}/\varepsilon_0 c)|^2}{|\eta_1 + \eta_2 + (\sigma^{2D}/\varepsilon_0 c)|^2} \quad \text{and} \quad A = 1 - R - T = \frac{4\eta_1(\text{Re}\,\sigma^{2D}/\varepsilon_0 c)}{|\eta_1 + \eta_2 + (\sigma^{2D}/\varepsilon_0 c)|^2}$$

(22)

Here T, R, and A are the transmission, reflection, and absorption, respectively, and η_1 and η_2 are the refractive indices before and after the active layer. For $\sigma^{2D} \ll \varepsilon_0 c = (377\Omega)^{-1}$ and $\eta_1 = \eta_2 = \eta$ we get $A = \text{Re}\,\sigma^{2D}/\varepsilon_0 c\eta$. We show later that this is identical to the preceding α_{2D}, in Eq. (20), when $\sigma(\omega)$ is calculated assuming a Drude–Lorentz oscillator.

It is generally advisable to work with the dielectric function $\varepsilon(\omega)$ or the conductivity $\sigma(\omega)$ instead of the absorption coefficient, whenever the electromagnetic properties of a sample are to be calculated consistently (see Section IV) or many-body effects in the absorption are to be taken into account (see Section VII). The relation

$$\varepsilon(\omega) = 1 + i\frac{\sigma}{\varepsilon_0 \omega} \tag{23}$$

(for now neglecting the contribution from the background dielectric constant) known from three dimensions has to be modified in two dimensions to

$$\varepsilon(\omega) = 1 + i\frac{\sigma^{2D}}{\varepsilon_0 \omega L_{\text{eff}}} \tag{24}$$

where σ^{2D} is again the 2D (areal) conductivity and L_{eff} is some effective thickness of the quasi-2D system. Using the relation between the optical constants

$$\text{Im}(\varepsilon) = 2\,\text{Re}(\eta)\,\text{Im}(\eta) \tag{25}$$

and

$$\alpha_{3D} = 2\frac{\omega}{c}\,\text{Im}(\eta)$$

where η is the refractive index, we obtain

$$\alpha_{3D} = \frac{\omega}{c\,\text{Re}(\eta)}\text{Im}(\varepsilon) = \frac{\text{Re}(\sigma^{2D})}{\varepsilon_0 c\,\text{Re}(\eta) L_{\text{eff}}} =: \frac{\alpha_{2D}}{L_{\text{eff}}} \qquad (26)$$

Inserting here a Drude–Lorentz oscillator with oscillator strength f_{12} and damping constant $\gamma(=\Gamma/\hbar)$,

$$\sigma^{2D}(\omega) = \frac{n_s e^2 f_{12}}{m^*}\frac{-i\omega}{\omega_{21}^2 - \omega^2 - 2i\gamma\omega} \qquad (27)$$

and approximating the denominator with the near-resonance expression $(\omega_{21} - \omega)(\omega_{21} + \omega) \approx 2\omega(\omega_{21} - \omega)$, we obtain

$$\text{Re}\,\sigma^{2D}(\omega) = \frac{n_s e^2 f_{12}}{2m^*}\frac{\gamma}{(\omega_{21} - \omega)^2 + \gamma^2} \qquad (28)$$

which is exactly equivalent to Eq. (20). For more detailed discussion, see Section IV.

In the discussion up to here we have pretended that one can simply shine light through a 2D system and get intersubband absorption according to Eq. (19). However, the polarization selection rule dictates that there must be a component of the electric field that is perpendicular to the 2D layers (z direction) to get coupling of the radiation to the intersubband transition. Obviously this is not the case for normal-incidence radiation. Therefore, experimentalists have used other geometries to obtain a finite coupling and thus a finite absorption. These geometries and their theoretical description will be discussed in more detail in Section IV. Here we consider only the simplest method, namely, oblique incidence of the radiation (see Fig. 2). Denoting the angle between the growth axis (z direction) and the propagation direction of the optical beam with θ, the electric-field component interacting with the intersubband transition is $E_z = E_0 \sin\theta$, which gives a factor $\sin^2\theta$ for the absorbed intensity. However, the effective interaction length of the radiation with the quantum well is then increased by $1/\cos\theta$, so α_{2D} in Eqs. (19)–(21) must be multiplied by $\sin^2\theta/\cos\theta$. Finally, if we are dealing with a multi-quantum-well (MQW) system of N quantum wells, the transmitted intensity will be proportional to $\exp(-N\alpha_{2D}\sin^2\theta/\cos\theta)$. Experimentally, one usually derives the absorption ("absorbance") from the transmission by *absorbance* $= -\log(transmission)$, and proper normalization to obtain a flat baseline.

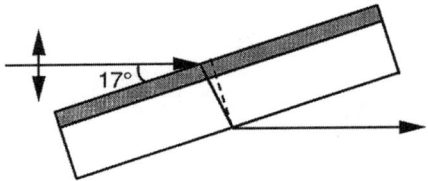

FIG. 2. Oblique-incidence geometry. Here the radiation is incident in Brewster's angle (73° for GaAs with $\eta = 3.3$), resulting in $\theta = 17° = 90° - 73°$.

In MQWs it is common to define a three-dimensional absorption coefficient α_{3D} with dimensions in inverse centimeters. This can be achieved by simply dividing α_{2D} by d, the MQW period. Yet we must recognize that α_{3D} then becomes dependent on the barrier thickness. This concept becomes more useful (and even necessary) when the barriers are thin (for strongly coupled superlattices). In this case, one can actually proceed just like for bulk material and extend the integration in Eqs. (17) and (18) over k_x, k_y, and k_z, leading to a three-dimensional absorption coefficient in the most natural way. We come back to this point in Section VI on superlattices. A third alternative definition of an α_{3D} that is sometimes used in the literature is obtained through normalizing α_{2D} by the quantum well thickness.

III. The Symmetric Quantum Well

In this section, we discuss in more detail the properties of a single, symmetric quantum well, with both infinitely high barriers and finite barriers. Especially the case of the infinite quantum well is very instructive, since all the energy levels, wave functions, and matrix elements can be easily calculated analytically.

For an infinitely deep quantum well (i.e., with a potential that is zero from $z = 0$ to $z = L$ and infinity everywhere else), the following eigenvalues and envelope wave functions are obtained

$$E_{n,\mathbf{k}_\perp} = \frac{n^2 \pi^2 \hbar^2}{2m^* L^2} + \frac{\hbar^2 \mathbf{k}_\perp^2}{2m^*} \tag{29}$$

$$\varphi_n(z) = \sqrt{\frac{2}{L}} \sin\left(\frac{n\pi z}{L}\right) \tag{30}$$

The momentum matrix element between the ground state ($n = 1$) and the $n = 2$ excited subband can be calculated explicitly to yield

$$|\langle 1|p_z|2\rangle| = \frac{8\hbar}{3L} \tag{31}$$

which results in an oscillator strength

$$f_{12} = \frac{256}{27\pi^2} = 0.96 \tag{32}$$

where Eq. (29) must be used for the energy difference $E_2 - E_1 = \hbar\omega_{21}$. The oscillator strength has the nice feature of being independent of the quantum well width. The same expression for the oscillator strength can be obtained by using the dipole matrix element

$$|\langle 1|z|2\rangle| = \frac{16L}{9\pi^2} \tag{33}$$

It is even possible to derive a general expression for all allowed transitions in an infinite quantum well:

$$f_{mn} = \frac{64}{\pi^2} \frac{n^2 m^2}{(n^2 - m^2)^3} \tag{34}$$

Only parity changing (odd–even or even–odd) transitions are allowed due to the inversion symmetry of the potential. As further examples, $f_{14} = 0.03$, $f_{23} = 1.87$, and so on. Thus, we see that by far the strongest transition is the one with $n = m + 1$. The higher ones carry a smaller and smaller oscillator strength. We can also see that transitions between excited states are much stronger than transitions from the ground state, a fact well known from atomic physics and necessary to full the oscillator sum rule. For large n, it can actually be shown using Eq. (34) that $f_{n,n+1}$ increases linearly with n. In Fig. 3 an infinite square quantum well is sketched together with the allowed intersubband transitions.

A more realistic description must include the finite depth of the quantum well. This can be done either analytically (in the case of a symmetric square well) or numerically with the transfer matrix method, as mentioned in Section II. In general, the subband energies will be somewhat lower than for a structure having the same width, but infinite barriers and the wave functions penetrate into the barrier. But no matter how low the barriers and

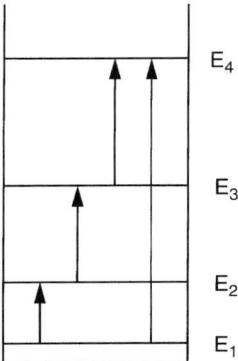

Fig. 3. Possible intersubband transitions in an infinite quantum well; only transitions with a change in parity are allowed.

how thin the quantum well, for a one-dimensional binding potential there will always be at least one confined state, which is well-known from quantum mechanics. Figure 4 shows a state-of-the-art absorption spectrum of an $In_{0.53}Ga_{0.47}As$–$In_{0.52}Al_{0.48}As$ quantum well (barrier height approximately 500 meV) with two bound states. The measurement here has been performed with the sample prepared in a multipass waveguide geometry (see Section IV) and taking the ratio of the transmission signals of two orthogonal polarizations, where only one polarization has an intersubband-active component (z direction).

In the remainder of this section we look at transitions from the ground state to the continuum in a finite, symmetric QW, which has some practical relevance for infrared detectors. For simplicity, we assume that the QW is reasonably deep so that we can approximate the wave function of the first subband by the infinite-well wave function. Taking the center of the well to be at $z = 0$, we have

$$\varphi_1(z) = \sqrt{\frac{2}{L}} \cos\left(\frac{\pi z}{L}\right) \qquad (35)$$

and the continuum wave function is simply written as

$$\varphi_c(z) = \frac{1}{\sqrt{\ell}} \exp(ikz) \qquad (36)$$

where we have neglected the effect of the boundaries between the well and

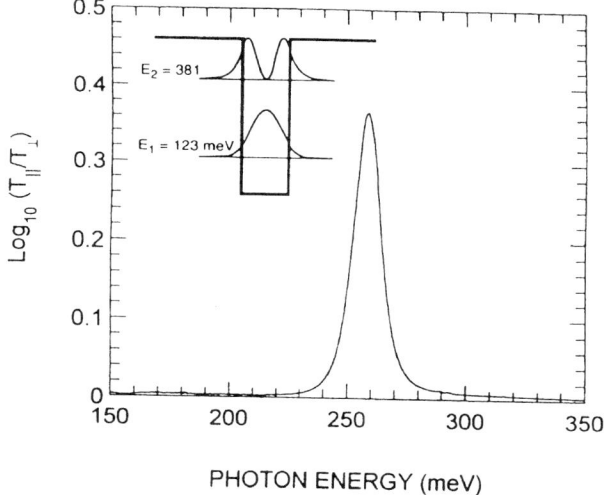

FIG. 4. Absorption spectrum of an $In_{0.53}Ga_{0.47}As–In_{0.52}Al_{0.48}As$ quantum well sample with two bound states (width 52 Å) obtained by taking the log of the transmission ratio of the two orthogonal polarizations. The subband levels and squared wave functions are shown in the inset (from Sirtori et al., 1994).

the barrier, which normally give rise to standing-wave (Fabry–Pérot) effects in the continuum, and ℓ is the normalization length of the continuum. With these wave functions, the momentum matrix element can be calculated to give (Bastard, 1988)

$$\langle \varphi_1 | p_z | \varphi_c \rangle = \sqrt{\frac{2}{L\ell}} \cos\left(\frac{kL}{2}\right) \left[\frac{1}{k + \pi/L} - \frac{1}{k - \pi/L} \right] \hbar k \qquad (37)$$

Summing over the continuum states (k-integration) leads to the 2D absorption coefficient

$$\alpha_{2D} = \frac{n_s e^2}{2\varepsilon_0 c\eta\omega} k_0 L \cos^2(k_0 L/2) \left[\frac{1}{k_0 L + \pi} - \frac{1}{k_0 L - \pi} \right]^2 \qquad (38)$$

with $k_0^2 = (2m^*/\hbar^2)(\hbar\omega - V_b + E_1)$ and V_b is the barrier height. This function yields a steep rise at the ionization threshold $\hbar\omega = V_b - E_1$, and a more slowly decaying tail on the high-frequency side. Liu (1993) performed an exact analytical calculation of the absorption of a finite, symmetric QW. He showed that the inclusion of a broadening parameter Γ removes all divergences and there is a smooth transition between bound-to-bound and bound-to-continuum transition. This is shown in Fig. 5, where the absorption is plotted for a QW with varying thicknesses (with $\Gamma = 5\,\text{meV}$). For $L < 49\,\text{Å}$ the QW binds only one state, and the three curves for $L = 47$, 41, and 35 Å corresponds to continuum transitions. Only a decrease of the peak absorption and an asymmetric broadening is observed. Liu (1993) compared this calculation with experimental data of Asai and Kawamura (1990) and found good agreement. Figure 6 shows the bound-continuum absorption spectrum of a narrow (30 Å) $\text{In}_{0.53}\text{Ga}_{0.47}\text{As}$–$\text{In}_{0.52}\text{Al}_{0.48}\text{As}$ multiquantum well that has only one bound state.

Standing-electron-wave effects in the continuum can be used intentionally to drastically modify the absorption of the quantum well structure. For example, so-called Bragg confinement (Lenz and Salzman, 1990; Capasso *et al.*, 1992, Sirtori *et al.*, 1992) can be achieved by embedding a quantum well with one bound stated is into a superlattice, where the superlattice has to act like a quarter-wave dielectric stack for electrons. This is discussed more extensively in Chapter 2 of the present volume.

Dupont *et al.* (1995) have observed a photocurrent due to transitions into the continuum of a multi-QW structure, which was excited by simultaneous

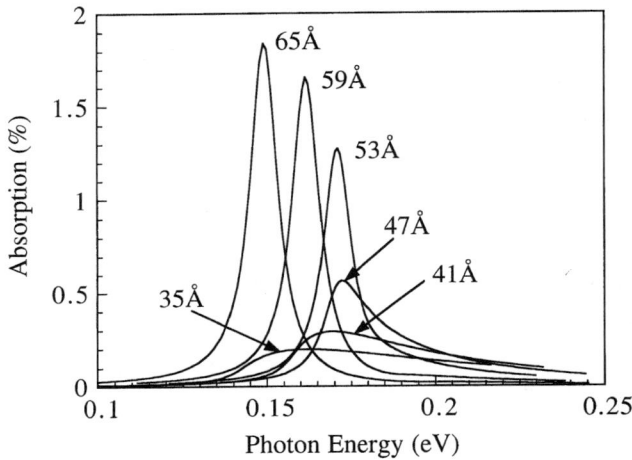

FIG. 5. The calculated intersubband absorption spectrum of a single GaAs–$\text{Al}_{0.3}\text{Ga}_{0.7}\text{As}$ QW plotted for different quantum well thicknesses as indicated. The second subband is bound only for thicknesses greater than 49 Å, otherwise the absorption is to the continuum (after Liu, 1993).

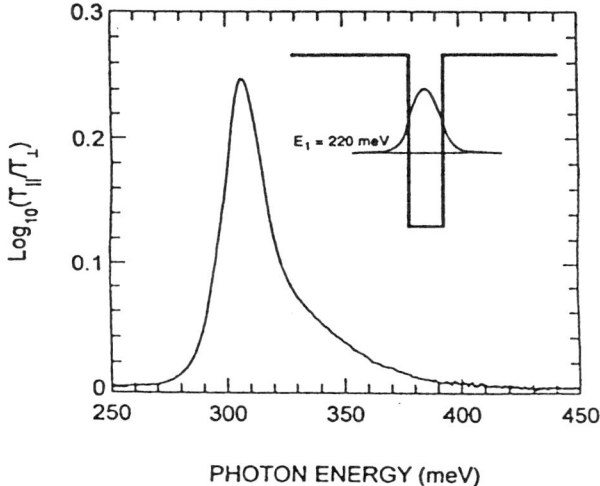

FIG. 6. Bound-to-continuum absorption spectrum of a 30-Å-wide $In_{0.53}Ga_{0.47}As$–$In_{0.52}Al_{0.48}As$ quantum well sample with one bound state obtained by taking the log of the transmission ratio of the two orthogonal polarizations. The bound state with its squared wave function is shown in the inset (from Sirtori et al., 1994).

one-photon and two-photon absorption at a wavelength of 5.3 and 10.6 μm, respectively. Since the 5.3-μm radiation was generated through frequency doubling from the 10.6-μm radiation, the two waves had a constant but tunable phase relation. They were able to coherently control the direction of the induced photocurrent by changing the phase difference.

Before we proceed further and discuss other types of quantum wells, it seems appropriate to devote a section to some experimental techniques commonly employed for the study of intersubband transitions.

IV. Experimental Geometries and Their Electromagnetics

The specific nature of intersubband transitions — that is, the selection rule requiring an electric-field component perpendicular to the QW layers — necessitates the use of nonstandard geometries to perform absorption experiments, which in turn require a careful consideration of the electromagnetics (i.e., the spatial distribution of the electromagnetic field in the sample). These two issues are the topic of this section.

In a standard transmission geometry, where the light is incident perpendicular to the sample surface, the electric field has components only in the QW plane and intersubband transitions cannot be induced (at least in the simple band structure discussed so far; exceptions are discussed in Sections VIII and IX). The simplest way to overcome this problem is to shine the light on the sample at an oblique angle of incidence. A frequent choice is Brewster's angle (73° for GaAs with a refractive index of $\eta = 3.3$) (see Fig. 2 in Section II), at which the reflection on the surfaces also vanishes (West and Eglash, 1985; Levine et al., 1987). However, due to the high refractive index of most semiconductors, the angle of incidence within the sample is still rather small ($\theta = 17°$ for GaAs) leading to a small intersubband-active electric-field component E_z. The effective coupling factor $\sin^2\theta/\cos\theta$ in this case is given by $1/\eta\sqrt{\eta^2 + 1} \approx 0.09$ for GaAs. Due to this small coupling, the Brewster-angle geometry is normally used only for relatively highly doped multi-QW systems. An advantage, however, is the very well defined electric-field strength and intensity at the QW position.

For samples exhibiting weaker absorption (e.g., for lower doping, single QWs or large linewidths), several types of waveguide geometries have been devised by various authors, where the incident radiation undergoes several total internal reflections in the sample (Fig. 7). In these geometries the light is coupled into the sample at the edges, which are wedged at a certain angle α. The number of passes through the active layer, M (= twice the number of reflections), is determined by the length (L_{substr}) and thickness (D_{substr}) of the substrate, $M = L_{substr}/(D_{substr} \cdot \tan\theta)$. For a sample with 0.5 mm thickness, 5 mm length, and an angle of $\theta = 45°$ one obtains five total internal reflections at each surface ($M = 10$), leading to a coupling strength of $10 \times \sin^2 45°/\cos 45° \approx 7$. Two frequently employed types of waveguides are shown in Figs. 7a and 7b. The geometry (Fig. 7a) (Levine et al., 1987; Kane et al., 1988; Wieck et al., 1988) has the advantage that the surface reflection loss of the incident light is the same for both orthogonal polarizations, which is convenient for taking reference spectra. (The TM or p polarization is the one that couples to the intersubband transition, since the electric field contains a z component, whereas the TE or s polarization does not couple, since the electric field lies in the QW plane.) A slight drawback is the displacement of the beam after traversing the sample. In geometry (Fig. 7b) (Hertle et al., 1991; Fromherz et al., 1994) the surface reflections are different for both polarizations, but the beam is not displaced. The internal angle θ is given by $\theta = \alpha + \arcsin[\sin(90° - \alpha)/\eta]$, and when α is chosen so that $\theta = 90° - \alpha$ ($\alpha = 38°$, $\theta = 52°$ for GaAs), the propagation direction inside the sample is parallel to one of the wedged facets. Variations of these geometries contain a two-pass arrangement, where either the sample is so short (e.g., an etched mesa structure, Capasso et al., 1994, see Fig. 7c), or the

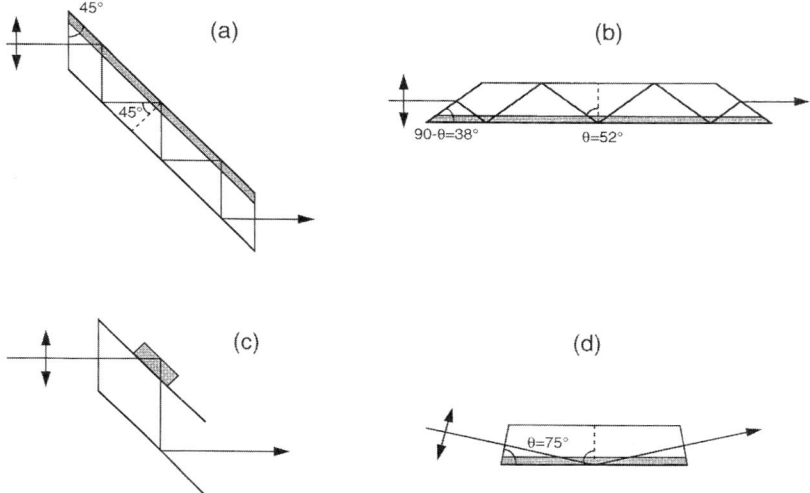

FIG. 7. Several common waveguide geometries for the measurement of intersubband absorption are shown. Top (a) and (b): two types of multipass waveguides. Bottom (c) and (d): Two types of two-pass waveguides; (c) shows an etched mesa structure. For a discussion, see text.

wedge angle α is so large (nearly 90°, Seilmeier et al., 1987, see Fig. 7d) that the light undergoes only one total reflection. This is especially useful when the intersubband transition is pumped with a laser and one wants to know the effective light intensity in the QWs accurately.

To calculate the absorption of an electromagnetic wave in such a multipass waveguide structure, it is essential to consider what is happening at the boundary of the semiconductor slab, that is, at the semiconductor–air interface. Due to the interference of the incident and reflected wave, a standing-wave intensity pattern of the field component E_z will form along the z direction having a periodicity $\lambda/2\eta \cos\theta$ (Vodopyanov et al., 1997). Thus it is clear that only as long as the total thickness of the active layer ($=Nd$) is larger than this standing-wave period (or essentially the resonant wavelength in the sample, λ/η), the nodes and antinodes are largely averaged out and the preceding simple approach for calculating the absorption will work. As an example, if the resonant wavelength is 10 μm, the corresponding intensity pattern has a period of 2.15 μm (for $\eta = 3.3$, $\theta = 45°$). On the other hand, a multiquantum well system with $N = 100$ periods of $d = 300$ Å thickness (well plus barrier width) yields a total thickness of 3 μm, and the preceding condition is satisfied. If this condition is not fulfilled, as is obviously the case in a single quantum well, the absorption strongly depends on the position of the QW relative to the intensity nodes and antinodes.

Let us now consider a QW near the surface — that is, near the semiconductor–air interface (see Fig. 8) — and the corresponding electromagnetic boundary conditions for light incident from inside the semiconductor at an angle of $\theta = 45°$. According to Fresnel's formulas, the totally reflected wave will undergo a phase shift of 168° (for $\eta = 3.3$), which corresponds to an almost complete phase reversal, and produces a node in the perpendicular field component E_z. Therefore there is virtually no interaction of the light with the intersubband transition, when the active layer is thin in the preceeding sense. This can, of course, be remedied, when the QW is moved away from the surface to the next antinode; that is, by half a period of the intensity pattern. There is, however, another method, which was first discussed by Kane et al. (1988) for quantum wells, but has been routinely used in the seventies in connection with Si inversion and accumulation layers (Kamgar et al., 1974; Kneschaurek et al., 1976). When a metallic layer is deposited on the sample surface, the boundary conditions are changed, leading to a maximum of the field component E_z at the semiconductor–metal interface, as illustrated in Fig. 8.

Quantitatively, the transmission through such a multipass waveguide structure can be approximated by

$$T \cong \exp(-C \cdot M \cdot N \cdot \alpha_{2D} \cdot \sin^2\theta/\cos\theta) = \exp\left(-C \frac{L_{\text{substr}}}{D_{\text{substr}}} N \cdot \alpha_{2D} \cdot \sin\theta\right) \quad (39)$$

Here L_{substr} and D_{substr} are the length and thickness, respectively, of the sample including substrate, N is the number of multiquantum well periods,

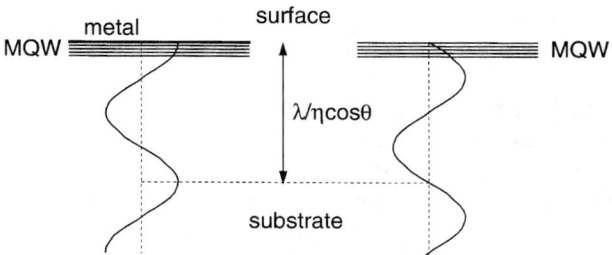

FIG. 8. Sketch of the electric-field distribution for a multiquantum well structure on a substrate, with (left) or without (right) a metal coating of the surface. A standing-wave pattern with period $\lambda/\eta \cos\theta$ is formed that has a crest or node at the surface, respectively.

M is the number of passes through the active layer, and C is a coupling factor between 0 and 2, which depends on the active layer thickness Nd relative to the wavelength λ/η and on the position of the QWs with respect to the standing-wave pattern.

For a thick MQW layer ($Nd > \lambda/\eta$), where the standing-wave effect is averaged out, $C = 1$, identical to what one gets in a simple traveling-wave picture. For a thin MQW layer ($Nd \ll \lambda/\eta$), C depends on the location of the active layer relative to the electric-field nodes and crests. If the MQW is in a crest of the electric-field pattern (for example, near the metal-coated surface), $C = 2$, whereas $C \ll 1$ if the MQWs are located at an electric-field node (for example, near the uncoated surface).

In summary, at a semiconductor–air interface the E_z component is nearly zero and the in-plane component E_{xy} is finite, whereas at a semiconductor–metal interface, the E_z component is finite (even enhanced) and the E_{xy} components are shorted out to zero. Therefore, intersubband absorption experiments with a small number of QW periods located near the surface require a metal coating of the surface.

Note that for *nonlinear* experiments (such as saturation measurements) even for a thick MQW layer, the local peak intensity is crucial, which can be a factor of four larger than in the traveling-wave picture (Vodopyanov et al., 1997).

For a consistent treatment of the waveguide transmission, the electromagnetic wave propagation through the multilayer stack of the sample should be calculated using proper expressions for the dielectric functions for each layer as well as the electromagnetic boundary conditions between the layers. The quantity of interest describing the absorption is then the reflectivity at the semiconductor–air or semiconductor–metal interface. The numerics can be done, for example, using a transfer-matrix method (Harbecke, 1986; Terzis et al., 1990). In this way, the substrate and (un)doped buffer or cap layers can be included in addition to the active MQW layer as well as a possible metallic layer on the surface, which may result in Fabry–Pérot-type fringes. For the substrate, buffer and cap layers, a classical Drude expression for $\varepsilon(\omega)$ can be employed (plus a phonon term, if any phonon resonances are close in frequency). The metal can be described in a similar way. The active MQW layers can be treated as a single layer (effective medium approximation), since the individual quantum well thicknesses are two orders of magnitude smaller than the wavelength of the light, and can be described with a simple anisotropic plasma model with an oscillator in the z direction (compare Eq. (27)) (Chen et al., 1976; Kane et al., 1988)

$$\varepsilon_{zz} = \varepsilon_\infty - \frac{\varepsilon_\infty f_{12} \omega_p^2}{\omega^2 - \omega_{21}^2 + 2i\omega\gamma} \quad \text{and} \quad \varepsilon_{xx} = \varepsilon_{yy} = \varepsilon_\infty - \frac{\varepsilon_\infty \omega_p^2}{\omega^2 + i\omega/\tau} \quad (40)$$

Here γ is the HWHM of the intersubband absorption, τ is a in-plane scattering time, and ω_p is the two-dimensional effective plasma frequency, given by $\omega_p = \sqrt{(n_s e^2/\varepsilon_\infty \varepsilon_0 m^* L_{\text{eff}})}$, where L_{eff} is an effective quantum well thickness (see also Section VII, Eq. (58)).

Dahl and Sham (1977) and Nakayama (1995, 1977) carefully considered the nonlocal electromagnetics of a quasi-2D electron system. Liu (1994) has even included retardation effects. In a nonlocal effective-medium approach for multiquantum well systems, Zaluzny and Nalewajko (1997, 1998) derive the dielectric tensor component ε_{zz} from the two-dimensional dynamic conductivity of each individual quantum-well layer. The conductivity σ^{2D} can be calculated microscopically, including depolarization shift and all many-body effects (see Section VII) properly, and with the resulting dielectric tensor, the transmission can be calculated. It turns out that not only the absorption strength but also the absorption frequency and the line shape may depend on the electromagnetic boundary conditions (semiconductor–air or semiconductor–metal); a metallic surface shifts the absorption to higher frequencies (related to this is the so-called Berreman effect, an absorption occurring in thin films at the LO-phonon frequency (see Berreman, 1963; Harbecke *et al.*, 1985). In general, the MQW absorption is usually proportional to $\text{Im}(-1/\varepsilon_{zz})$, and not to $\text{Im}(\varepsilon_{zz})$. In many cases, however, especially if the absorption of each individual QW is small, the transmission can be well approximated by Eq. (39).

Note that a detailed analysis of intersubband absorption strength, energy, and lineshape must include both many-body effects (see Section VII) and electromagnetics. The level of sophistication of the theoretical description must be chosen depending on the sample structure, the geometry, and the absorption strength.

A number of other geometries have been used to excite intersubband transitions, including a type of transmission line arrangement. Here the light is coupled directly into a cleaved facet of the semiconductor slab, which must be covered by a metal for the reasons already discussed. This technique has mostly been employed in the far-infrared spectral range ($\lambda > 50\ \mu\text{m}$), for example, in the earliest experiments on Si-MOS (metal-oxide semiconductor) accumulation and inversion layers (Kamgar *et al.*, 1974; Kneschaurek *et al.*, 1976), and theoretically discussed by Nakayama (1977) and Zaluzny (1996). It has also been used in intersubband absorption experiments performed with a far-infrared free-electron laser (Heyman *et al.*, 1994). Note that an experimentally relevant 3D absorption coefficient can be obtained here by dividing α_{2D} by the sample thickness, which is identical to the second expression in Eq. (39) for $\theta = 90°$. In another method, a Si or Ge prism is attached to the substrate and the radiation is coupled into the sample through the prism, thus avoiding refraction at the semiconductor–

air interface and retaining a large E_z component (McCombe et al., 1979). A particularly elegant and useful way is the so-called "critical incidence coupling" proposed and demonstrated by Keilmann (1994), where radiation is coupled into the sample at the critical angle for total reflection, which forces the p-polarized light to be exactly perpendicular to the QW layer.

Finally, a technique that has proven very useful for many applications is the use of a grating coupler, either by etching a grating into the surface of the sample or by depositing a metallic grating. This enables coupling of normal-incidence radiation to the intersubband transitions. Two regimes can be distinguished: When the period of the grating is of the order of the wavelength in the semiconductor, the first diffracted order propagates nearly parallel to the surface, thus providing a large electric-field component E_z (Fig. 9, right). This has been applied in mid-infrared detectors based on intersubband transitions (Goossen and Lyon, 1985; Goossen et al., 1988; Hasnain et al., 1989; Andersson et al., 1991; Andersson and Lundqvist, 1991; Yu et al., 1992; Ralston et al., 1992), and here the light is often incident from the substrate side (reflection grating mode). If, in contrast, the grating period is much smaller than the wavelength of the radiation (Fig. 9, left), which corresponds to a quasi-static rather than the optical regime (the diffracted modes are evanescent and not propagating), the electric-field components in the near field of the (metallic) grating get scrambled and, in particular, a finite component E_z results (Heitmann et al., 1982; Heitmann and Mackens, 1986; Batke et al., 1989; Helm et al., 1991). Nonvertical intersubband transitions with a finite wavevector can also be excited (Heitmann and Mackens, 1986). This regime has mostly been employed for the far-infrared spectral region. Coupling efficiencies up to 30% in the FIR region (Li et al.,

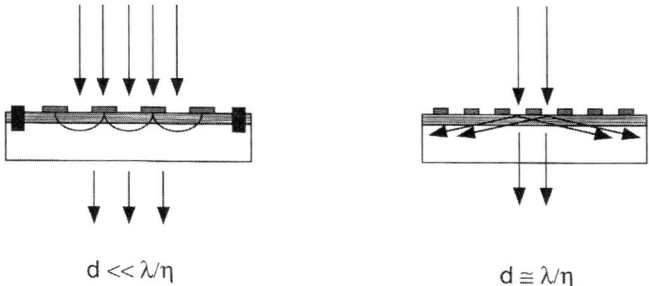

FIG. 9. Grating coupling for intersubband absorption: quasi-static regime (left), where the grating period is much smaller than the wavelength, and diffraction regime (right), where the grating period is of the same order as the wavelength. For the quasi-static regime the electric-field lines in the near field of the grating are sketched.

1990a, 1990b), and up to near unity in the midinfrared (MIR) region (Hasnain *et al.*, 1989; Andersson *et al.*, 1991; Andersson and Lundqvist, 1991) have been reported. The drawback of the grating coupler is that the exact coupling strength and field distribution are not very well known, so the light intensity and absorption strength cannot be quantified very accurately. Furthermore, a reliable theoretical treatment requires a nontrivial electromagnetic calculation (Zheng *et al.*, 1990; Li and McCombe, 1992; Andersson and Lundqvist, 1992; Duboz, 1996; Xu and Hu, 1997; Wendler *et al.*, 1997). Wendler *et al.* (1997) also studied the influence of the grating period on the absorption line shape. In some experiments, the metallic grating coupler was used as a Schottky gate to modulate the electron concentration, which results in increased sensitivity and in absorption spectra free of system artifacts (Helm *et al.*, 1991). The same can be done in waveguide geometry using the metal coating as a Schottky gate (Helm *et al.*, 1993).

V. Asymmetric Quantum Wells

In this section, we discuss quantum wells where the inversion symmetry with respect to the quantum well center is broken by some means. This leads to a relaxation of the selection rules (i.e., transitions between all subbands become allowed). The symmetry breaking can be achieved in several different ways:

1. Applying an electric field along the growth direction.
2. Varying the material (alloy) composition in the QW (e.g., "step quantum well").
3. Two quantum wells with different thicknesses separated/coupled by a thin barrier (asymmetric coupled QW; ACQW).

Of course, all of these three methods can be incorporated at the same time in a single structure, making possible true band-structure engineering. In addition, the symmetry can be broken through asymmetric doping profiles, generating an internal electric field in a quantum well or a heterostructure. An important example for an asymmetric quasi-two-dimensional system is, of course, also the inversion or accumulation layer at the interface of a heterostructure or at the surface in a MOS structure. In this case, the potential near the band edge has an asymmetric triangular shape.

The case of a square quantum well in an electric field F can be treated by second-order perturbation theory, as long as eFL is small compared to the confinement energy E_1. In this case, the $n = 2$ level remains nearly unperturbed, since it still "feels" mainly the vertical energy barriers, whereas the $n = 1$ level is shifted down by an amount $\Delta E_1 = -C(m^* e^2 F^2 L^4/\hbar^2)$; that is, it is quadratic in the electric field. Here $C = 0.0022$ is a numerical constant (Bastard et al., 1983). This results in a *blue* shift of the intersubband absorption and corresponds to the quantum confined Stark effect. (Note, however, that this term was originally coined for interband transitions, where the electric field causes a *red* shift of the absorption edge.) This effect can be used for electrooptic devices such as modulators. Experimental results were reported by Harwit and Harris (1987). In the nonperturbative regime of high electric fields and/or large well widths the energy levels can be calculated variationally (Bastard et al., 1983) or by the usual numerical methods such as the transfer matrix (Ahn and Chuang, 1986, 1987). Figure 10 shows the calculated dependence of the subband separation E_{21} on the electric field for a GaAs–Al$_{0.4}$Ga$_{0.6}$As QW with three different thicknesses. The initial quadratic behavior turns into linear at higher fields. For extremely high fields ($eFL \gg E_{21}$), E_{21} finally becomes independent of QW thickness, corresponding to the situation of a triangular potential. The conduction-band profile of a 120-Å-wide QW with $F = 100$ kV/cm is shown in Fig. 11. The deformation of the ground-state wave function can clearly be seen, giving rise to a partial transfer of oscillator strength from the $1 \rightarrow 2$ to higher transitions ($1 \rightarrow 3$, $1 \rightarrow 4$, etc.).

A simple way of breaking the symmetry with varying alloy composition is to introduce a potential step into the QW (Yuh and Wang, 1989).

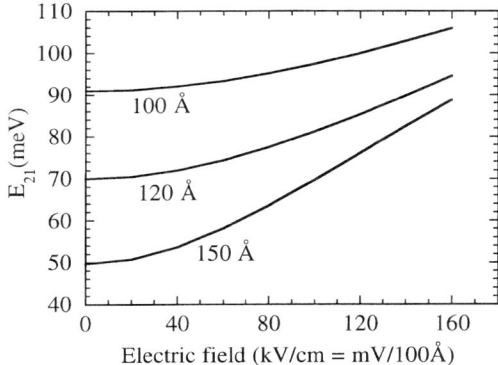

FIG. 10. Stark shift of the intersubband absorption. The energy separation of the two lowest subbands for GaAs–Al$_{0.4}$Ga$_{0.6}$As quantum wells with different thicknesses (as indicated) is plotted vs electric field. Note the transition from quadratic to linear behavior.

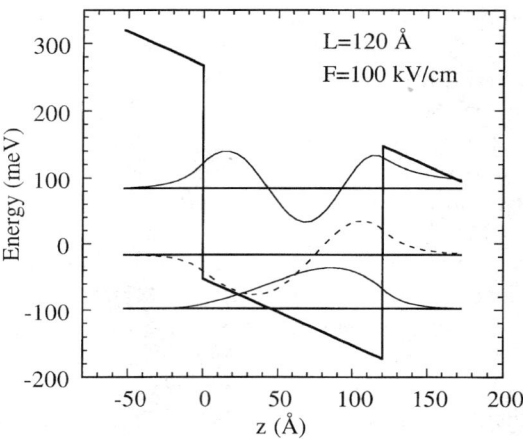

FIG. 11. Potential profile, energy levels, and wave functions of a 120-Å-wide GaAs–Al$_{0.4}$Ga$_{0.6}$As quantum well with an electric field of 100 kV/cm.

Absorption experiments in such structures were first performed by Mii *et al.* (1990a), also including an additional electric field (Mii *et al.*, 1990b; see Fig. 12). The absorption spectrum (Fig. 12, right panel) shows two strong peaks, which correspond to the $1 \rightarrow 2$ and $1 \rightarrow 3$ transitions. Depending on the polarity of the applied electric field, a red or blue shift can occur (Yuh *et al.*, 1990). This Stark shift is much stronger than for a square well and nearly linear in the electric field, and is therefore useful for electrooptic modulators. A particularly valuable feature of step quantum wells is that the energy differences $E_3 - E_2$ and $E_2 - E_1$ can be made equal, which opens up the possibility of observing doubly resonant nonlinear optical phenomena (Rosencher and Bois, 1991; Rosencher *et al.*, 1992) such as second-harmonic generation. Such applications are discussed in great detail in Chapter 6 by Sirtori, *et al.*, in the present volume.

The same can be achieved with asymmetric coupled quantum wells. The additional application of an electric field enables one to observe anticrossing between different subbands (Yuh and Wang, 1988) and tuning of the oscillator strengths of the various transitions (Yuh and Wang, 1988; Faist *et al.*, 1993a). Such structures have again been used for giant nonlinear optical effects (Capasso *et al.*, 1994) and they also serve as the active cell in quantum cascade lasers (see Chapter 5 in the present volume). In Figs. 13 and 14 examples for the intersubband absorption in such structures is presented. The InGaAs–InAlAs structure in Fig. 13 has been optimized for doubly resonant nonlinear effects ($E_3 - E_2 \cong E_2 - E_1$), and in the

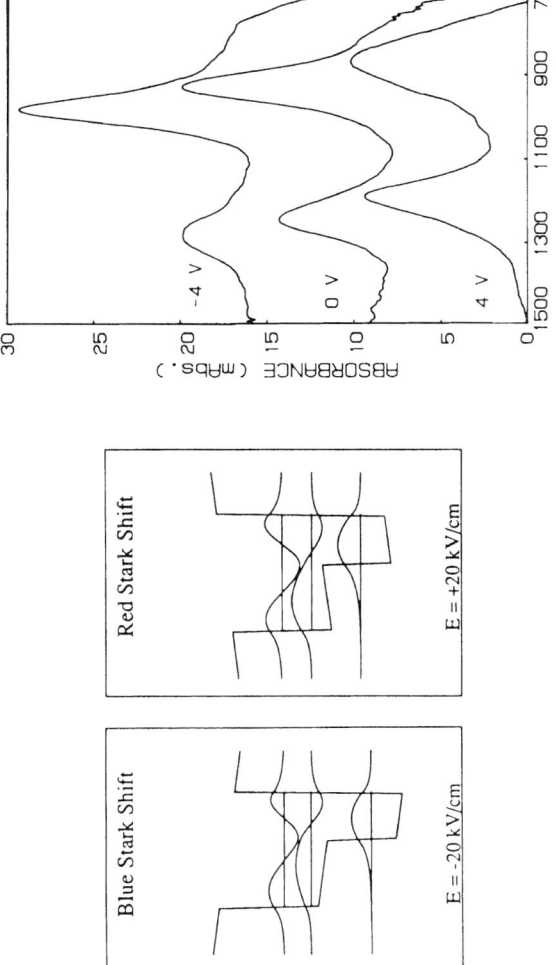

FIG. 12. Stark effect in a GaAs–Al$_{0.18}$Ga$_{0.82}$As–Al$_{0.44}$Ga$_{0.56}$As step-quantum well (width of the GaAs part 60 Å, width of the step 90 Å). Left: Potential profile and wave functions for positive and negative bias. Right: Measured absorption spectrum for positive, zero, and negative bias (from Mii et al., 1990b).

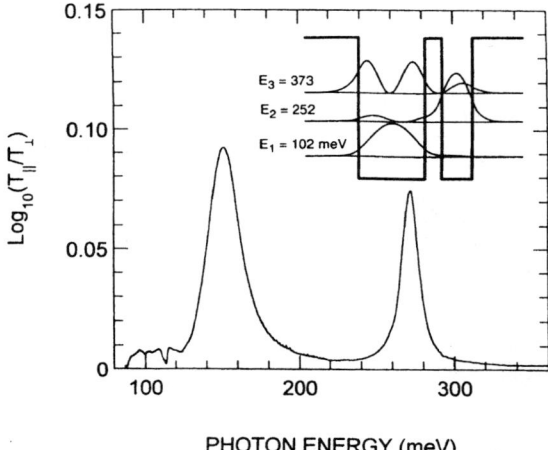

FIG. 13. Absorption spectrum of a strongly asymmetric InGaAs–InAlAs coupled quantum well structure with three bound states (thicknesses are 59, 13, and 24 Å for the first well, barrier, and second well, respectively). The subband levels and squared wave functions are shown in the inset (from Sirtori *et al.*, 1994).

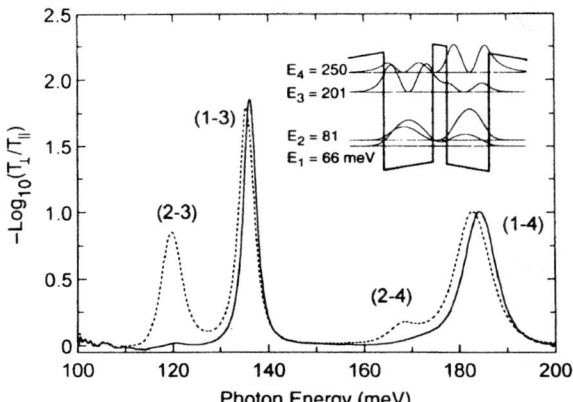

FIG. 14. Absorption spectrum of a weakly asymmetric GaAs–$Al_{0.33}Ga_{0.67}As$ coupled quantum well structure with four bound states (well, barrier, and well thicknesses are 70, 20, and 60 Å). The subband levels and squared wave functions are shown in the inset. The solid curve is recorded at $T = 10$ K, the dashed curve at $T = 60$ K. The relevant transitions are indicated; note the sharpness of the 1–3 transition (from Faist *et al.*, 1994b).

GaAs–AlGaAs structure of Fig. 14 the energy difference $E_2 - E_1$ is so small that the second subband can be populated with increasing temperature, allowing for the observation of the transitions $2 \rightarrow 3$ and $2 \rightarrow 4$.

Finally, we briefly discuss the "oldest" two-dimensional system, the inversion or accumulation layer in a heterostructure or MOSFET. Such a structure is also asymmetric due to the built-in electric field, and intersubband transitions from the ground state to several excited states were observed long before the investigation of quantum wells began. As examples, note the work of Heitmann and Mackens (1986) on Si inversion layers and of Batke *et al.* (1989) and Batke (1991) on GaAs inversion layers; the latter article contains many references to previous investigations. A sketch of a modulation doped GaAs–AlGaAs heterostructure is depicted in Fig. 15 and the intersubband absorption spectrum for such a structure is shown in Fig. 16 for different electron densities. Transitions from the ground state (labeled 0 in Fig. 16) to three excited states can be observed.

To summarize, by introducing several symmetry-breaking elements into a QW structure, the subband energy levels and the resulting absorption properties can be specifically tailored with great design freedom. Such structures can, in some sense, be regarded as artificial, man-made atoms or molecules, although they contain two-dimensional subbands and not really discrete energy levels such as, for instance, quantum dots. But because the joint density of states for intersubband transitions is essentially a δ function (when nonparabolic effects are neglected; see Section VII), quantum wells behave like atoms as far as their intersubband absorption properties are concerned. (One big difference are the short relaxation and dephasing times, as discussed in Section X). This can be exploited for novel, efficient devices like harmonic generators (see Chapter 2 on nonlinear optics in Volume 66) or infrared lasers (Chapter 1 of Volume 66).

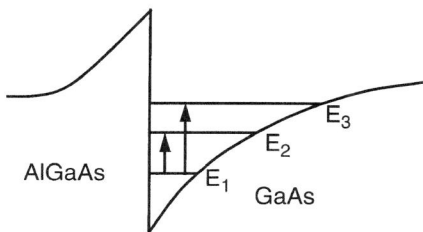

FIG. 15. Conduction-band edge of a modulation doped GaAs–AlGaAs heterostructure. An inversion layer with a two-dimensional electron gas is formed at the interface. Three subbands and possible transitions from the ground state are indicated.

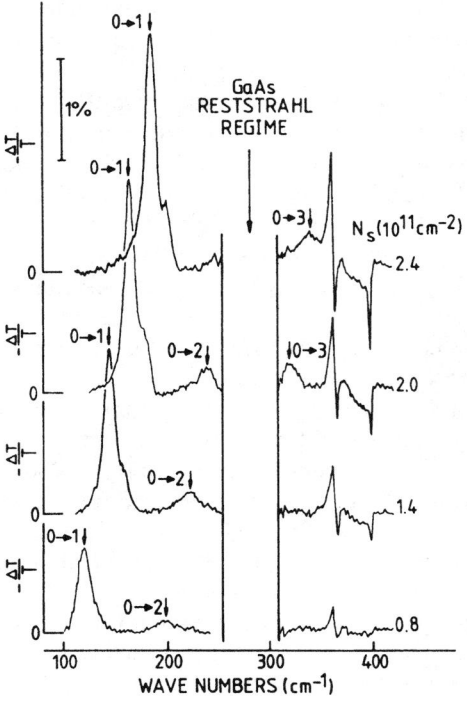

FIG. 16. Intersubband absorption in a gated GaAs–AlGaAs heterostructure. Plotted is the transmission change (relative to the depleted sample) for different electron concentrations as indicated. Transitions to three excited states are observed. (Note that the level numbering starts with 0 in the figure.) In the reststrahlen regime the sample is opaque; the sharp features above 350 cm^{-1} are related to AlAs photons (from Batke et al., 1989).

VI. Multiquantum Wells and Superlattices

From a theoretical point of view, there are some fundamental differences between an isolated system of one or several quantum wells and a periodic sequence of an infinite number of quantum wells (with periodic boundary conditions). In the first case, the energy spectrum always consists of a set of bound states (subbands) and a continuum, where the bound states become more and more closely spaced, when the number of quantum wells becomes larger. For a large number of identical quantum wells, the system is better described by periodic boundary conditions. In this case, one obtains energy bands $E_n(k_z)$, where k_z is the wavevector component parallel to the growth direction, which is now a good quantum number. This description is especially adequate if the barriers are so thin that strong coupling between different QWs occurs. Such a system is normally called a superlattice and the resulting energy bands are called minibands. There is no essential difference between minibands below and above the top of the barriers. When the barriers are thick (multiquantum well system), the minibands below the top of the barriers are extremely narrow and we have degenerate states localized in each quantum well. Above the top of the barriers we still obtain minibands, as in the case of strong-coupling superlattices, but they are very dense, forming a quasi-continuum. The energy spectrum of a MQW system and a superlattice are schematicaly shown in Fig. 17. For a review article on superlattices, see Helm (1995).

A superlattice with, ideally, an infinite number of periods, can be described by the Kronig–Penney model of a one-dimensional lattice with periodic boundary conditions. The z-dependent part of the envelope wave function can then be written as the product of a Bloch part, which is periodic in the superlattice period, and a slowly varying plane-wave part

$$\varphi_n(z) = e^{ik_z z} u_n(z) \tag{41}$$

where $u_n(z) = u_n(z + d)$ and d is the superlattice period. Note that $u_n(z)$ is different from the $u_v(\mathbf{r})$ in Eq. (1), but there should be no confusion, since

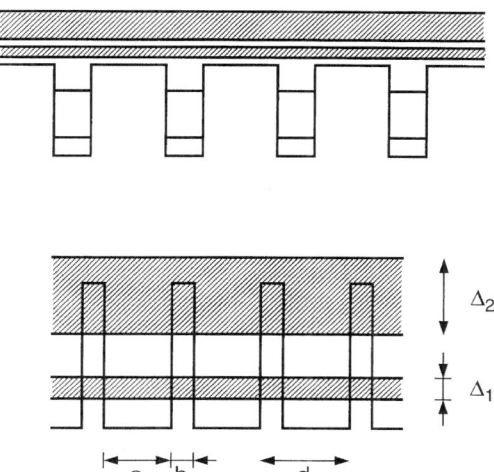

FIG. 17. Schematic conduction band profile and minibands of a multiquantum well structure (top) and a strongly coupled superlattice (bottom).

the two functions never appear in the same equation in this article. The resulting transcendental equation which yields the dispersion $E_n(k_z)$ reads

$$\cos(k_z d) = \cos(\alpha a)\cos(\beta b) - \frac{1}{2}\left(\xi + \frac{1}{\xi}\right)\sin(\alpha a)\sin(\beta b) \qquad (42)$$

where

$$\xi = \frac{m_w^* \beta}{m_b^* \alpha} \quad \alpha = \sqrt{2m_w^* E/\hbar^2} \quad \text{and} \quad \beta = \sqrt{2m_b^*(E-V)/\hbar^2}$$

where a and b are the superlattice well and barrier widths, and m_w^* and m_b^* are the well and barrier effective masses, respectively. In Fig. 18, the first three minibands of a GaAs–Al$_{0.3}$Ga$_{0.7}$As superlattice with $a = 75$ Å and $b = 25$ Å are shown. (The 1s and 2p_z impurity levels, also shown in Fig. 18, are discussed in Subsection 3 of Section XI.) When the widths of the minibands are smaller than the gaps between them (which are, of course, not real gaps, since they contain a constant, two-dimensional density of

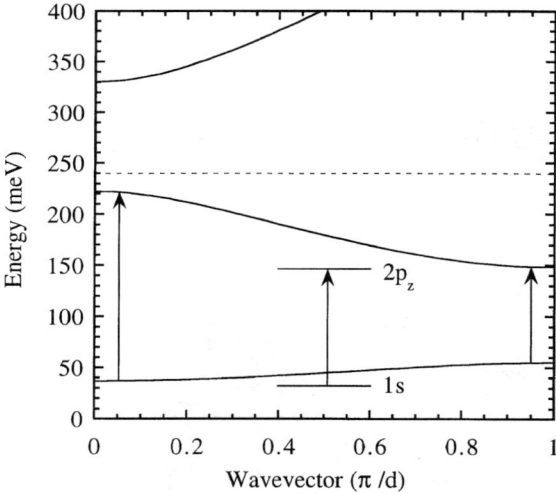

FIG. 18. Calculated miniband dispersions of the three lowest minibands of a GaAs–Al$_{0.3}$Ga$_{0.7}$As superlattice with $a = 75$ Å and $b = 25$ Å. The horizontal dashed line indicates the top of the barriers. The 1s and 2p_z impurity states are also included schematically (for a discussion, see Subsection 3 in Section XI). The interminiband transitions at the center and the edge of the mini-Brillouin zone are indicated as well as the impurity transition.

states), the miniband dispersions can be approximated by the explicit analytic expression

$$E_n(\mathbf{k}) = \varepsilon_n + \frac{\Delta_n}{2}(1 \pm \cos k_z d) + \frac{\hbar^2(k_x^2 + k_y^2)}{2m^*} \quad (43)$$

where the minus sign holds for odd minibands ($n = 1, 3, \ldots$) and the plus sign for even ones ($n = 2, 4, \ldots$). We no longer distinguish between well and barrier effective mass. Equation (43) corresponds to a tight-binding approximation, taking into account the nearest neighboring wells.

The calculation of the absorption coefficient α proceeds in a manner similar to that in Section II for quantum wells, with the difference that we are now dealing with a three-dimensional problem. Therefore α is obtained through integration over k_x, k_y, and k_z, leading to (Helm et al., 1993)

$$\alpha = \frac{e^2 kT}{\varepsilon_0 c \eta \hbar^2 \pi m^* \omega} \int_0^{\pi/d} dk_z |\langle 1|p_z|2\rangle|^2 \ln\left[\frac{1+\exp([E_F - E_1(k_z)]/kT)}{1+\exp([E_F - E_2(k_z)]/kT)}\right]$$

$$\times \left(\frac{\Gamma/\pi}{(E_2(k_z) - E_1(k_z) - \hbar\omega)^2 + \Gamma^2}\right) \quad (44)$$

The $k_x k_y$ integration has already been performed, assuming parabolic bands. The integral over k_z must be evaluated numerically, using the miniband dispersions $E_i(k_z)$ in the Lorentzian. Note that also the matrix elements are k_z-dependent.

Further note that in a periodic system with unbounded wave functions the $\mathbf{A} \cdot \mathbf{p}$ interaction for the electron–photon coupling is preferable over the $e\mathbf{E} \cdot \mathbf{r}$ interaction, since the latter can lead to wrong results (edge terms related to the choice of the unit cell would have to be included; Peeters et al., 1993).

It is now instructive to analyze the different contributions in the preceding formula, namely, the oscillator strength (or the squared matrix element), the Fermi–Dirac thermal occupation factor, and the Lorentzian. Together with the k_z integration the latter is nothing less than the joint density of states (JDOS), which has two singularities at the center ($k_z = 0$) and the edge ($k_z = \pi/d$) of the mini-Brillouin zone with a $1/\omega$ divergence, characteristic of its one-dimensional character. The JDOS for transitions between the two lowest minibands of the above superlattice is shown in Fig. 19 (dotted curve). The singularities are smoothed out by a broadening parameter of $\Gamma = 10$ meV. Note that due to the different curvatures of the bands near $k_z = 0$ and $k_z = \pi/d$, respectively, the shape is not symmetric. (In the tight-binding approximation of (Eq. (43)) the shape can be described

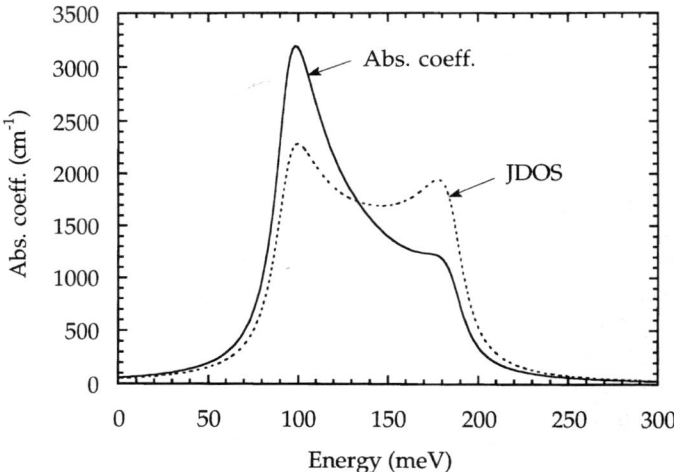

FIG. 19. Calculated JDOS (dotted) and absorption coefficient (solid line) for the above superlattice with $n = 6 \times 10^{17}$ cm^{-3} at $T = 5$ K. The units for the JDOS are arbitrary (from Helm et al., 1993).

analytically and turns out to be symmetric (Helm et al., 1991). The full curve in Fig. 19 reflects the total absorption coefficient according to Eq. (44) using an electron concentration of 6×10^{17} cm^{-3} and a temperature of $T = 5$ K. (At this doping level the Fermi energy lies above the top of the first miniband, i.e., the first miniband is "full.") Now the asymmetry is further enhanced so that the low-frequency peak, resulting from transitions at the edge of the mini-Brillouin zone, becomes much stronger than the high-frequency peak (corresponding to transitions near $k_z = 0$), which is merely visible as a shoulder. (Note that naive use of the $e\mathbf{E} \cdot \mathbf{r}$ interaction leads just to a reversed asymmetry; Kim et al., 1990). The reason for this is the variation of the oscillator strength, $f_{12}(k_z)$, across the Brillouin zone (Helm, 1995) illustrated in Fig. 20 for GaAs–Al$_{0.3}$Ga$_{0.7}$As superlattices with 75-Å well width and barrier widths of $b = 15$, 25, and 40 Å, corresponding to miniband widths Δ_1 of 36, 18, and 7 meV, respectively. For comparison, the oscillator strength f_{12} of an infinite single QW is also shown. Evidently, for wider minibands, more and more oscillator strength is concentrated near the zone edge. In the opposite limit of very narrow minibands, the oscillator strength is independent of k_z and of the order unity. Note that the average of $f_{12}(k_z)$ across the minizone remains of the order unity even for wide bands. This can be understood with the help of the extension of the oscillator sum rule (Eq. (16)) for energy bands (and not only discrete levels),

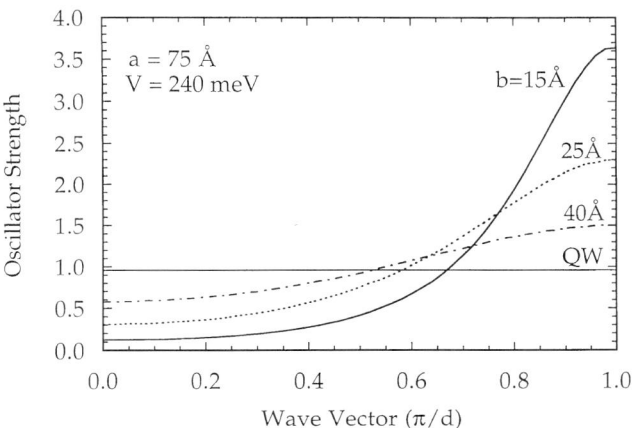

FIG. 20. Oscillator strength f_{12} as a function of k_z for superlattices with the same well width (75 Å), but different barrier widths b, as indicated. The horizontal line represents the case of an infinite QW (from Helm, 1995).

which reads

$$\sum_j f_{ij}(k_z) + \frac{m^*}{\hbar^2}\frac{\partial^2 E_i(k_z)}{\partial k_z^2} = 1 \quad \text{or} \quad \sum_j f_{ij}(k_z) = 1 - \frac{m^*}{m_{SL}^{(i)}} \quad (45)$$

Here $m_{SL}^{(i)}$ is the effective mass along the z direction, which is related to the curvature of the miniband at a certain point along k_z. In the preceding equation, the first term describes transitions between different minibands (interminiband transitions), whereas the second term corresponds to free-carrier type of transitions within one miniband (Helm, 1995). From Eq. (45) it is clear that a large curvature (or a small effective miniband mass along the z direction) will strongly influence the values of $f_{ij}(k_z)$. Since the miniband curvature is positive near $k_z = 0$, the oscillator strength must be reduced there, whereas it must be enhanced due to the negative curvature near $k_z = \pi/d$. Thus, near $k_z = 0$, the sum rule is completed through free-carrier absorption, whereas near $k_z = \pi/d$ it is accomplished through free-carrier emission processes (Helm, 1995).

Experiments on intersubband absorption in strongly coupled superlattices (or interminiband absorption) are rather scarce. After some initial reports in connection with infrared detectors (Byungsung et al., 1990; Gunapala et al., 1991), Helm et al. (1991, 1993) presented absorption spectra clearly showing the predicted asymmetric line shape resulting from the singularities in the JDOS. As can be seen in Fig. 21, when compared to the calculated spectrum (Fig. 19) the miniband absorption is well described by the above theory. In this superlattice ($n = 6 \times 10^{17}$ cm^{-3}) the Fermi energy at low temperatures lies above the top edge of the first miniband, and thus the spectra are basically temperature independent (apart from some additional line broadening at high temperature). The situation is different for a lower doping level ($n = 6 \times 10^{16}$ cm^{-3}), when the Fermi energy lies approximately in the middle of the first miniband (Fig. 22). In this case the top edge of the first miniband can be populated with electrons by increasing the temperature, and thus the absorption spectrum becomes strongly temperature dependent. The additional line appearing at low temperature at $\hbar\omega = 125$ meV is due to the 1s–$2p_z$ donor transition, as discussed in more detail in Section XI, Subsection 3.

At the end of this section, we mention another experiment in which Streibl et al. (1996) observed the interminiband absorption in a finite superlattice, which was embedded in a modulation doped, wide parabolic quantum well (see Section XI, Subsection 2). In this way, the donor impurities are spatially separated from the entire superlattice (Jo et al., 1990), leading to a higher electron mobility. The authors observed an indication of a line narrowing due to the collective effects (see Section VII), which was predicted by Zaluzny (1992b).

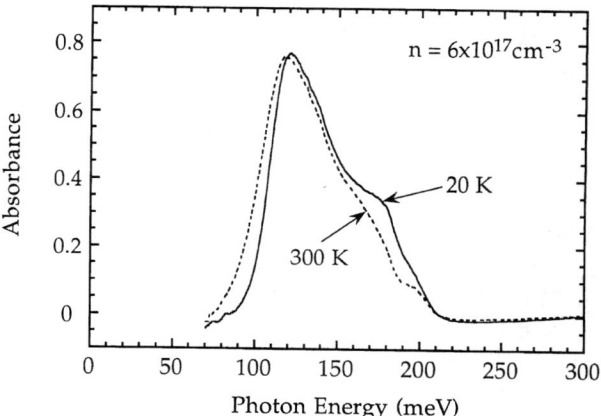

FIG. 21. Measured absorption spectrum for a GaAs–Al$_{0.3}$Ga$_{0.7}$As superlattice ($a = 75$ Å, $b = 25$ Å) with $n = 6 \times 10^{17}$ cm^{-3} at $T = 20$ and 300 K. This should be compared with the theoretical curve in Fig. 19 (from Helm, 1995).

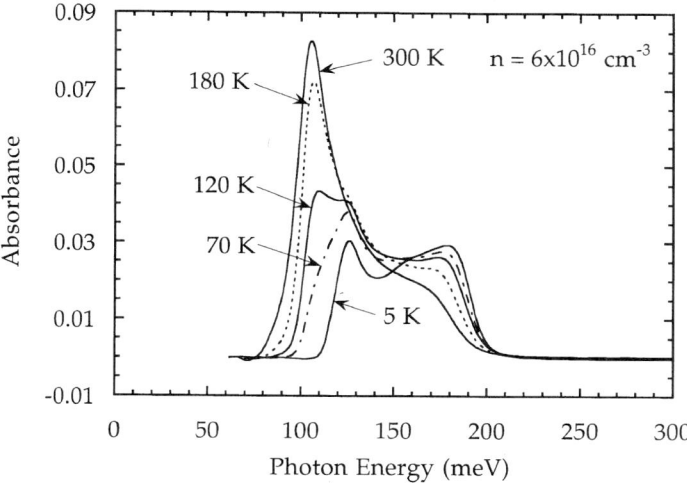

FIG. 22. Measured absorption spectrum for a GaAs–Al$_{0.3}$Ga$_{0.7}$As superlattice ($a = 75$ Å, $b = 25$ Å) with $n = 6 \times 10^{16}$ cm^{-3} for different temperatures as indicated. For an explanation, see text (from Helm et al., 1993).

VII. Nonparabolicity and Many-Body Effects

So far we have discussed the basic energy level structure of quantum wells and superlattices (SLs). In reality, however, it turns out that the exact energy levels and the measured absorption line positions can significantly differ from these simple predictions. There are a number of physical effects that can cause such energy shifts, all of which were studied in detail over the past decade. They can be divided into

1. Band structure effects: nonparabolicity in the z direction (Nelson et al. 1987; Yoo et al., 1989), nonparabolicity in the QW plane (Ekenberg, 1987, 1989), and its effect on the intersubband absorption line shape (Newson and Kurobe, 1988; von Allmen et al., 1988)
2. Effects of Coulomb interaction on the energy levels (Hartree self-consistent potential and exchange-correlation energy)
3. Effects of Coulomb interaction on absorption frequency (depolarization shift, exciton shift) (Ando et al., 1982; Batke, 1991).

In the following, we briefly summarize all these effects, but we see that not all of them can simply be separated.

For many experimental situations (such as GaAs QWs with thicknesses between 50 and 100 Å and sheet electron density of 10^{11} to 10^{12} cm^{-2}), all of the preceding effects contribute not more than around 10% to the absorption peak position, so they often can be neglected as long as only a crude estimate is desired. A detailed understanding, however, requires inclusion of all of them, since they are of the same order of magnitude and partly compensate each other. For wide quantum wells, on the other hand, the Coulomb effects 2 and 3 become very important, since the Coulomb energy becomes of the same order of magnitude as the subband separation.

1. NONPARABOLICITY

The parabolic approximation for the band structure is valid only at low energies, close to the relevant conduction band minimum, which is the Γ point in most cases. At higher energy, coupling to different bands (the valence and higher conduction bands) plays a significant role, especially for narrow-gap materials, which makes the effective mass energy-dependent. The theoretical standard approach for, for example, the electrons in the conduction band of GaAs–AlGaAs quantum wells, is the evaluation of the full 8 × 8 or 14 × 14 **k**·**p** matrix Hamiltonian by applying Kane's theory to heterostructures (Bastard, 1981, 1982, 1988). Simpler **k**·**p** models (such as the Kane–Bastard two-band model), which are desired for less cumbersome calculations, put severe restrictions on the effective masses involved (e.g., masses of electrons and light holes are equal, as are the well and barrier masses at the same energy). Therefore some simple, more semiempirical methods have been proposed, which use a larger number of independent input parameters. Nelson et al. (1987) proposed an "empirical two-band model," which is based on Kane's two-band model, but three independent parameters are taken either from experiment or from a more accurate (14 × 14) **k**·**p** calculation for the bulk material. Within this model the energy dispersion can be written in the following form:

$$\frac{\hbar^2 k^2}{2m^*} = E(1 + E/E'_g) \quad \text{or} \quad E = \frac{\hbar^2 k^2}{2m^*}(1 - \gamma k^2) \qquad (46)$$

A relation of this type is valid for the well and barrier materials, leading to totally four parameters: the band-edge masses m^* in the well and the barrier, and the nonparabolicity parameter γ in both materials (or, equivalently, an "effective" energy gap E'_g). The band-edge masses in the well and barrier materials can be taken from experiment and are known with high accuracy in most materials. The value of γ (or E'_g) is chosen to achieve the best agreement with the experiments, or can be obtained from a more extensive $\mathbf{k} \cdot \mathbf{p}$ calculation with the results fitted by Eq. (46). Then for a given set (m_w^*, m_b^*, γ_w), γ_b is uniquely determined through the condition $\gamma_w/\gamma_b = (m_b^*/m_w^*)^2$ (or $m_w^*/m_b^* = E'_w/E'_b$), when the interband matrix elements are assumed to be the same in both materials. Thus, this scheme uses the simplest possible functional form to describe nonparabolicity and cures the remaining inaccuracy by determining the relevant adjustable parameters by some other means. Another way to put it is that the usual parameters in $\mathbf{k} \cdot \mathbf{p}$ theory, the (real) energy gap E_g and the Kane matrix element P are eliminated in favor of m^* and γ (Zaluzny, 1991). For example, the non-parabolicity parameter for GaAs is $\gamma = 4.9 \times 10^{-19}$ m^2 (Nelson et al., 1987). A detailed description of this procedure can be found in Nelson et al. (1987) and Yoo et al. (1989). For further discussions, see Eppenga et al. (1987), Persson and Cohen (1988), Winkler and Rössler (1993), Burt (1992), and Meney et al. (1994). A simple approximate description was also presented by Altschul et al. (1992).

One interesting result of this model concerning subbands in QWs is that the position of the lowest subband is virtually uninfluenced by non-parabolicity, no matter how narrow the QW and how large the confinement energy. This is so because the strong nonparabolicity in the well material (increasing effective mass with energy) is compensated due to the strong wave function penetration into the barriers, where the nonparabolicity acts the opposite way (decreasing mass deeper in the barriers). The largest energy shifts (toward lower energy) are observed for subbands with high quantum numbers, since they are high in energy with respect to the QW band edge, but very close to the barrier band edge. This behavior obviously gives rise to a red shift of the intersubband absorption.

Nonparabolic effects, of course, not only shift the subband edges, but also increase the effective mass for electrons moving in the quantum well planes. Ekenberg (1987, 1989) presented a thorough discussion of the effect of nonparabolicity on the perpendicular mass, m_z^* (which influences the subband energies) and the in-plane mass, $m_\parallel^*$ (which influences the in-plane dispersion), starting from a fourth-order expansion of $E(\mathbf{k})$, which is based on the $14 \times 14 \, \mathbf{k} \cdot \mathbf{p}$ model (Rössler, 1984; Braun and Rössler, 1985)

$$E(\mathbf{k}) = \alpha_0 k_z^4 + \left[\frac{\hbar^2}{2m_1} + (2\alpha_0 + \beta_0)(k_x^2 + k_y^2) \right] k_z^2 + \frac{\hbar^2}{2m_1}(k_x^2 + k_y^2)$$
$$+ (2\alpha_0 + \beta_0) k_x^2 k_y^2 + \alpha_0(k_x^4 + k_y^4) \quad (47)$$

Here the spin splitting is neglected and the k_z terms are collected separately.

Note that α_0 and β_0 are negative. For heterostructures, k_z must be replaced by $-id/dz$ and the QW potential $V(z)$ must be added. He obtains approximate expressions for m_z^* and $m_\parallel^*$, which are

$$m_z^*(E) = m^*(1 + \alpha E) \tag{48}$$

and

$$m_\parallel^*(E) = m^*[1 + (2\alpha + \beta)E] \tag{49}$$

Here m^* is the band-edge effective mass, and α and β are related to the fourth-order coefficient of the $E(\mathbf{k})$ expansion ($\alpha = -(2m_1/\hbar^2)^2\alpha_0$, $\beta = -(2m_1/\hbar^2)^2\beta_0$, given by $\alpha = 0.64\,\text{eV}^{-1}$ and $\beta = 0.70\,\text{eV}^{-1}$, respectively, for GaAs. Therefore, the nonparabolicity enhancement in the QW planes is about three times larger than along the growth direction. Ekenberg's result also confirm the weak influence on the ground-state binding energy. He further discusses the effect of different boundary conditions for matching the wave functions at the interfaces (Ekenberg, 1989; Burt, 1992).

The influence of nonparabolicity on the intersubband absorption (Newson and Kurobe, 1988) is, from a theoretical point of view, a crucial one, since the integration over k_x and k_y cannot be performed analytically and the JDOS is no longer a δ function. Thus, the absorption coefficient must be calculated through a two (for QWs) or three (for SLs) dimensional integration over k-space using some form of energy dispersion $E(\mathbf{k}_\perp k_z)$, as in Eq. (18) of Section II. The intersubband absorption line shape then acquires some additional, asymmetric broadening on the low-frequency side, which should become relevant at high doping levels, when the Fermi energy is large. Recently it has been shown, however, that this broadening can be entirely compensated by many-body effects (see below in Subsection 4; also Zaluzny, 1991; von Allmen, 1992; Warburton et al., 1996).

2. Self-Consistent Coulomb Potential

To provide the electrons or holes necessary to observe intersubband transitions, quantum wells have to be doped. Since the electron charge distribution, which is determined by the wave functions of all occupied subband levels, will never be identical to the spatial distribution of the donor host ions, the positive and negative contributions do not cancel and there will be a remaining electrostatic potential. Although this effect is relatively weak, when the doping is placed directly in the QWs, it becomes very large when the doping is placed into the barrier material. In this case, the electrons are transferred to the well and are spatially separated from the dopants, leading to a strong modification of the potential. This is called modulation doping and is known to increase the electron mobility and decrease the intersubband absorption line broadening, since impurity scattering becomes suppressed due to the spatial separation of the electrons from the impurities.

This contribution to the potential energy is called Hartree potential V_H and can be calculated via Poisson's equation (with $V_H = -e\Phi$)

$$\frac{\partial^2 \Phi}{\partial z^2} = \frac{e}{\varepsilon \varepsilon_0}[n(z) - N_D(z)] \tag{50}$$

the solution to which is obtained by two integrations, but it can be brought into a form containing only one integration, which is more convenient for numerical evaluation

$$\Phi(z) = \frac{e}{\varepsilon \varepsilon_0} \int_{-\infty}^{z} (z - z')[n(z') - N_D(z')] \, dz' \tag{51}$$

In these equations, $N_D(z)$ is the donor doping profile (ionized donors only) and $n(z)$ is the three-dimensional electron density, given by

$$n(z) = \sum_i n_i |\varphi_i(z)|^2 = \frac{m^* kT}{\pi \hbar^2} \sum_i \ln\left[1 + \exp\left(\frac{E_F - E_i}{kT}\right)\right] \cdot |\varphi_i(z)|^2 \tag{52}$$

for finite temperatures and multiple subband occupancy. If different effective masses m_i^* are used for the different subbands, m_i^* must be taken under the sum. Here n_i is the areal electron concentration in the ith subband and $\varphi_i(z)$ is the respective wave function. The Fermi energy E_F must also be determined self-consistently so that $\int n(z) = \Sigma_i n_i = n_s$, where n_s is the total areal electron concentration. Equations (50) and (52) must be solved self-consistently together with Schrödinger's equation with V_H added to the Hamiltonian. Both are coupled via the appearance of the wave function in Poisson's equation.

In the usual picture, where electron energies are counted positive, negative charge gives rise to negative curvature of the potential profile, while positive charge results in positive curvature. As an example, let us consider an 80-Å-wide GaAs–$Al_{0.3}Ga_{0.7}As$ quantum well. Using the bare confining potential (barrier height 240 meV) leads to two bound states, $E_1 = 42.3$ meV and $E_2 = 153.5$ meV, corresponding to a subband separation of $E_{21} = 111.1$ meV.

Now the self-consistent Hartree potential is taken into account for two doping profiles. Figure 23a shows a modulation doped structure, where the barriers are doped with $N_D = 5 \times 10^{18}$ cm^{-3} over 10 Å, 40 Å away from the quantum well. This yields an areal electron density of $n_s = 1 \times 10^{12}$ cm^{-2} in the QW. The strong band bending reduces the subband separation to $E_{21} = 104.4$ meV. The second structure (Fig. 23b) is doped over the center 50 Å of the well with $N_D = 2 \times 10^{18}$ cm^{-3}, resulting in the same areal electron concentration. The band bending is, however, much weaker, and the subband separation is slightly increased to 112.3 meV with respect to the bare potential.

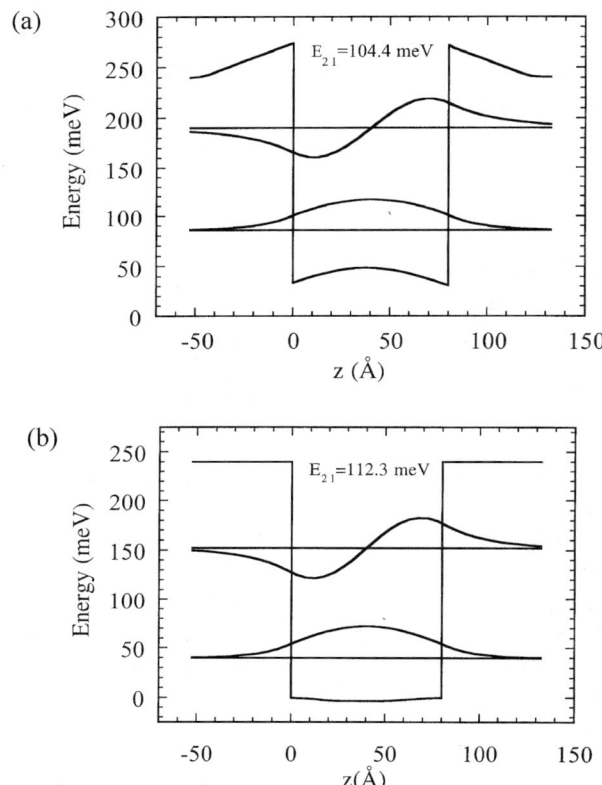

FIG. 23. Conduction-band profile, energy levels, and wave functions of an 80-Å-wide GaAs–Al$_{0.3}$Ga$_{0.7}$As quantum well with an areal electron concentration of $n_s = 1 \times 10^{12}$ cm^{-2}: (a) doping in the barriers (modulation doping) and (b) doping in the well. The energy separation $E_2 - E_1$ is given in the figure.

Some complication arises, when a nonparabolic variation of the effective mass is taken into account, since due to the band bending m^* becomes z-dependent even within in the same material. In this case, a weighted spatial average of the effective mass can be used (Bastard, 1988), defined through

$$\frac{1}{m_i^*} = \int_{-\infty}^{\infty} \frac{|\varphi_i(z)|^2}{m_i^*(z)} dz \qquad (53)$$

3. MANY-BODY EFFECTS ON THE ENERGY (EXCHANGE AND CORRELATION)

At the electron densities usually encountered in semiconductor QWs and heterostructures, many-body effects on the energy of the electron gas also play a nonnegligible role (i.e., the exchange and correlation energies). Two methods for their calculation have been frequently used for the quasi-two-dimensional electron gas in a quantum well: The Hartree–Fock method, which yields an explicit expression for the exchange energy but neglects correlations, and the local-density approximation (LDA) within the Kohn–Sham density functional theory (Kohn and Sham, 1965).

The exchange integral in the Hartree–Fock equation, which represents a nonlocal potential, has been evaluated in an approximate way by Bandara et al. (1988, 1989) for electrons in the ground subband using infinite QW wave functions. This leads to

$$E_{\text{exch}}(k) \cong \frac{-e^2}{4\pi\varepsilon\varepsilon_0} k_F \left[\frac{2}{\pi} \mathbf{E}(k/k_F) - 0.32(k_F/k_L) \right] \qquad (54)$$

for $k < k_F$. Here k_L is an inverse length characteristic of the QW (e.g., $k_L = \pi/L$ for an infinite QW), and $\mathbf{E}(k/k_F)$ is a complete elliptical integral of the second kind. For $k > k_F$, a similar expression results (Bandara et al., 1988, 1989; Manasreh et al., 1991; Szmulowicz et al., 1994); also Zaluzny (1992a) provided a simple expression.

Within the Kohn–Sham density functional theory using the LDA, the combined exchange-correlation potential can be written as a functional of the electron density $V_{xc}(n(z))$ and added as a (now local) potential in the single-particle Schrödinger equation, in a way similar to the Hartree potential $V_H(n(z))$. (Note that for the LDA to hold strictly, the electron density is required to exhibit only slow spatial variations, which is, in fact, not well fulfilled in QWs.) Various forms for $V_{xc}(n(z))$ have been suggested in the literature. The following expression given by Hedin and Lundqvist (1971) has also been used frequently for electrons in quantum wells (Bloss, 1989):

$$V_{xc} = -\left(\frac{9\pi}{4}\right)^{1/3} \frac{2}{\pi r_s}\left[1 + \frac{B}{A}r_s \ln\left(1 + \frac{A}{r_s}\right)\right] \cdot \frac{e^2}{8\pi\varepsilon\varepsilon_0 a^*} \qquad (55)$$

Here r_s is the dimensionless parameter characterizing the electron gas, given by $r_s = [(4\pi/3)a^{*3}n(z)]^{-1/3}$ and corresponding to the mean electron distance normalized to the effective Bohr radius $a^* = (\varepsilon/m^*)a_B$. The constants A and B were quoted as $A = 21$ and $B = 0.7734$ (in a later paper, Gunnarsson and Lundqvist, 1976, give values of $A = 11.4$ and $B = 0.6213$, see also Chuang et al., 1992; Luo et al., 1993); both values probably lie within the uncertainty of the theory with the approximations made). This expression is claimed by its originators to be valid in a rather wide range of r_s.

In contrast to Eq. (54), Eq. (55) does not exhibit any k-dependence and leads to a correction that is two to three times smaller than in the Hartree–Fock approximation (Zaluzny, 1992a), which tends to overestimate the energy correction. For a discussion of the use of the Hartree–Fock equation vs density-functional theory, compare Jogai (1991) and the comment by Szmulowicz and Manasreh (1992).

The main effect of including the exchange (and correlation) energies is a lowering of the total electron energy (or the effective potential) in regions of high electron density. This results in a lowering of the subband levels. Since the ground subband is mostly influenced, this leads to a blue shift of the intersubband absorption. Note that in cases where the doping is placed into the QW (no charge separation), the exchange effect can be larger than the effect of the direct Coulomb term (Hartree potential). Especially in systems with a large effective mass, the exchange energy becomes appreciable when compared to the confinement energy.

Thus finally, using the preceding parameterization of the exchange-correlation energy, the following Schrödinger equation must be solved together with Eqs. (50) and (52):

$$\left[\frac{-\hbar^2}{2m^*}\frac{d^2}{dz^2} + V(z) + V_H(n(z)) + V_{xc}(n(z))\right]\varphi_n(z) = E_n\varphi_n(z) \qquad (56)$$

neglecting all complications that arise from a nonconstant effective mass for simplicity.

As an illustration, the preceding GaAs–Al$_{0.3}$Ga$_{0.7}$As QW (with doping in the well) is shown again in Fig. 24, now including the exchange-

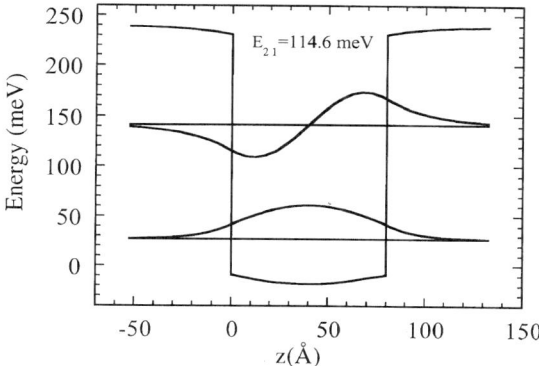

FIG. 24. The GaAs–Al$_{0.3}$Ga$_{0.7}$As quantum well of Fig. 23 doped in the well, but with the exchange-correlation energy added to the potential.

correlation potential according to Eq. (55). The potential lowering in the regions of high electron density is clearly observed, and the subband separation is slightly increased to $E_{21} = 114.6$ meV.

4. Collective Effects on the Absorption (Depolarization and Exciton Shift)

Many-body effects are relevant not only for the energy of the electron gas but also for its electric conductivity (or polarizability), which must be calculated to obtain the absorption spectrum in a many-body theory. The single-particle approach presented in Section II is not sufficient in this case. It has been shown that there are two main terms that cause a shift of the absorption maximum with respect to the bare energy levels; that is, the absorption does not occur at the energy difference $E_{21} = E_2 - E_1$, but rather at an energy given by

$$\tilde{E}_{21}^2 = E_{21}^2(1 + \alpha - \beta) \tag{57}$$

The frequency shifts represented by α and β (both >0) are known as the depolarization shift and the exciton or final-state interaction, respectively, in the many-body calculation of the conductivity or polarizability (Vinter, 1976, 1977; Ando et al., 1982; Batke, 1991, and references therein). Here α and β should not be confused with the nonparabolicity parameters in the preceding section.

The depolarization shift essentially comes from a (time-dependent) Hartree term related to the high-frequency field inducing the absorption. Each electron feels an effective field that is different from the external field by the mean Hartree field of the other electrons polarized by the external field. Thus, the external field is screened by the quasi-two-dimensional electron gas (corresponding to the random phase approximation for the dynamic conductivity). In this spirit, Allen et al. (1976) calculated the depolarization shift using time-dependent perturbation theory by introducing a self-consistent AC potential in Schrödinger's equation and evaluating the induced oscillating charge density and AC current self-consistently from Poisson's equation. In this treatment, we restrict ourselves to a two-level system, which requires that there is significant oscillator strength for transitions to only one excited state. A generalization for several levels was performed by Allen et al. (1976) and Chun et al. (1993).

The frequency shift α (depolarization shift) is given by

$$\alpha = \frac{2e^2 n_s}{\varepsilon \varepsilon_0 E_{21}} S \quad \text{with} \quad S = \int_{-\infty}^{\infty} dz \left[\int_{-\infty}^{z} dz' \varphi_2(z') \varphi_1(z') \right]^2 \quad (58)$$

where S (in the literature often called S_{11}) has the units of length, and α is proportional to the electron density, in fact, the squared frequency shift is of the order of the three-dimensional plasma frequency and the absorption frequency can be written in the form $\tilde{\omega}_{21}^2 = \omega_{21}^2 + f_{12} \omega_p^2$, as was recognized by Chen et al. (1976). Here ω_p is a 3D plasma frequency defined by $\omega_p^2 = (n_s e^2/\varepsilon \varepsilon_0 m^* L_{\text{eff}})$, with $L_{\text{eff}} = (\hbar^2 f_{12}/2m^* S E_{21})$. This is the L_{eff} that must be used in Eq. (40) for consistency. Note that the oscillator strength f_{12} cancels in the final shift. For an infinite QW of thickness L, the preceding can be calculated analytically to give the values $S = (5/9\pi^2) \cdot L = 0.056 \cdot L$ and $L_{\text{eff}} = f_{12} \cdot (3/5) \cdot L$.

The physical content of the depolarization shift can be understood through the collective nature of the intersubband absorption. Exposing the system to external radiation not only excites electrons into the higher subband but also modulates their charge density. The restoring Coulomb force gives rise to a kind of plasma oscillation. The combined intersubband–Coulomb problem can thus be regarded as two coupled oscillators with frequencies ω_{21} and ω_p (Sherwin et al., 1995).

As evident from the preceding formulas, the depolarization shift is important only at high electron densities and/or small energy separations. For quantum wells with E_{21} around 100 meV and $n_s < 10^{12} \text{cm}^{-2}$ it is a minor correction (see example following).

The same is true for the second term, β, which acts to reduce the absorption frequency, and is due to the Coulomb interaction between the excited electron and the quasi-hole left behind in the ground subband. Since this is similar to an excitonic electron–hole pair, the effect is called the exciton correction, or the final-state interaction. It corresponds to a reduction of the mean Hartree field around each electron due to exchange-correlation effects. In a systematic many-body theory, it can be shown that β represents a local field or vertex correction to the depolarization shift (Ando et al., 1982; Bloss, 1989). Ando (1977a, 1977b) calculated both terms, α and β, using the density functional theory in the local density approximation and obtained

$$\beta = -\frac{2n_s}{E_{21}} \int_{-\infty}^{\infty} dz \varphi_2(z)^2 \varphi_1(z)^2 \frac{\partial V_{xc}[n(z)]}{\partial n(z)} \tag{59}$$

where V_{xc} is the exchange-correlation energy of Eq. (55). Since dV_{xc}/dn is negative (V_{xc} becomes more negative, when n is increased), β is a positive number. For GaAs QWs, β is usually much smaller than α. However, in the two-dimensional electron gas in Si accumulation or inversion layers, they can be of the same order of magnitude (resulting mostly from the higher effective mass), so that both effects nearly compensate each other.

Ando (1977a, 1977b, 1978) presented a treatment using the two-dimensional conductivity σ_{zz} and modified (observable) conductivity $\tilde{\sigma}_{zz}$ that reflect the response to the total electric field and external electric field, respectively. The induced current can be written as

$$j_z = \sigma_{zz} E \tag{60}$$

The external field E_{ext} is related to the total field through

$$E_{ext} = \varepsilon_{zz} E \tag{61}$$

where ε_{zz} is the dielectric function,

$$\varepsilon_{zz} = 1 + i \frac{\sigma_{zz}}{\varepsilon_0 \varepsilon_{st} \omega L_{eff}} \tag{62}$$

and ε_{st} the static background dielectric constant. The current can be expressed as

$$j_z = \tilde{\sigma}_{zz} E_{ext} \tag{63}$$

with

$$\tilde{\sigma}_{zz} = \frac{\sigma_{zz}}{\varepsilon_{zz}} = \frac{\sigma_{zz}}{1 + i(\sigma_{zz}/\varepsilon_0 \varepsilon_{st} \omega L_{eff})} \qquad (64)$$

As a consequence, the absorption of the 2D layer occurs not at the poles, but at the zeroes of the dielectric function. If σ_{zz} is the conductivity of a two-dimensional electron gas with parabolic energy dispersion (i.e., E_{21} is independent of $\mathbf{k}_\perp$), the resulting $\tilde{\sigma}_{zz}$ has the same Drude-form as σ_{zz} (see Eq. (27)), however, with a renormalized resonance frequency $\tilde{\omega}_{21}$:

$$\tilde{\sigma}_{zz} = \frac{n_s e^2 f_{12}}{m^*} \frac{-i\omega}{\tilde{\omega}_{21}^2 - \omega^2 - 2i\gamma\omega} \qquad (65)$$

The absorption A (if $A \ll 1$) is then again given by $A = (\text{Re}\tilde{\sigma}_{zz}/\varepsilon_0 c \eta)$ (compare Eq. (26)).

To determine some numerical values let us go back to the earlier well-doped, 80-Å-wide GaAs–Al$_{0.3}$Ga$_{0.7}$As QW. For this structure, we obtain $\alpha = 0.16$ and $\beta = 0.058$, which corresponds to a depolarization shift of 8.8 meV and exciton correction of -3.4 meV, if each is considered alone, and in a combined shift (according to Eq. (57)) of 5.7 meV. We can calculate now the position of the resonance absorption $\tilde{E}_{21}$ when the various correction terms are added successively:

	Resonance position
bare potential	111.2 meV
+Hartree	112.3 meV
+exchange correlation	114.6 meV
+depolarization	123.4 meV
+exciton	120.3 meV

The total shift is less than 10% and is dominated by the depolarization effect. It is clear, however, that for wide QWs with a bare subband separation of less than 20 meV the collective energies can become as large as E_{21}, and an intersubband transition cannot be described in a single-particle picture anymore.

For systems with nonparabolic energy dispersions (which require integration over k_x and k_y) the conductivity takes the more general form (Ando,

1978; Zaluzny, 1991, 1992a; Warburton et al., 1996)

$$\tilde{\sigma}_{zz}(\omega) = \frac{n_s e^2 f_{12}}{m^*} \frac{-i\omega}{\omega_{21}^2(0)} \frac{G(\omega)}{1 + (\alpha - \beta)G(\omega)} \quad (66)$$

with

$$G(\omega) = \frac{1}{n_s} \frac{2}{(2\pi)^2} \int d^2 k_\perp \frac{\omega_{21}^2(0)}{\omega_{21}^2(\mathbf{k}_\perp) - \omega^2 - 2i\omega\gamma} \quad (67)$$

The integration goes over all occupied states up to the Fermi wavevector. The change of the oscillator strength with $\mathbf{k}_\perp$ has been neglected. For superlattices, one proceeds similarly (Zaluzny, 1992b), except that the integration is over k_z. At finite temperature, $G(\omega)$ would contain the difference between the initial and final state distribution functions in addition.

Zaluzny (1991, 1992a, b) noted that under such conditions, the depolarization effect not only shifts the absorption resonance but can also induce a significant modification of the line shape. For a highly doped QW with large nonparabolicity, one expects a broad absorption line due to the $\mathbf{k}_\perp$ dependence of the transition energy $E_{21}(\mathbf{k}_\perp)$. When the depolarization effect is included, however, it turns out that the absorption spectrum consists of a narrow line near the high-energy side of $E_{21}(\mathbf{k}_\perp)$; thus nonparabolicity is compensated by the depolarization field (Zaluzny, 1991). This effect was also confirmed through experiments on InAs–AlSb QWs (Warburton et al., 1996). Figure 25 shows the calculated absorption for a 150-Å-wide InAs–AlSb quantum well with an electron concentration of $n_s = 1 \times 10^{12}$ cm^{-2}. The small effective mass of InAs results in a large Fermi energy (over 50 meV) and a high nonparabolicity. Note that E_{21} varies by nearly 20 meV between $k_\perp = 0$ and $k_\perp = k_F$, as can be seen from the dotted curves, where the depolarization effect is neglected. When it is included (full curves), the absorption is shifted above the single-particle excitations (which are still visible as a low-energy shoulder) and its width is determined by the Lorentzian broadening parameter γ (in Fig. 26, the HWHMs are $\hbar\gamma = 0.5$ and 2.5 meV, respectively) and not by the variation of $E_{21}(\mathbf{k}_\perp)$. The corresponding experiment (Fig. 26) clearly confirms the validity of the collective-excitation picture: the absorption consists of one narrow, nearly Lorentzian peak with no obvious sign of nonparabolicity broadening.

According to Zaluzny (1992a), the nonparabolicity due to the k-dependence of the exchange energy (Eq. (54)) is also eliminated in the same way. A similar mechanism is predicted to narrow interminiband absorption in

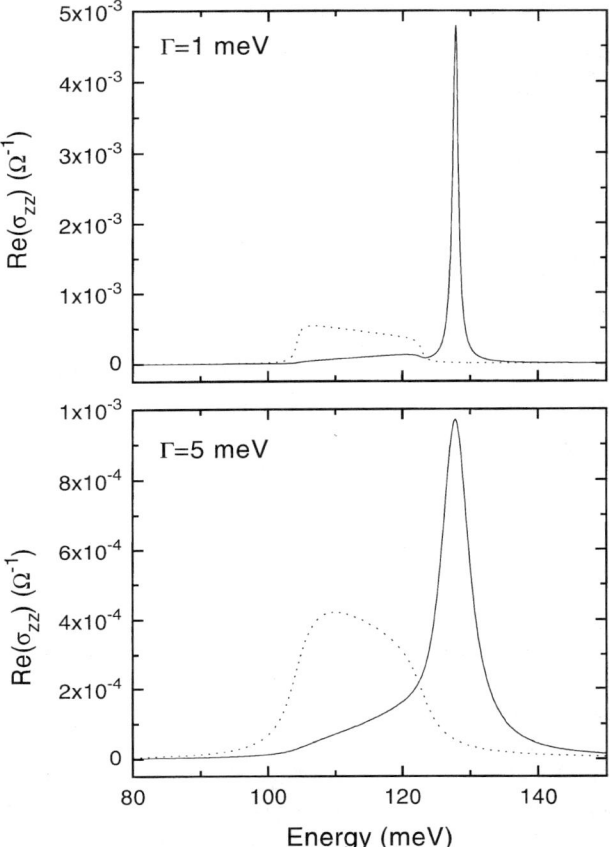

FIG. 25. Calculated real part of the conductivity (proportional to the absorption) of a 150-Å-wide InAs–AlSb quantum well. For the dotted curves, the depolarization effect was neglected; in the solid curves it is included. Spectra for two values of the broadening parameter Γ are shown. Note that here Γ is the FWHM, whereas in the main text we have used Γ (or γ) for the HWHM (from Warburton *et al.*, 1996).

strongly coupled superlattices (Zaluzny, 1992b). Nikonov *et al.* (1997) calculated the intersubband absorption in the framework of the semiconductor Bloch equations (Haug and Koch, 1993) in the Hartree–Fock approximation. In particular, they study the situation of two different effective masses in the first and second subbands, and find a similar line narrowing induced by the Coulomb and exchange interactions.

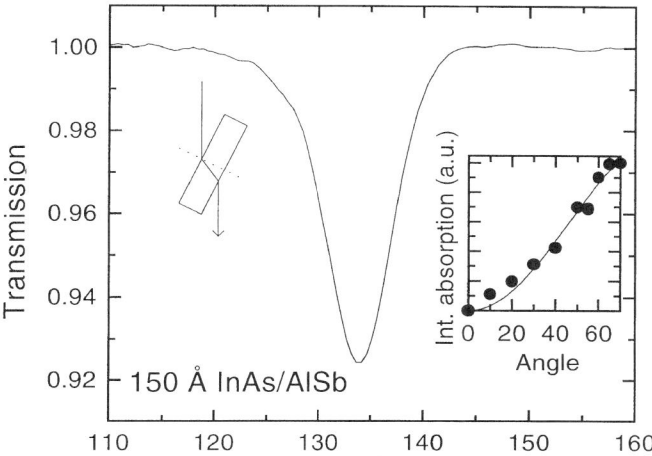

FIG. 26. Measured transmission spectrum of a 150-Å-wide InAs–AlSb quantum well. The experimental geometry is schematically shown as well as the angle (θ) dependence of the integrated absorption (from Warburton et al., 1996).

A systematic treatment of many-body effects in intersubband transitions was presented by Chuang et al. (1992) and Luo et al. (1993) on the basis of the Bethe–Salpeter equation, and also by Huang et al. (1995). Other related approaches are due to Jiang (1992) and Boykin and Chui (1997).

It is interesting to note that in Raman scattering experiments (for a review see Pinczuk and Abstreiter, 1989), both the bare and dressed intersubband resonance can be measured (Pinczuk and Worlock, 1982; Pinczuk et al., 1989; Ramsteiner et al., 1990; Chuang et al., 1992; Luo et al., 1993). When the pump and scattered light have crossed polarizations, one measures the spin-density excitation (SDE), which is (nearly) coincident with the single-particle subband separation, whereas in crossed polarizations the charge-density excitation (CDE) is measured, which is identical to the depolarization shifted intersubband resonance. Intersubband excitations with a finite wavevector q have also been discussed (Heitmann et al., 1982; Heitmann and Mackens, 1986; Yi and Quinn, 1983; Batke et al., 1991).

The conclusion of this section is that a proper description of intersubband absorption has to proceed via a full quantum mechanical calculation of the conductivity or polarizability including band-structure and many-body effects on the same footing, and even treat the electromagnetics dictated by the sample geometry as outlined in Section IV. In other words, strictly the collective modes of the interacting system in response to the total (not external) field have to be evaluated, taking into the correct boundary conditions and mode propagation. Still, in many realistic cases, the approach of Section II, which is based on the single-particle calculation of the absorption coefficient, can lead to satisfactory results within 10% accuracy.

VIII. Mechanisms for In-Plane Absorption

We have seen in Section II that the main intersubband absorption occurs for an electromagnetic wave polarized along the growth direction (i.e., perpendicular to the layer planes). This selection rule holds very well for many cases (such as, e.g., a typical n-type GaAs QW), but there are several mechanisms that can lead to relaxing this rule. In fact, theorists have actually searched for such mechanisms, since it would be useful for applications in infrared detectors (see Chapters 3 and 4 of this volume) to obtain absorption of in-plane polarized light, which can then be utilized in a normal-incidence geometry.

Strong normal-incidence intersubband absorption has been predicted and observed for holes in valence-band quantum wells, which is due to the mixing of the various hole bands. This is discussed in detail in the next section. In this section, we focus on mechanisms that enable in-plane polarized absorption in the simpler band structure of the conduction band.

1. INDIRECT-GAP SEMICONDUCTORS

In semiconductors with an indirect bandgap, the conduction-band minimum is located away from the Brillouin-zone center either near the X point (in the (001) direction) or near the L point (in the (111) direction). The constant-energy surfaces are then ellipsoids characterized by a longitudinal and a transverse effective mass. If the principal axis of such an ellipsoid is tilted with respect to the QW growth and confinement direction, the effective-mass tensor provides coupling between the parallel and perpendicular motion of the electrons. As a consequence, normally incident light (polarized in the plane of the layers) light leads to an electronic polarization component perpendicular to the layers and thus to intersubband transitions (Yang et al., 1989; Brown and Eglash, 1990; Xu et al., 1993, 1994). This concept was recognized many years ago in connection with Si-MOSFETs on (110) and (111) Si substrates (Yi and Quinn, 1983; Cole and McCombe, 1984). Interest has revived in this subject due to the prospect of realizing normal-incidence infrared detectors based on QWs.

The main candidates for this effect are of course Si and Ge, the former with the conduction-band minimum along the (001) direction (Δ) near the X point, the latter in the (111) direction at the L point. In addition, some III–V semiconductors are indirect as well: For an Al content $x > 0.45$ $Al_xGa_{1-x}As$ becomes indirect with the minimum at the X point, and $Al_xGa_{1-x}Sb$ is indirect for $0.25 > x > 0.55$ with the minimum at the L point (Xie et al., 1991a), among others.

For electrons in an elliptical valley of a semiconductor heterostructure the Hamiltonian in the effective-mass approximation has the form

$$H = \tfrac{1}{2}\mathbf{p} \cdot \bar{\mathbf{w}} \cdot \mathbf{p} + V(z) \tag{68}$$

where $\bar{\mathbf{w}}$ is the 3×3 inverse effective mass tensor (Stern and Howard, 1967). The perturbation Hamiltonian for the electron–photon coupling is given by

$$H' = -\frac{e}{2}(\mathbf{A} \cdot \bar{\mathbf{w}} \cdot \mathbf{p} + \mathbf{p} \cdot \bar{\mathbf{w}} \cdot \mathbf{A}) \tag{69}$$

The absorption coefficient is now calculated in the same way as in Section II, by using the transition matrix elements $H'_{ij} = \langle \psi_i | H' | \psi_j \rangle$ and the dipole approximation. Since intersubband absorption is induced by the z component of the electronic polarization, only terms that contain a p_z component give a finite contributions. (Mathematically, the operators p_x and p_y have no net effect on the wave function and the matrix elements vanish due to the orthogonality of the wave functions.) Making also use of the symmetry of the inverse effective mass tensor, $w_{mn} = w_{nm}$, we obtain the transition rate

$$W_{if} = \frac{2\pi}{\hbar}e^2(A_x w_{xz} + A_y w_{yz} + A_z w_{zz})^2 |\langle i|p_z|f\rangle|^2 \delta(E_f - E_i - \hbar\omega) \tag{70}$$

where A_n are the Cartesian components of the amplitude of the vector potential. The absorption coefficient follows analogous to that in Section II. Xu et al. (1993, 1994) showed that with a proper coordinate transformation very general expressions for different material systems, substrate orientations, and polarizations can be obtained.

In the late 1970s it was predicted that the two-dimensional electron gas (2DEG) in Si-MOSFETs on (110) and (111) Si substrates should exhibit intersubband absorption for in-plane polarized radiation. Both theoretical (Ando et al., 1977; Yi and Quinn, 1983) and extensive experimental (Cole and McCombe, 1984; Nee et al., 1984; Wieck et al., 1984) investigations have been carried out. One important result was the absence of the depolarization shift for parallel excitation (Nee et al., 1984). Chun and Wang (1992) (also Wang and Karunasiri, 1993; Wang et al., 1994a) extensively discussed the intersubband absorption in n-type strained Si/SiGe quantum wells for various growth directions. Here, conduction-band QWs occur in the Si layers, when grown on a relaxed SiGe buffer. Experimental work include a study of n-type δ-doped Si/SiGe QWs on Si (110) (Lee and Wang, 1992) and Ge/SiGe QWs on Si (001) (Lee and Wang, 1994).

AlAs–$Al_xGa_{1-x}As$ X-valley QWs were investigated theoretically by Xie et al. (1992a, 1992b) and experimentally by Katz et al. (1992) for several substrate orientations other than (001). Here, an AlAs QW is embedded between AlGaAs barriers with $x \cong 0.4$. Wang et al. (1993) and Zhang et al. (1994) have fabricated a normal-incidence infrared detector using this scheme, the latter, however, grown on a Si substrate.

L-valley QWs and SLs based on the $Al_xGa_{1-x}Sb$–AlSb system were discussed by Xie et al. (1991a) and by Shaw and Jaros (1994). The latter work includes a microscopic calculation of the band structure and the linear and nonlinear susceptibilities. The normal-incidence absorption of an $Al_{0.09}Ga_{0.91}Sb$ QWs was experimentally demonstrated by Brown et al. (1992). Abramovich et al. (1994) studied the photoinduced intersubband absorption in GaSb–AlSb superlattices. Zhang et al. (1993) reported normal-incidence absorption and photoresponse in GaSb QWs, where the ground state is an L-type subband due to the confinement and the larger effective mass at the L point as compared to the Γ point. Normal-incidence electrooptic modulators were also proposed on this basis (Xie et al., 1993, 1994).

Finally, another approach worth mentioning are QWs oriented vertically with respect to the wafer plane. Berger et al. (1995) fabricated such vertically oriented AlGaAs QWs (with low Al content) through growth on a patterned substrate and reported normal-incidence intersubband absorption of Γ point electrons. This is just the usual intersubband absorption, which is made possible here through the vertical orientation of the wells.

2. SPATIAL VARIATION OF THE EFFECTIVE MASS

Yang (1995a, 1995b) has proposed a mechanism for in-plane polarized intersubband absorption, which applies even at the Γ point in a spherical conduction band, and relies on the spatial variation of the effective mass (i.e., m^* takes a different value in each of the material layers). When the variation of the effective mass along the z direction is taken into account, the proper form of the electron-photon Hamiltonian is

$$H' = \frac{e\mathbf{A}}{2} \cdot \left(\frac{1}{m^*(z)} \mathbf{p} + \mathbf{p} \frac{1}{m^*(z)} \right) \tag{71}$$

When the polarization of the perturbing electromagnetic field is taken in the QW plane, for instance, in the x direction, the induced transition rate according to Fermi's golden rule is then given by

$$W_{if} = \frac{2\pi}{\hbar} e^2 A_0^2 \hbar^2 k_x^2 \left| \left\langle i \left| \frac{1}{m^*(z)} \right| f \right\rangle \right|^2 \delta(E_f - E_i - \hbar\omega) \tag{72}$$

If the effective mass were constant throughout the quantum well system, it is clear that this rate would identically vanish due to the orthogonality of the wave functions. A z-dependent effective mass, however, leads to a finite value of the matrix elements and thus, of the absorption. For a symmetric quantum well, the matrix element is finite only when initial and final state have the same parity, which implies the selection rule $\Delta n = n' - n = 2, 4, \ldots$, in contrast to usual intersubband transitions. Candidates for observation of this effect are, for instance, step quantum wells, but so far no clear observation has been reported. Note also that in a multiband model, this effect actually tends to come out weaker, since the effective masses are energy dependent and become more alike in different materials at the same energy.

3. Coupling to the Valence Band

There is another mechanism that can give rise to in-plane polarized intersubband absorption of electrons at the Γ point. If one goes beyond the effective mass approximation (i.e., taking into account nonparabolicity), then not only the energy dispersions but also the electronic wave functions are modified. That is, they are not simple products of the conduction-band Bloch functions and an envelope function, but due to the band mixing in $\mathbf{k} \cdot \mathbf{p}$ theory, they acquire a contribution from the valence band. The wave functions are then linear combinations of electron as well as light, heavy, and split-off hole Bloch functions. The complete optical matrix elements within the 8×8 $\mathbf{k} \cdot \mathbf{p}$ model can thus be written (Yang et al., 1994) (cf. Eq. (11))

$$\langle \psi_n | \mathbf{e} \cdot \mathbf{p} | \psi_{n'} \rangle = \sum_{j,j'=1}^{8} [\langle e^{i\mathbf{k}_\perp \cdot \mathbf{r}} \varphi_{nj} | \mathbf{e} \cdot \mathbf{p} | e^{i\mathbf{k}_\perp \cdot \mathbf{r}} \varphi_{n'j'} \rangle \langle u_j | u_{j'} \rangle + \langle \varphi_{nj} | \varphi_{n'j'} \rangle \langle u_j | \mathbf{e} \cdot \mathbf{p} | u_{j'} \rangle] \tag{73}$$

where φ_{nj} is the z-dependent part of the envelope function (n is the subband index) and u_j is the rapidly varying, periodic Bloch function of the jth band (with $\langle u_j | u_{j'} \rangle = \delta_{jj'}$). The first term is the envelope-function matrix element, which is responsible for the "usual," z-polarized intersubband absorption. The second term, which reflects the admixture from the other bands, can give rise to transitions for both orthogonal polarizations, and also transitions with and without a change of parity become allowed in principle. This issue was discussed in detail by Yang et al. (1994), Lew Yan Voon et al. (1995), Warburton et al. (1996), and Flatté et al. (1996). In most common quantum wells, the by far strongest transition is still the one with z polarization and parity change (Δn odd; e.g., the 1–2 transition). The matrix element for xy polarization with odd Δn reaches at maximum a few percent of this value, the ratio between the two can be approximated as (Yang et al., 1994)

$$R_{xy/z} \approx \frac{1}{3} \frac{(E_{n'} - E_n)\Delta}{(E_n + E_g + \Delta)(E_{n'} + E_g)} \tag{74}$$

where E_n and $E_{n'}$ are the subband energies, E_g is the band gap, and Δ is the spin–orbit splitting of the valence band. Evidently the spin–orbit interaction is crucial here. The transitions with even Δn are still weaker (Warburton et al., 1996) for both polarizations (of course not for asymmetric structures; there z-polarized transitions with Δn even can be very strong, cf. Section V). These effects should be (relatively) strongest in narrow-gap semiconductors such as InAs (small E_g, large Δ), where the band mixing is largest and also large k values are usually involved due to the small effective mass (Warburton et al., 1996), but most likely still too small to be observed. Note that in the late 1970s and early 1980s in-plane polarized absorption was observed in surface layers of the narrow-gap semiconductors InSb (Beinvogl and Koch, 1977; Wiesinger et al., 1982) and InAs (Reisinger and Koch, 1981), and corresponding calculations were performed by Zawadzki (1983), however on the basis of a 4×4 $\mathbf{k \cdot p}$ model completely neglecting spin. In view of today's knowledge (Yang et al., 1994; Warburton et al., 1996), these calculations as well as that of Shik (1988, 1992) appear to overestimate the strength of the normal-incidence absorption. To date, only a few reports exist on the observation of in-plane polarized intersubband absorption in quantum wells ascribed to this mechanism. Peng et al. (1992, 1993), Peng and Fonstad (1993), Hirayama et al. (1993), and Smet et al. (1994) published a series of papers reporting both TM- and TE-polarized intersubband absorption of similar strength in InGaAs–InAlAs and InGaAs–AlAs structures prepared in slab waveguide structures down to 40 μm thin and also in direct normal-incidence geometry. The peaks of the TM and TE mode were reported to be shifted from each other (polarization splitting) by a magnitude proportional to the strain in the structure. The authors have claimed to substantiate their experiments with a 14 × 14 $\mathbf{k}\cdot\mathbf{p}$ calculation (Peng and Fonstad, 1995), but this in turn was questioned by Lew Yan Voon et al. (1996). To our knowledge, the controversy on this topic has not yet been completely settled. In case the theoretical analysis by Yang et al. (1994), Warburton et al. (1994), and Lew Yan Voon et al. (1996) is correct, the origin of the observations of Peng et al. (1993) would still not be clear. The same is true for another work by Li et al. (1993) and Karunasiri et al. (1995), who have reported similar experimental results. Liu et al. (1998) conducted a systematic study of TE vs TM absorption and found the ratio to be less than 0.2% for a GaAs QW and still less than 3% for an InGaAs QW.

Finally, note that the band mixing just described as a mechanism for in-plane intersubband absorption is much more important in the valence band, where, in fact, it leads to strong, observable, normal-incidence intersubband absorption. The valence band is the topic of the next section.

IX. Intersubband Absorption in the Valence Band

So far we have considered only intersubband transitions of electrons in the conduction band of quantum wells. In a similar way, of course, valence-band intersubband transitions can be observed in quantum wells that are p-type doped. The main difference arises from the complexity of the valence band, which makes reasonably accurate calculations significantly more difficult, but also gives rise to some new phenomena, especially related to normal-incidence absorption. This is the reason why p-doped quantum wells are mainly being studied for applications in infrared detectors.

The valence-band maximum of most common semiconductors is located at the Γ point of the Brillouin zone, where it is fourfold degenerate (heavy holes (HHs) and light holes (LHs), each being doubly spin-degenerate). The degeneracy between heavy and light holes is lifted for nonzero values of the wavevector $\mathbf{k}$. In addition, a doubly degenerate band is split off by an amount Δ through the spin-orbit interaction (spin-orbit split-off band, SO). At $\mathbf{k} = 0$ all bands are decoupled, whereas for $\mathbf{k} \neq 0$ all bands are coupled (except in the (001) direction, where there is only a LH–SO coupling). In the case of a large spin-orbit splitting Δ, the interaction of the HH and LH bands with the SO band can be neglected. In Fig. 27 the valence-band structure is depicted schematically as it occurs in GaAs, Si, Ge, and other

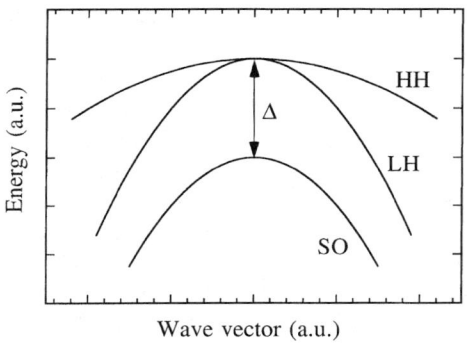

Fig. 27. Schematic of the valence-band structure of most common semiconductors at the Γ point. HH, LH, and SO indicate heavy-hole, light-hole, and spin-orbit split-off band, respectively. The splitting due to the spin-orbit interaction is Δ.

semiconductors. In a quantum well the degeneracy of the HH and LH bands at $\mathbf{k}_\perp = 0$ is removed due to the different effective masses, which results in different binding energies (see Fig. 28). Yet (for QWs grown in the (001) direction) they are still decoupled from each other at $\mathbf{k}_\perp = 0$ and thus the energetic positions of the subband edges can be obtained from a simple one-band calculation for HH and LH separately (as long as the SO band is neglected; for more details, see below).

The nonparabolicity of the conduction band is usually described with Kane's model, where the six (including spin) valence bands and two (or more) conduction bands are taken into account (8×8 $\mathbf{k} \cdot \mathbf{p}$ model). This procedure, however, leads to an free-electron like heavy-hole mass (with the wrong curvature) and is therefore not well suited to describe the valence band. Instead, the Luttinger–Kohn (LK) model (Luttinger and Kohn, 1955)

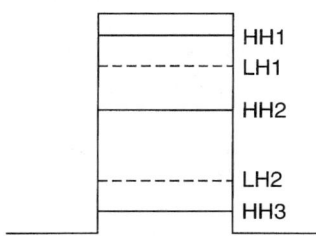

Fig. 28. Subbands of a typical valence-band quantum well (e.g., in GaAs). The energetical order of the HH and LH subbands depends on the QW thickness (see later in the text).

is appropriate; here, only the six valence bands are treated exactly, but the interaction with remote bands is taken into account perturbatively to second order in $\mathbf{k}\cdot\mathbf{p}$ (for the energies; to first order for the wave functions). This interaction gives rise to a finite heavy-hole mass and to the warping of the constant-energy surfaces.

Due to its crucial importance for calculating the valence-band structure and, of course, also intersubband transitions in the valence band, we briefly discuss the Luttinger–Kohn Hamiltonian, although good descriptions can be found in several textbooks. First, the theory is outlined for bulk semiconductors and then extended to quantum wells. The description closely follows the textbook by Chuang (1995).

The Schrödinger equation for the complete Bloch wave function including the spin-orbit interaction is given by (neglecting higher relativistic terms)

$$\left[\frac{p^2}{2m_0} + V(\mathbf{r}) + \frac{\hbar}{4m_0^2 c^2}(\nabla V \times \mathbf{p})\cdot\sigma\right]\psi_{v\mathbf{k}}(\mathbf{r}) = E_v(\mathbf{k})\psi_{v\mathbf{k}}(\mathbf{r}) \qquad (75)$$

where σ are the Pauli spin matrices. The relevant quantum numbers are the band index v and the wavevector $\mathbf{k}$. According to the Bloch theorem, the wavefunction $\psi_{v\mathbf{k}}(\mathbf{r})$ is written as $\psi_{v\mathbf{k}}(\mathbf{r}) = e^{i\mathbf{k}\cdot\mathbf{r}} u_{v\mathbf{k}}(\mathbf{r})$, which is introduced in Eq. (75). The lattice-periodic basis functions $u_{v\mathbf{k}}(\mathbf{r})$ are expanded as a linear combination of the six zone-center Bloch functions of the hole bands $u_{v0}(\mathbf{r})$ and the remote bands (called set A and set B, respectively):

$$u_{v\mathbf{k}}(\mathbf{r}) = \sum_i^A a_i(\mathbf{k}) u_{i0}(\mathbf{r}) + \sum_j^B a_j(\mathbf{k}) u_{j0}(\mathbf{r}) \qquad (76)$$

The six valence bands contained in set A are included exactly, whereas the remote bands in set B are treated perturbatively.

The wave functions near the top of the valence band are p-like, and the bands can be best classified using the total angular momentum $j = l \pm s = 1 \pm 1/2$ and its projection m_z. The basis functions $u_{n0}(\mathbf{r}) = |j, m_z\rangle$ in set A are then explicitly given by (Chuang, 1995)

$$u_{10}(\mathbf{r}) = \left|\frac{3}{2}, \frac{3}{2}\right\rangle = \frac{-1}{\sqrt{2}} |(X + iY)\uparrow\rangle$$

$$u_{20}(\mathbf{r}) = \left|\frac{3}{2}, \frac{1}{2}\right\rangle = \frac{-1}{\sqrt{6}} |(X + iY)\downarrow\rangle + \sqrt{\frac{2}{3}} |Z\uparrow\rangle$$

$$u_{30}(\mathbf{r}) = \left|\frac{3}{2}, \frac{-1}{2}\right\rangle = \frac{-1}{\sqrt{6}}|(X - iY)\uparrow\rangle + \sqrt{\frac{2}{3}}|Z\downarrow\rangle$$

$$u_{40}(\mathbf{r}) = \left|\frac{3}{2}, \frac{-3}{2}\right\rangle = \frac{1}{\sqrt{2}}|(X - iY)\downarrow\rangle \qquad (77)$$

$$u_{50}(\mathbf{r}) = \left|\frac{1}{2}, \frac{1}{2}\right\rangle = \frac{1}{\sqrt{3}}|(X + iY)\downarrow\rangle + \sqrt{\frac{1}{3}}|Z\uparrow\rangle$$

$$u_{60}(\mathbf{r}) = \left|\frac{1}{2}, \frac{-1}{2}\right\rangle = \frac{1}{\sqrt{3}}|(X - iY)\uparrow\rangle - \sqrt{\frac{1}{3}}|Z\downarrow\rangle$$

The order for the $j = 3/2$ bands is with descending quantum number m_z. Thus u_{10} and u_{40} represent the HH bands, u_{20} and u_{30} the LH bands, and u_{50} and u_{60} the SO bands.

Now the coupling of the lattice-periodic functions $u_{v\mathbf{k}}(\mathbf{r})$ ($v = 1, \ldots, 6$) to the remote bands of set B can be removed by $\mathbf{k} \cdot \mathbf{p}$ perturbation theory (Löwdin method) via the transformed basis

$$u_{v\mathbf{k}}(\mathbf{r}) = u_{v0}(\mathbf{r}) + \frac{\hbar}{m_0} \sum_{j}^{B} \frac{\langle u_{j0}|\mathbf{k} \cdot \mathbf{p}|u_{v0}\rangle}{E_{v0} - E_{j0}} u_{j0}(\mathbf{r}) \qquad (78)$$

Here $v \in A$ and $j \in B$. The E_{v0} and E_{j0} are the respective band-edge energies at the Γ point. There is no sum over set A, since the $\mathbf{k} \cdot \mathbf{p}$ coupling within the valence band vanishes. When this expression is substituted into the Schrödinger equation, one obtains a 6 × 6 matrix Schrödinger equation

$$\sum_{v'=1}^{6} H_{vv'}^{LK}(\mathbf{k}) a_{v'}(\mathbf{k}) = E(\mathbf{k}) a_v(\mathbf{k}) \qquad (79)$$

where H^{LK} is the Luttinger–Kohn Hamiltonian, which is of the form

$$H_{vv'}^{LK} = E_{v0}\delta_{vv'} + \sum_{\alpha,\beta} D_{vv'}^{\alpha\beta} k_\alpha k_\beta \qquad (80)$$

The matrix $D^{\alpha\beta}$ ($\alpha, \beta = x, y, z$), which represents the nondiagonal elements of the LK Hamiltonian and comes from the interaction with the remote bands, plays the role of an inverse effective-mass tensor and is given by

$$D_{vv'}^{\alpha\beta} = \frac{\hbar^2}{2m_0}\left[\delta_{vv'}\delta_{\alpha\beta} + \sum_{j}^{B} \frac{p_{vj}^\alpha p_{jv'}^\beta + p_{vj}^\beta p_{jv'}^\alpha}{m_0(E_0 - E_j)}\right] \qquad (81)$$

1   THE BASIC PHYSICS OF INTERSUBBAND TRANSITIONS   63

Here E_0 is the average value of the band-edge energies in set A (valence band); its use instead of E_{v0} is an excellent approximation, when the spin-orbit splitting Δ is much smaller than the energy separation from the remote bands.

The total Luttinger–Kohn Hamiltonian matrix then reads

$$H_{vv'}(\mathbf{k}) = -\begin{pmatrix} P+Q & -S & R & 0 & -S/\sqrt{2} & \sqrt{2}R \\ -S^* & P-Q & 0 & R & -\sqrt{2}Q & \sqrt{3/2}S \\ R^* & 0 & P-Q & S & \sqrt{3/2}S^* & \sqrt{2}Q \\ 0 & R^* & S^* & P+Q & -\sqrt{2}R^* & -S^*/\sqrt{2} \\ -S^*/\sqrt{2} & -\sqrt{2}Q^* & \sqrt{3/2}S & -\sqrt{2}R & P+\Delta & 0 \\ \sqrt{2}R^* & \sqrt{3/2}S^* & \sqrt{2}Q^* & -S/\sqrt{2} & 0 & P+\Delta \end{pmatrix} \quad (82)$$

where

$$P = \frac{\hbar^2}{2m_0} \gamma_1 (k_x^2 + k_y^2 + k_z^2)$$

$$Q = \frac{\hbar^2}{2m_0} \gamma_2 (k_x^2 + k_y^2 - 2k_z^2)$$

$$R = -\frac{\hbar^2 \sqrt{3}}{2m_0} \left[\frac{\gamma_2 + \gamma_3}{2} (k_x - ik_y)^2 + \frac{\gamma_2 - \gamma_3}{2} (k_x + ik_y)^2 \right]$$

$$= -\frac{\hbar^2 \sqrt{3}}{2m_0} [-\gamma_2 (k_x^2 - k_y^2) + 2i\gamma_3 k_x k_y]$$

$$S = \frac{\hbar^2 \sqrt{3}}{m_0} \gamma_3 (k_x - ik_y) k_z$$

where γ_1, γ_2, and γ_3 are the so-called Luttinger parameters (related to the elements of the matrix D) and describe the influence of the remote bands on the valence band. They can be expressed as a sum over momentum matrix elements between the valence band and the remote bands (compare Eq. (81)). Usually, however, they are not explicitly calculated, but taken from experiments and used as empirical parameters (e.g., to fit the experimentally obtained hole masses), in the same way as the effective mass is introduced in the single-band $\mathbf{k} \cdot \mathbf{p}$ perturbation theory. A closer inspection of the preceding shows that γ_2 and γ_3 give rise to an anisotropic dispersion (warping) of the valence bands. When the coupling to the SO band is neglected, diagonalization of the Hamiltonian leads to a simple analytical result (see for example Bastard, 1988, p. 54, or other semiconductor physics textbooks).

When $k_x = k_y = 0$, the HH + LH part of the matrix (SO neglected) becomes diagonal. The diagonal elements $P + Q$ and $P - Q$ are proportional to $\gamma_1 - 2\gamma_2$ and $\gamma_1 + 2\gamma_2$, respectively. Thus, these two expressions can be viewed as the inverse effective masses for the heavy and light holes, respectively (note that this cannot be seen so easily when setting, e.g., $k_y = k_z = 0$, since the z direction has been used as quantization axis for the angular momentum).

To solve not the bulk but the quantum well problem, k_z must be replaced by the differential operator $-id/dz$ in Eq. (82), which, however, must be symmetrized before with respect to k_z to ensure Hermiticity. The resulting system of six coupled differential equations has to be solved for the six-component envelope wavefunction F_v (with proper boundary conditions, see, e.g., Altarelli et al., 1985; Altarelli, 1986):

$$\sum_{v'=1}^{6} [H_{vv'}^{LK} + V(z)\delta_{vv'}] F_{v'n\mathbf{k}_\perp}(\mathbf{r}) = E_{vn\mathbf{k}_\perp} F_{vn\mathbf{k}_\perp}(\mathbf{r}) \tag{83}$$

Here $V(z)$ is the quantum well potential (the proper offsets for each hole band—HH, LH, and SO—must be taken). Instead of the wavevector component k_z we now have the subband index n as a new quantum number, and $\mathbf{k}_\perp$ is the in-plane vector (k_x, k_y). Thus $E_{vn\mathbf{k}_\perp}$ gives the in-plane energy dispersion of the nth subband of the vth band. The envelope function $F_{vn\mathbf{k}}(\mathbf{r})$ can be separated into a z-dependent and an in-plane part:

$$F_{vn\mathbf{k}_\perp}(\mathbf{r}) = \frac{1}{\sqrt{A}} e^{i(\mathbf{k}_\perp \cdot \mathbf{r})} g_{vn\mathbf{k}_\perp}(z) \tag{84}$$

The total wave function is a linear superposition of the six basis functions, where the Fourier transform of $g_{vn\mathbf{k}_\perp}(z) = \Sigma_{k_z} g_{vn\mathbf{k}_\perp}(k_z) \cdot e^{ik_z z}$ is introduced for convenience (Altarelli, 1986; Chang and James, 1989):

$$\psi_{n\mathbf{k}}(\mathbf{r}) = \sum_{v,k_z} g_{vn\mathbf{k}_\perp}(k_z) e^{i\mathbf{k}\cdot\mathbf{r}} u_{v\mathbf{k}}(\mathbf{r}) \tag{85}$$

with $u_{v\mathbf{k}}(\mathbf{r})$ from Eq. (78). With this Fourier transform Eq. (83) is conveniently reduced to an algebraic matrix equation. The probability density (of the envelope function, i.e., averaged over one lattice constant) along the quantization direction z for a state $(n, \mathbf{k}_\perp)$ is then given by

$$\sum_{v=1}^{6} g^*_{vn\mathbf{k}_\perp}(z) \cdot g_{vn\mathbf{k}_\perp}(z) \qquad (86)$$

Several factors should be noted about the Luttinger–Kohn Hamiltonian applied to quantum wells:

1. When the spin-orbit splitting Δ is much larger than all energies of interest, the coupling to the SO band can be neglected. Then the Hamiltonian of Eq. (82) reduces to a 4×4 matrix. This is the case for the lowest subbands in not too narrow GaAs–AlGaAs quantum wells ($\Delta = 340$ meV), but never for Si–SiGe quantum wells, since Δ is as small as 45 meV for Si.
2. As a further simplification, one can put $\gamma_2 = \gamma_3 =: \bar{\gamma} = (\gamma_2 + \gamma_3)/2$ in the expression for R. This is called the axial approximation and removes the valence band warping (i.e., the in-plane dispersion becomes spherical).
3. As a main effect of the confinement, the degeneracy at $\mathbf{k}_\perp = 0$ of the HH and LH bands is lifted. In the 4×4 Hamiltonian (neglecting the SO band), the band-edge energies of the heavy and light holes can be determined separately like in the one-band model, since the Hamiltonian becomes diagonal for $k_x = k_y = 0$. The corresponding effective masses (which are relevant for the confinement energies) are given by

$$\frac{m^z_{HH}}{m_0} = \frac{1}{\gamma_1 - 2\gamma_2} \quad \text{and} \quad \frac{m^z_{LH}}{m_0} = \frac{1}{\gamma_1 + 2\gamma_2} \qquad (87)$$

Approximate effective masses in the quantum well plane can be obtained by setting $k_z = 0$ and an expansion for small k_x and k_y. The result is

$$\frac{m^{xy}_{HH}}{m_0} = \frac{1}{\gamma_1 + \gamma_2} \quad \frac{m^{xy}_{LH}}{m_0} = \frac{1}{\gamma_1 - \gamma_2} \qquad (88)$$

This approximation is never really good, since a finite confinement energy already corresponds to a nonzero value of k_z. Note that these expressions predict a light in-plane mass for the heavy-hole band and vice versa, which inevitably leads to an anticrossing of the light- and heavy-hole subbands at some finite $\mathbf{k}_\perp$ value. Interaction between light and heavy holes also often leads to a negative (electron-like) in-plane effective mass of some of the subbands. Figure 29 shows the hole subband dispersion along the (100) and (110) directions for a 50-Å-wide GaAs–AlAs quantum well, calculated with the full 6×6 LK Hamiltonian. The negative mass of the LH subbands is clearly observed. Near $k_\perp = 0.035\,\text{Å}^{-1}$ there is an anticrossing between the HH1 and LH1 states. The LH2 and HH3 states interact strongly near $k_\perp = 0$.

The energetic order of the HH and LH bands is determined by the confinement effective masses and the quantum well width. In GaAs QWs, the LH1 subband is below the HH2 subband for narrow wells (since in the extreme quantum limit, only one HH and one LH state are bound), while

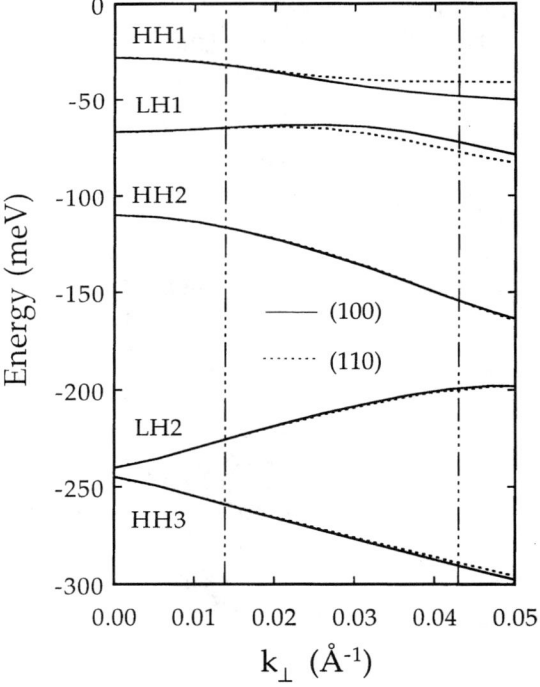

FIG. 29. In-plane hole–subband dispersions of a 50-Å-wide GaAs–AlAs quantum well calculated with the 6×6 Luttinger-Kohn theory along the (100) (solid curves) and (110) (dotted curves) directions. The vertical dash-dotted lines indicate the Fermi wavevector for $p = 3 \times 10^{11}\,\text{cm}^{-2}$ and $p = 3 \times 10^{12}\,\text{cm}^{-2}$, respectively. The character (HH, LH) of each subband at $\mathbf{k}_\perp = 0$ is indicated.

there is a crossover around $L = 190$ Å (Schulman and Chang, 1985). In asymmetric QW potentials, one would observe the lifting of the twofold spin degeneracy for $k_\perp \neq 0$.

Many quantum well systems are built from strained heterostructures, such as Si–Si$_{1-x}$Ge$_x$ or In$_x$Ga$_{1-x}$As–Al$_y$Ga$_{1-y}$As. In this case, strain terms must be added to the LK Hamiltonian following Pikus and Bir (1960) and Bir and Pikus (1974), which leads to a lifting of the HH–LH degeneracy at $\mathbf{k} = 0$. For biaxially strained QWs this yields additional terms for the diagonal elements of the HH and LH bands, leading to a shift of the respective band edges (which can change the order of HH and LH subbands) as well as a nondiagonal term, coupling the LH and SO bands (People and Sputz, 1990; Fromherz et al., 1994; Chuang, 1995). In tensile strained QWs the ground state can be a light–hole subband (Xie et al., 1991b, Stoklitsky et al., 1994, 1995).

Now we proceed to the next step, the calculation of the optical matrix elements for intersubband transition in the valence band (Chang and James, 1989; People et al., 1992a, 1992b, Fromherz et al., 1994; Szmulowicz and Brown, 1995; Kim and Majerfeld, 1995; Tsang and Chuang, 1995). Here we restrict ourselves to the 4×4 LK Hamiltonian neglecting the spin-orbit interaction. The matrix elements of the momentum operator between the complete wavefunctions $\psi_{n\mathbf{k}}$ have to be taken ($\mathbf{k} = \mathbf{k}'$ during a transition)

$$\frac{\hbar}{m_0}\mathbf{e}\cdot\mathbf{p}_{nn'} = \left\langle \psi_{n\mathbf{k}} \left| \frac{\hbar}{m_0}\mathbf{e}\cdot\mathbf{p} \right| \psi_{n'\mathbf{k}} \right\rangle \tag{89}$$

Due to the formal analogy of the expression $\mathbf{e}\cdot\mathbf{p}$ with $\mathbf{k}\cdot\mathbf{p}$, the matrix element can be expressed through the inverse-mass tensor elements $D^{ij}_{vv'}$ of the LK Hamiltonian (Chang and James, 1989; Kim and Majerfeld, 1995):

$$\frac{\hbar}{m_0}\langle u_{v\mathbf{k}}|\mathbf{e}\cdot\mathbf{p}|u_{v'\mathbf{k}}\rangle = \sum_{i,j}(D^{ij}_{vv'} + D^{ji}_{vv'})e_i k_j \tag{90}$$

Separating terms containing k_z (which actually becomes $-i d/dz$) from the rest it can be shown that the preceding can be written in the form

$$\frac{\hbar}{m_0}\mathbf{e}\cdot\mathbf{p}_{nn'} = \mathbf{e}\cdot\sum_{v,v'=1}^{4}(\mathbf{I}_{vv'}Q^{nn'}_{vv'} + \mathbf{J}_{vv'}R^{nn'}_{vv'}) \tag{91}$$

Here the indices v and v' go over the four (or six) valence bands, whereas n and n' indicate the respective subband levels. Note that $Q^{nn'}_{vv'}$ and $R^{nn'}_{vv'}$ have nothing to do with Q and R of Eq. (82). The preceding quantities are given by

$$\mathbf{e} \cdot \mathbf{I}_{vv'} = \sum_{i=x,y,z} e_i (D_{vv'}^{ix} + D_{vv'}^{xi}) k_x + (D_{vv'}^{iy} + D_{vv'}^{yi}) k_y$$

$$\mathbf{e} \cdot \mathbf{J}_{vv'} = \sum_{i=x,y,z} e_i (D_{vv'}^{iz} + D_{vv'}^{zi})$$

and

$$Q_{vv'}^{nn'} = \int dz F_v^*(n, \mathbf{k}_\perp, z) F_{v'}(n', \mathbf{k}_\perp, z)$$

$$R_{vv'}^{nn'} = \int dz F_v^*(n, \mathbf{k}_\perp, z) \left(-i \frac{d}{dz}\right) F_{v'}(n', \mathbf{k}_\perp, z) \tag{92}$$

The index i extends over the Cartesian components and represents the dot product. These terms of the transition matrix elements $\mathbf{p}_{nn'}$ can be written in the form of the following matrix, which contains the Luttinger parameters, γ_i, explicitly (Kim and Majerfeld, 1995):

1. x polarization

	HH(3/2)	HH(−3/2)	LH(1/2)	LH(−1/2)
HH(3/2)	$\hbar k_x(\gamma_1+\gamma_2)Q^{nn'}$	0	$-\sqrt{3}\hbar\gamma_3 R^{nn'}$	$-\sqrt{3}\hbar(\gamma_2 k_x - i\gamma_3 k_y)Q^{nn'}$
HH(−3/2)	0	$\hbar k_x(\gamma_1+\gamma_2)Q^{nn'}$	$-\sqrt{3}\hbar(\gamma_2 k_x + i\gamma_3 k_y)Q^{nn'}$	$\sqrt{3}\hbar\gamma_3 R^{nn'}$
LH(1/2)	$-\sqrt{3}\hbar\gamma_3 R^{nn'}$	$-\sqrt{3}\hbar(\gamma_2 k_x - i\gamma_3 k_y)Q^{nn'}$	$\hbar k_x(\gamma_1-\gamma_2)Q^{nn'}$	0
LH(−1/2)	$\sqrt{3}\hbar(\gamma_2 k_x + i\gamma_3 k_y)Q^{nn'}$	$\sqrt{3}\hbar\gamma_3 R^{nn'}$	0	$\hbar k_x(\gamma_1-\gamma_2)Q^{nn'}$

$$(93)$$

The y polarization or any linear superposition would be analogous.

2. z polarization

	HH(3/2)	HH(−3/2)	LH(1/2)	LH(−1/2)
HH(3/2)	$\hbar(\gamma_1-2\gamma_2)R^{nn'}$	0	$-\sqrt{3}\hbar\gamma_3(k_x+ik_y)Q^{nn'}$	0
HH(−3/2)	0	$\hbar(\gamma_1-2\gamma_2)R^{nn'}$	0	$\sqrt{3}\hbar\gamma_3(k_x-ik_y)Q^{nn'}$
LH(1/2)	$-\sqrt{3}\hbar\gamma_3(k_x-ik_y)Q^{nn'}$	0	$\hbar(\gamma_1+2\gamma_2)R^{nn'}$	0
LH(−1/2)	0	$\sqrt{3}\hbar\gamma_3(k_x+ik_y)Q^{nn'}$	0	$\hbar(\gamma_1+2\gamma_2)R^{nn'}$

$$(94)$$

For easier reading, the lower indices (vv') are suppressed in $R_{vv'}^{nn'}$ and $Q_{vv'}^{nn'}$, since the corresponding band (HH, LH) is already clear from the respective

1 The Basic Physics of Intersubband Transitions

row and column of the matrix. Note that now the bands have not been ordered by descending angular momentum component, but have been grouped into heavy and light holes.

Let us now try to get some understanding of the transition matrix elements and selection rules. According to Eqs. (93) and (94) there are two types of contributions:

> The terms containing $\mathbf{J}R^{nn'}$: These contain a dipole matrix element of the envelope function ($R^{nn'}$). Due to the dipole matrix element only transitions between states of different parity are allowed. Since they are not proportional to $\mathbf{k}_\perp$, they are allowed at $\mathbf{k}_\perp = 0$. That these terms correspond to "usual" intersubband transitions, like in the conduction band, can be seen in the following way: $\mathbf{J}$ contains only elements of the D tensor with a z component (Eq. (92)), so it couples z- or xy-polarized radiation to an intersubband transition, analogous to the situation in an ellipsoidal conduction valley (compare Eq. (70)). Looking up the HH–HH transition in z polarization (Eq. (94)), one sees that the transition matrix element is proportional to $(\gamma_1 - 2\gamma_2) \cdot R^{nn'}$, which is just the inverse heavy-hole confinement mass (Eq. (87)) times the dipole matrix element, identical to intersubband absorption in a spherical conduction band. The same is true for LH–LH transitions. In addition, however, HH–LH transitions are allowed in xy polarization.

> The terms containing $\mathbf{I}Q^{nn'}$: They contain an overlap integral between envelope wave functions, and thus only transitions between states of the same parity are allowed. Furthermore, they are proportional to $\mathbf{k}_\perp$, and are thus allowed only for $\mathbf{k}_\perp \neq 0$. Looking at Eqs. (93) and (94) one can see that HH–HH, LH–LH, and HH–LH transitions arise from these terms in xy polarization, but only HH–LH transitions in z polarization. These transitions are a specific feature of the valence band, and they originate from interband coupling to the remote bands (the direct interband coupling between different valence bands vanishes).

Finally the 2D absorption coefficient is given by analogue to Eq. (18)

$$\alpha_{2D} = \frac{\pi e^2}{\varepsilon_0 c \eta \omega m_0^2 A} \sum_{n,n'} \sum_{\mathbf{k}_\perp} |\mathbf{e} \cdot \mathbf{p}_{nn'}(\mathbf{k}_\perp)|^2 \cdot [f(E_n(\mathbf{k}_\perp)) - f(E_{n'}(\mathbf{k}_\perp))]$$

$$\cdot \frac{\Gamma/\pi}{[E_n(\mathbf{k}_\perp) - E_{n'}(\mathbf{k}_\perp) - \hbar\omega]^2 + \Gamma^2} \quad (95)$$

From this discussion, we see that both the dipole and the overlap terms give rise to normal-incidence intersubband absorption. For illustration, the calculated absorption spectrum of the previously discussed 50-Å-wide GaAs–AlAs QW is shown in Fig. 30 for both polarizations and for two different hole concentrations of 3×10^{11} and 3×10^{12} cm^{-2}. The relevant transitions can be identified with the help of Fig. 29, where the corresponding Fermi wavevectors have been indicated by the dash-dotted vertical

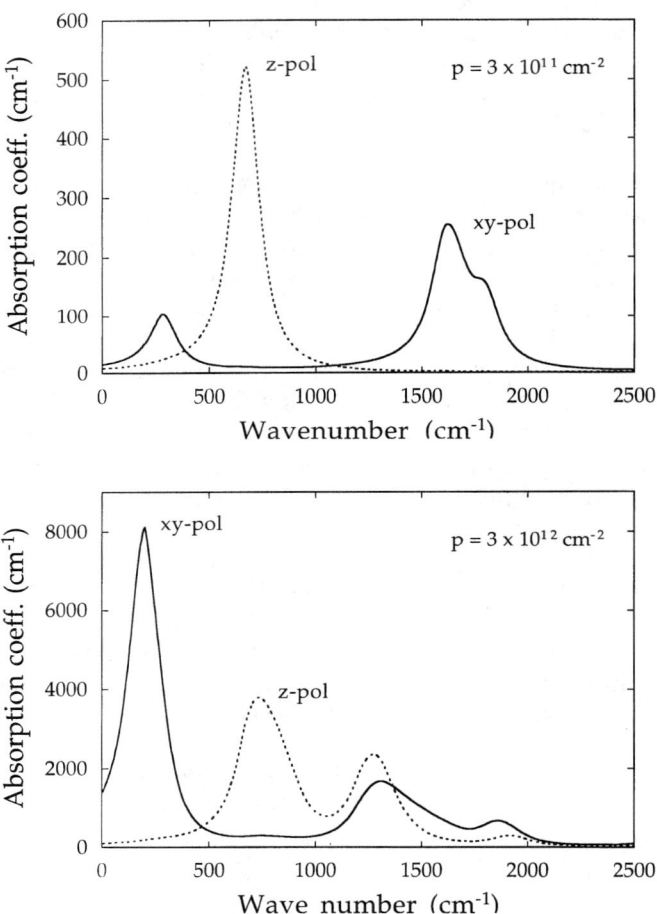

FIG. 30. Calculated absorption spectrum of the GaAs–AlAs quantum well of Fig. 29 for two different hole densities, $p = 3 \times 10^{11}$ cm^{-2} (upper panel) and $p = 3 \times 10^{12}$ cm^{-2} (lower panel). The spectra for z and xy polarization are represented by the dotted and full curves, respectively. For a discussion, see text.

lines. (For ease of calculation, k-space integration was not performed over the complete 2D space, but only along several representative directions with proper averaging). The z-polarized peak at $700\,\text{cm}^{-1}$ (87 meV) results from the HH1–HH2 transition. At 10 times higher hole concentration, it is (nearly) 10 times stronger, since the corresponding matrix element is independent of $\mathbf{k}_\perp$. This is the "usual" intersubband transition. The broadening at higher density is due to the nonparabolic dispersion. The xy-polarized HH1–LH1 transition occurs at $280\,\text{cm}^{-1}$ (35 meV). Since its matrix element is proportional to $\mathbf{k}_\perp$, its size increases stronger than linear with the hole concentration, and in the lower panel it has become stronger than the HH1–HH2 transition. The doublet around $1700\,\text{cm}^{-1}$ (or 210 meV, upper panel) is due to the HH1–LH2, HH3 transition, which latter are strongly mixed due to their proximity. The HH1–LH2 transition is allowed at $\mathbf{k}_\perp = 0$ for xy polarization, but in z polarization its strength is proportional to $\mathbf{k}_\perp$, so it appears only at high density (lower panel). The strength of the HH1–HH3 transition should be proportional to $\mathbf{k}_\perp$, but due to the strong mixing of the HH3 with the LH2 state, it can already be observed at low hole density. One has to keep in mind that a clear assignment of a specific subband to a band type (HH, LH, or SO) can strictly be done only at $\mathbf{k}_\perp = 0$. For $\mathbf{k}_\perp \neq 0$ the wave functions become strong mixtures of all basis states and thus basically all transitions become allowed to some degree.

Similar calculations were carried out for various material systems by Man and Pan (1992), Xie *et al.* (1991b, 1992c), and Stoklitsky *et al.* (1994, 1995). Chun *et al.* (1993) included many-body effects such as the depolarization shift properly into the multiband model. Corbin *et al.* (1994) employed a full-scale pseudopotential calculation for computing band structure and absorption of p-type QWs.

There have been a number of intersubband absorption studies in p-type quantum wells, interestingly mostly in Si–SiGe QWs (Park *et al.*, 1992; Fromherz *et al.*, 1994; Boucaud *et al.*, 1995; Zanier *et al.*, 1995). The reason is that in this material system p-type QWs are more easily achieved than n-type QWs (Wang and Karunasiri, 1993). Clear experimental identification of the various valence-band intersubband transitions has, however, remained rather scarce (Fomherz *et al.*, 1996), in part due to the large broadening one usually has in p-type quantum wells. Moreover, most experimental investigation were performed in structures where the final states of the strongest transitions already lie in the continuum, as desired for detector applications (Levine *et al.*, 1991; Wang *et al.*, 1994b; Szmulowicz and Brown, 1995), which leads to additional broadening. Often only photocurrent, but no transmission measurements have been performed on such samples. There have been two reports on far-infrared hole–intersubband absorption in relatively wide GaAs quantum wells (Shayasteh et al., 1996; Cole et al., 1996).

As an example, the transmission spectrum of a modulation doped Si–$Si_{0.71}Ge_{0.29}$ multiquantum well structure is shown in Fig. 31 (Helm et al., 1997). Both p- and s-polarized spectra (full curves) were obtained by normalizing the transmission against the transmission of an undoped Si substrate in the same polarization. By comparing with a calculation as already outlined, but including an electromagnetic simulation of the waveguide transmission (lines with open symbols), one can identify the main absorption features. In s polarization, the absorption at $400 \, cm^{-1}$ corre-

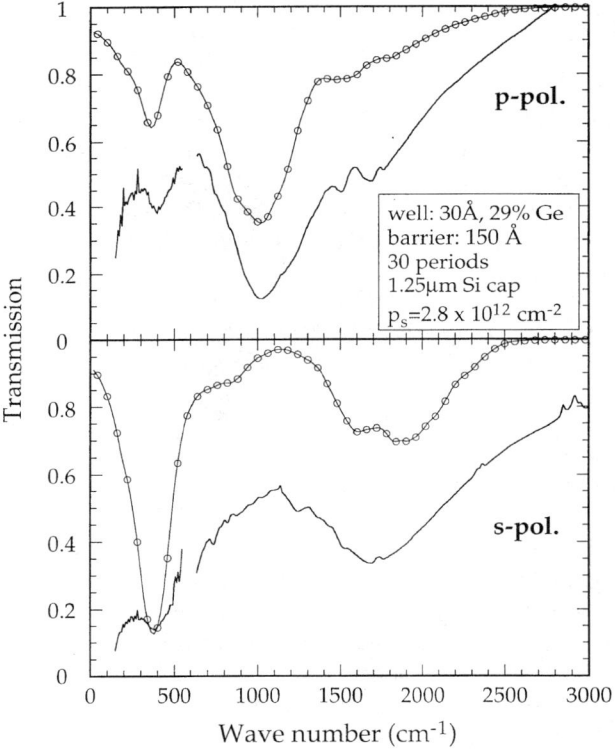

FIG. 31. Polarization dependent waveguide transmission spectrum (at $T = 10 \, K$) of a Si–SiGe multiquantum well, as described in the inset. Solid curves: experiment; lines with open symbols: calculation. Note that the s polarization contains only xy components, whereas the p polarization contains both z and xy components. For a discussion, see text (from Helm et al., 1997).

sponds to the HH1–LH1 transition, and the broad minimum around 1800 cm^{-1} stems from transitions to a continuum SO subband (with admixtures of HH and LH character). The p-polarized spectrum shows a strong absorption at 1000 cm^{-1}, which can be identified as the "usual" HH1–HH2 intersubband transition. For the interpretation of the additional features one has to keep in mind that p-polarized light contains both z and xy (in-plane) electric-field components, so these features are the same as in the s-polarized spectrum. Note that the spectra (s polarization in particular) contain a background that can be ascribed to free-carrier absorption.

X. Line Broadening and Relaxation

So far we have dealt with the mechanism of intersubband absorption in various systems and calculated the absorption coefficient on the basis of Fermi's Golden Rule or including many-body effects. We have, however, not discussed any dissipative processes, which give rise to line broadening and relaxation, but we have simply introduced a phenomenological line-broadening parameter Γ. In this section, we expand somewhat on which mechanisms give rise to the finite linewidth and discuss the experimental situation. The linewidth is an important parameter for many applications; in particular, narrow lines give rise to a larger peak absorption in intersubband detectors or to a larger gain in intersubband lasers. Also any experiments or applications making use of the coherence between quantum mechanical energy levels (see Chapter 2), such as dressed states (Sadeghi et al., 1995) or lasing without inversion (Imamoglu and Ram, 1994), require very narrow absorption lines. On the other hand, energy relaxation, which is described by an intersubband lifetime, is important for the electrical bandwidth of intersubband detectors (where a fast relaxation is desired) or for the possibility of achieving population inversion between subbands and hence for intersubband lasers.

In atomic physics, the relaxation processes in an optical transition are usually described in the framework of the density matrix. One can distinguish between a population decay time T_1, which describes the decay of the diagonal elements of the density matrix, and a polarization (or coherence) decay time T_2, which describes the decay of the nondiagonal elements (Milloni and Eberly, 1988). We assume a Lorentzian line shape

$$L(\hbar\omega) = \frac{\Gamma/\pi}{(E_{21} - \hbar\omega)^2 + \Gamma^2} \tag{96}$$

where Γ is the half width at half maximum, which is composed of two contributions

$$\Gamma = \hbar\left(\frac{1}{2T_1} + \frac{1}{T_2}\right) \qquad (97)$$

Note that T_1 contributes only half as strongly as T_2. In atomic physics, T_2 is normally related to elastic collisions, whereas T_1 relates to inelastic collisions and spontaneous photon emission. (Note that the total relaxation time $\hbar/\Gamma$ is sometimes expressed as T_2). In quantum well subbands, the situation is slightly more complicated, since due to the free-electron subband dispersion along k_x and k_y even elastic scattering processes can induce a transition to another subband (at a different $\mathbf{k}_\perp$, however).

Let us now take a look at the various scattering processes in a semiconductor. As for inelastic processes, we have acoustic and optical phonon scattering, the former with a typical time constant of a few 100 ps; the latter, about 1 ps. At low temperatures, when there is no optical phonon population in the crystal, only emission processes are relevant. To emit optical phonons, electrons require a threshold energy of $\hbar\omega_{op}$. So naturally there will be a quite different behavior depending on whether the subband separation E_{21} is larger or smaller than $\hbar\omega_{op}$. Both situations together with the relevant scattering processes for relaxation are sketched in Figs. 32a and 32b. Elastic processes can be, for example, scattering by ionized impurities or scattering from interface roughness. Impurity scattering can be quite effective with a time constant of 1 ps, but in modulation-doped QWs with a large setback of the dopants, the scattering rate is reduced to the order of 10 ps. Interface roughness scattering (with respect to the electron mobility) is known to exhibit a strong L^{-6} dependence on the well width L, becoming significant around $L < 100$ Å (Sakaki et al., 1987). If we now consider a typical GaAs–Al$_{0.3}$Ga$_{0.7}$As QW with 100 Å thickness (leading to $E_{21} \approx$ 100 meV), we see that optical phonon emission will be the dominant contribution to T_1 (see Fig. 32a), whereas impurity and interface roughness scattering will determine T_2. We discuss this situation, as well as the more intricate case of wide QWs, where $E_{21} < \hbar\omega_{op}$ (Fig. 32b), in the following.

In the first intersubband absorption experiments on GaAs–AlGaAs QWs with E_{21} of the order of 100 meV the measured full linewidths (FWHM) were of the order of $2\Gamma \approx 10$–20 meV and were attributed to a combined effect of optical phonon emission, impurity scattering (only for well-doped samples), interface roughness, and, for MQW systems, also thickness fluctuations from layer to layer. Later, the material quality was improved significantly, and in 1994 Faist et al. (1994b) reported a linewidth of $2\Gamma = 2.66$ meV. They claimed to have reached the intrinsic broadening limit due

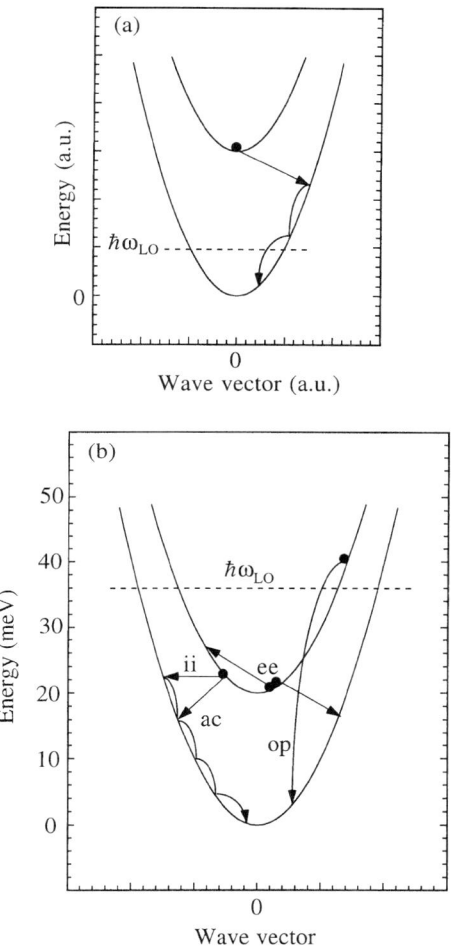

FIG. 32. Schematic view of intersubband relaxation for narrow (a) and wide (b) quantum wells, where the energy separation E_{21} is larger or smaller than the optical phonon energy, $\hbar\omega_{LO}$, respectively. In narrow QWs only optical-phonon emission is relevant for intersubband relaxation, whereas in wide wells several processes play a role: optical phonon emission (op), acoustic-phonon emission (ac), electron-electron scattering (ee), and elastic processes such as ionized-impurity scattering (ii).

to optical phonon emission. Remarkably, the sample was not a square well, but rather an asymmetric coupled MQW system (50 periods), where the 1–3 transition showed this narrow linewidth. Campman et al. (1996) conducted another study, showing that a linewidth of $2\Gamma = 2.5$ meV can also be achieved in a 100-Å-wide GaAs–AlGaAs square QW. They demonstrated that, under the same growth conditions, the linewidth increases when the well width gets smaller (to 4.4 meV for 75 Å QW in their study). Optical-phonon scattering, on the other hand, should get stronger for wider QWs, since its strength is proportional to $1/q^2$, with q the momentum transfer. This tendency, together with an evaluation of the optical-phonon scattering rate, led them to conclude that the intrinsic limit has not been reached yet, but should lie around $2\Gamma = 1.2$ meV for a 100-Å well. Interestingly, they found that the electron mobility varies much stronger with the well width than does the intersubband linewidth. Also, the linewidth is not too much affected by using an alloy QW ($In_xGa_{1-x}As$ or $Al_xGa_{1-x}As$ with $x < 10\%$, see Fig. 33). So the conclusion here seems to be that, although technology has driven the quality of QWs close to its intrinsic limit as far as

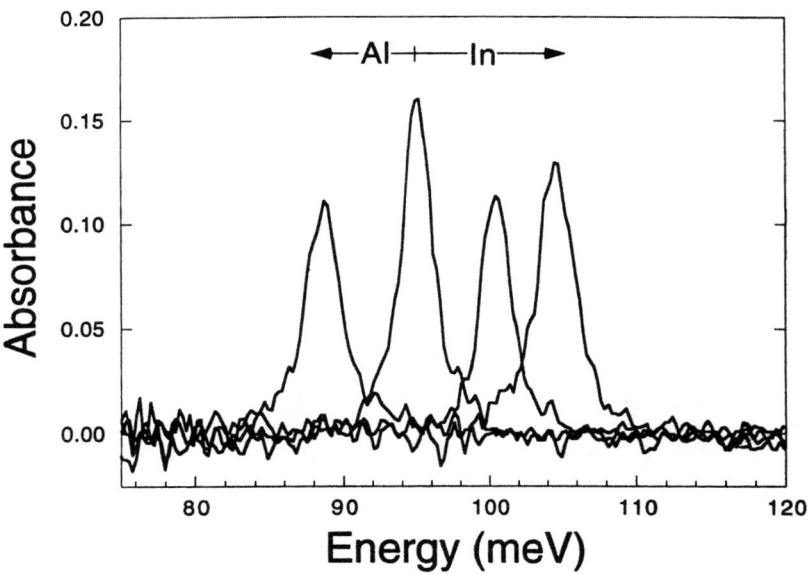

FIG. 33. Measured absorption spectrum of four single 100-Å-wide quantum wells with varying alloy composition. The well materials are (from left to right) $Al_{0.05}Ga_{0.95}As$, GaAs, $In_{0.05}Ga_{0.95}As$, and $In_{0.1}Ga_{0.9}As$, the barrier material is $Al_{0.3}Ga_{0.7}As$ for all samples (from Campman et al., 1996).

intersubband transitions are concerned, this limit has not been completely reached as yet, leaving some room for improvement. Finally, note that nearly the same as the record linewidth (namely, $2\Gamma = 2.8$ meV) has also been achieved in an MQW system containing 30 periods of 95-Å-wide QWs. This requires extremely accurate long-term control of the molecular beam epitaxy (MBE) growth (Gauer *et al.*, 1995; see also Fig. 35 in Section XI). Although intersubband absorption has been observed in many different material systems to date, the linewidths achieved in GaAs–AlGaAs are still the narrowest. An additional contribution to the linewidth can be due to nonparabolicity, but this has not been unambiguously identified, since it tends to be compensated by many-body effects (see also Section VII and Zaluzny, 1991; von Allmen, 1992; Warburton *et al.*, 1996).

An interesting issue is the question whether intersubband transitions are homogeneously or inhomogeneously broadened. Intrinsic processes such as phonon scattering give rise to homogeneous broadening, but the same is true for scattering by imperfections such as impurities or interface roughness, as long as their lateral length scale is smaller than the characteristic in-plane wavelength of the electrons. A lower bound for this is the Fermi wavelength, which is typically a few 100 Å for $n_s = 10^{11}-10^{12}$ cm^{-2}. Compared to this, the mean distance between impurities (in samples doped heavily in the wells) is somewhat smaller than 100 Å. Interface roughness has the same lateral length scale, but only when the interfaces are grown without interruption (for a review, see Herman *et al.*, 1991). Thus both processes will give rise to homogeneous broadening. When the growth is interrupted at the interfaces, however, the lateral length scale can be much larger, up to micrometers. In this case, one can expect some inhomogeneous broadening. The most important sources of inhomogeneous broadening are probably vertical, well-to-well thickness fluctuations in MQW samples with many periods. However, in high-quality MQW structures, the intersubband absorption could be fitted by a Lorentzian line shape, indicating homogeneous broadening (Faist *et al.*, 1994b).

Experiments specifically investigating this issue were reported by Beadie *et al.* (1997) and Vodopyanov *et al.* (1996). Beadie *et al.* (1997) analyzed the saturation behavior in strained, narrow (40-Å) $In_{0.45}Ga_{0.55}As$–$Al_{0.45}Ga_{0.55}As$ MQWs and concluded that both homogeneous and inhomogeneous processes contribute to the line shape. Vodopyanov *et al.* (1996) performed a two-color pump–probe experiment on a somewhat wider (58.5-Å) $In_{0.5}Ga_{0.5}As$–$Al_{0.45}Ga_{0.55}As$ MQW, but could not observe any spectral hole burning; this indicated homogeneous broadening despite the quite large linewidth (>20 meV) of the sample used. A direct measurement of the polarization decay time T_2 in a lattice-matched 60-Å-wide InGaAs/InAlAs QW has been performed by Kaindl *et al.* (1998) by time-resolved four-wave mixing in the mid-infrared. They found a dephasing time T_2 of a few 100 fs, caused by electron–electron scattering, and could determine the homogeneous contribution to the linewidth to be 4 meV, which was about 30% of the total linewidth.

Thus, this issue has apparently not been completely resolved, but as a preliminary conclusion one can probably say that for very high quality and not too narrow QWs, the intersubband absorption is homogeneously broadened, but for many other structures there is a significant part of inhomogeneous broadening. To conclude this discussion, let us note another fundamentally inhomogeneous (yet intrinsic) broadening mechanism, namely, subband nonparabolicity. This mechanism, however, is hard to track down due to its already mentioned (see Section VII, Subsection 4, Figs. 25 and 26, and Warburton et al., 1996) cancellation through manybody effects (Zaluzny, 1991; von Allmen, 1992; Warburton et al., 1996). Theoretical work (Nikonov et al., 1997) suggests that intersubband absorption in QWs with a large nonparabolicity (such as InAs–AlSb) may be mostly homogeneously broadened due to the collective character of the excitation. This view is also supported through experiments by Warburton et al. (1998), demonstrating the need for a microscopic theory of line broadening in intersubband absorption, which does not exist to date. The first step toward such a theory has, however, been published by Ullrich and Vignale (1998).

A lot of effort has also been put into the measurement of the intersubband relaxation time T_1. Experimental techniques reaching from interband (Oberli et al., 1987; Tatham et al., 1989; Levenson et al., 1990; Hunsche et al., 1994; Hartig et al., 1996) or intersubband (Seilmeier et al., 1987; Bäuerle et al., 1988; Elsaesser et al., 1989; Boucaud et al., 1996; Lutgen et al., 1996a, 1996b) time-resolved pump-and-probe measurements with short-pulse lasers to steady-state absorption saturation measurements (Julien et al., 1988, erratum 1993; Faist et al., 1993b; Cui et al., 1993; West and Roberts, 1994) have been employed. Whereas initially the relaxation time was somewhat overestimated (Seilmeier et al., 1987) due to the specific sample design, there seems to be agreement now that the intersubband relaxation time is of the order of 0.3 to 0.6 ps for GaAs QWs with E_{21} of the order of 100 meV. This would be in agreement with a intrinsic lifetime broadening of $2\Gamma = 2 \cdot \hbar/2T_1 = 1.1$ to 2.2 meV.

In wide quantum wells with $E_{21} < \hbar\omega_{op}$ the situation is totally different. Optical phonon emission is irrelevant for the linewidth, but also, monolayer thickness fluctuations play only a minor role (for a 400-Å QW, thickness fluctuations about one monolayer contribute only 0.1 meV to the linewidth). This is the reason that linewidths of $2\Gamma < 1.5$ meV (Helm et al., 1991) have been readily achieved even with samples that were doped in the QWs. Here, most likely impurity scattering was the dominant broadening mechanism. In such samples, the intersubband linewidth is also correlated with the electron mobility. Thus, it is obvious that the absorption lines can be made narrower by modulation doping with a large setback (thick spacer layer). In this way, linewidths of $2\Gamma = 0.8$ meV ($=6.5$ cm^{-1}) have been achieved in multiquantum wells (Fig. 34) and of $2\Gamma = 0.3$ meV in single-period QW systems (Craig et al., 1994, 1996). It is likely that remote-impurity scattering is responsible for this, but also interface roughness comes back into play in this regime.

As far as intersubband lifetime and energy relaxation in wide QWs are concerned, the situation has been far from well understood until recently, and a wide range of lifetimes from a few picoseconds to a few 100 ps were reported (Oberli et al., 1987; Levenson et al., 1990; Faist et al., 1994b; Heyman et al., 1995, 1996; Murdin et al., 1994; Luo et al., 1997; Hartig et al., 1998). These values were again achieved by a variety of time-resolved and steady-state experimental methods. The possible relaxation channels are sketched in Fig. 32b. In principle, acoustic-phonon emission is the only

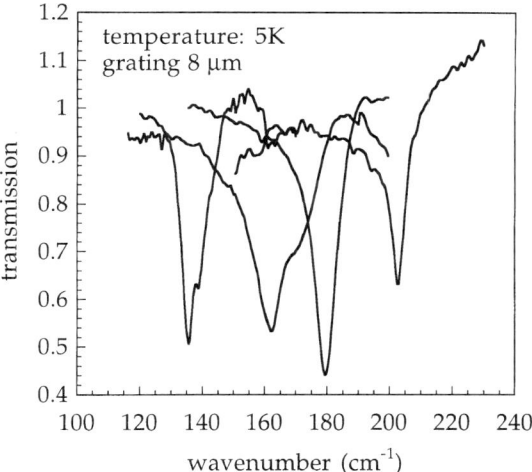

FIG. 34. Transmission spectra of wide modulation-doped GaAs–Al$_{0.3}$Ga$_{0.7}$As multiquantum wells with thicknesses of 250, 280, 300, and 320 Å and an electron concentration of 2–3 × 10^{11} cm^{-2}, measured with a grating coupler. The high-energy shoulders in the two low-energy curves are most likely due to well-thickness variations in the MQW, thus these two lines are inhomogeneously broadened, whereas the two lines at higher energy appear to be mostly homogeneously broadened (courtesy of G. Strasser).

energy-dissipating process in this regime at low temperatures and leads to a theoretical lifetime of a few 100 ps. Time-resolved pump-probe measurements conducted with the far-infrared free-electron laser (FELIX) as well as careful analysis of previous experiments have, however, shed new light on this problem (Murdin et al., 1997). It has been found that the measured lifetime depends strongly on the experimental situation, such as the lattice temperature T_L of the sample and the optical excitation intensity I. Both high T_L and high I lead to a strong reduction of the measured lifetime. This can be interpreted in terms of the heating and cooling of a hot electron distribution. For high T_L or strong pumping a high electron temperature is reached within less than a picosecond due to electron-electron scattering (Dür et al., 1996; Lee and Galbraith, 1997). Some electrons from the tail of this distribution can then relax by emitting optical phonons, thus giving rise to a shorter relaxation time (Lee et al., 1995). This cooling takes place until an electron temperature of around $T_e = 35$ K is reached. Below this temperature, there are not enough electrons in the high-energy tail to emit optical phonons and thus acoustic-phonon emission becomes dominant. For $T_e > 40$ K lifetimes of a few picoseconds and for $T_e < 30$ K of a few 100 ps are measured. A detailed analysis and discussion can be found in Murdin et al. (1997).

XI. Other Phenomena Related to Intersubband Transitions

1. Magnetic-Field Effects

When a magnetic field is applied perpendicular to the plane of a quantum well, the magnetic part of the Hamiltonian is decoupled from the electric confinement $V(z)$ and the energy spectrum consists of a ladder of Landau levels for each electric subband. The infrared absorption spectrum then shows cyclotron resonance for in-plane polarized light and intersubband absorption for z polarization. We won't discuss this case here further, since it brings no new aspects concerning the intersubband absorption. The situation is different when the magnetic field is applied in the layer plane or in some tilted direction. In these cases, the cyclotron and electric motions are coupled.

Let us first discuss the geometry, where the magnetic field lies in the plane of the QW layer. The confinement direction is still assumed to be the z axis, whereas the magnetic field points along the x direction. Using the gauge $\mathbf{A}_B = (0, -eBz, 0)$ (the subscript B is used to distinguish this from the AC vector potential) the Hamiltonian is still separable and its z-dependent part can be written as

$$H = -\frac{\hbar^2}{2m^*}\frac{\partial^2}{\partial z^2} + \frac{1}{2}m^*\omega_c^2(z-z_0)^2 + V(z) \qquad (98)$$

where $\omega_c = eB/m^*$ is the cyclotron frequency and $z_0 = -\hbar k_y/eB$ is the Landau level center coordinate. The electronic motion is clearly determined by two z-dependent effective potentials, the electric confinement, $V(z)$, and the parabolically shaped magnetic confinement. Depending on their relative strength, the motion and the energy levels are electric or magnetic in character.

For small magnetic fields, obeying the condition $\omega_c \ll \omega_{21}$, or equivalently $l \gg L$ (where $l = \sqrt{\hbar/eB}$ is the magnetic length) the electric subbands (and their energy separation) undergo only a small quadratic shift, the diamagnetic shift. When the magnetic field increases ($\omega_c \approx \omega_{21}$), the energy becomes position-dependent and one obtains hybrid magneto-electric subbands (Zawadzki, 1987). In the limit of large magnetic fields ($\omega_c \gg \omega_{21}$) the electron does not feel the electric confining potential anymore and one recovers three-dimensional Landau levels.

It is interesting to calculate the optical matrix elements for magnetic fields so small that the wave functions can be assumed to remain unchanged (Gauer et al., 1995). In the interaction term $(e/m^*)\mathbf{A}\cdot\mathbf{P}$, the canonical momentum $\mathbf{P} = \mathbf{p} - e\mathbf{A}_B$ must be used. The usual, z-polarized intersubband matrix element then reads

$$\frac{eA_z}{m^*}\langle 1|p_z|2\rangle = eA_z\omega_{21}\langle 1|z|2\rangle \qquad (99)$$

The matrix element for radiation polarized in the y direction (i.e., in the layer plane), but perpendicular to the magnetic field, is

$$\frac{eA_y}{m^*}\langle 1|p_y - eBz|2\rangle = eA_y\omega_c\langle 1|z|2\rangle \qquad (100)$$

since the term containing p_y vanishes. Thus, we get the remarkable result that the in-plane polarized absorption is proportional to the usual intersubband z matrix element, however reduced by a factor ω_c/ω_{21}. Physically, the magnetic field couples the y and z motions of the electrons, thus making normal-incidence absorption possible. This effect has been discussed and experimentally observed by Gauer et al. (1995), shown in Fig. 35. (Note the extremely narrow linewidth of $2\Gamma = 2.8$ meV in this sample). When the magnetic field is increased further, the oscillator strength is transferred from the z polarization (intersubband type) to the y polarization (cyclotron resonance), and the magnetic limit is reached. Note also that somewhat

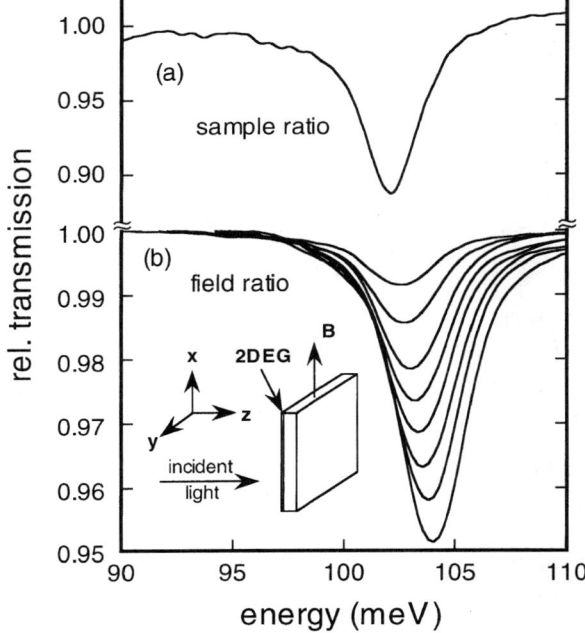

FIG. 35. (a) Intersubband absorption of a 30-period, 95-Å-wide modulation doped GaAs–Al$_{0.35}$Ga$_{0.65}$As multiquantum well sample measured under an oblique angle (electron concentration $n_s = 6 \times 10^{11}$ cm^{-2}). (b) The transmission spectra recorded in Voigt geometry with a magnetic field of $B = 6, 8, 10, 11, 12, 13, 14,$ and 15 T parallel to the layers (geometry sketched in the inset) (from Gauer et al., 1995).

related experiments were carried out already by Oelting et al. (1986) on InSb inversion layers.

For a parabolic potential $V(z)$ the Hamiltonian, Eq. (98), becomes particularly simple, since it consists of the sum of two harmonic oscillators. Then an exact solution can even be obtained for an arbitrarily tilted magnetic-field direction (Maan, 1984; Merlin, 1987). Other properties of parabolic quantum wells are discussed in the following subsection.

If the magnetic field is tilted from the surface-normal (the z direction) by only a small angle θ, a small coupling is induced between the electric subbands and the Landau levels. This leads to an anticrossing and an absorption line doublet near resonance (at $\hbar\omega_c = E_{21}$), which can be observed in a normal-incidence absorption experiment. This was the method that actually provided the first evidence of the subbands in a GaAs–AlGaAs heterostructure (Schlesinger et al., 1983). Later, this technique was applied by a number of authors for intersubband spectroscopy on 2D systems (Rikken et al., 1986; Wieck et al., 1987, 1989; Ensslin et al., 1989; Pillath et al., 1989). In Fig. 36 this anticrossing is observed at $B = 11.25$ T at an energy of $150\,\text{cm}^{-1}$ (Fig. 36a) and compared to a direct (grating-coupler induced) intersubband absorption measurement (Fig. 36b), where the resonance occurs at $168\,\text{cm}^{-1}$. This experiment was regarded as a proof that the anticrossing measures the bare energy separation, whereas the intersubband absorption includes the depolarization shift (and exciton correction) (Pillath et al., 1989). However, theory (Zaluzny, 1989) and later experiments (Wixforth et al., 1994) were in contradiction with this and showed that the anticrossing is also affected by the depolarization shift.

Away from the resonance (i.e., for $\hbar\omega_c \neq E_{21}$) again the diamagnetic shift of the intersubband absorption has been observed, but also combined intersubband–cyclotron resonances (Beinvogl and Koch, 1978; Wieck et al., 1988; Batke et al., 1991). These resonances involve intersubband transitions, which are accompanied by a change of the Landau quantum number, so

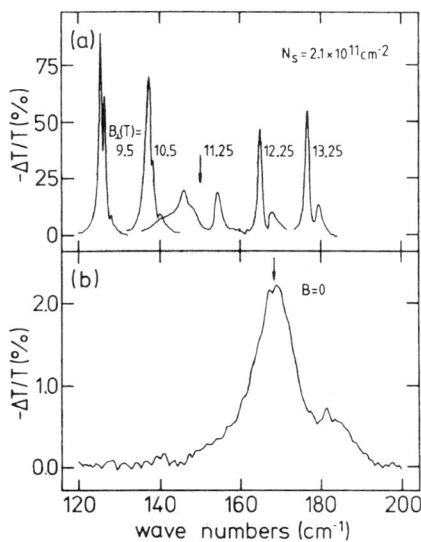

FIG. 36. (a) Cyclotron resonance absorption of a GaAs–AlGaAs heterostructure in a magnetic field slightly tilted from the surface normal. Spectra are plotted for different values of the normal magnetic-field projection, $B_\perp$. At $B_\perp = 11.25$ T a line splitting is observed due to the subband–Landau-level coupling (at $168\,\text{cm}^{-1}$). (b) In comparison, the direct (grating coupled) intersubband absorption peaked at $150\,\text{cm}^{-1}$. The frequency difference is ascribed to the depolarization shift (from Pillath et al., 1989).

they occur at energies of $E = E_{21} \pm n\hbar\omega_c$, where n is an integer. A review of these effects can be found in Batke (1991).

2. Parabolic Quantum Wells

At first sight, parabolically shaped quantum wells are not fundamentally different from any other QWs. It turns out, however, that they exhibit some very interesting properties, especially regarding their infrared (IR) absorption properties.

A parabolic potential shape (see Fig. 37) can be achieved by grading the Al content continuously from zero to a certain value; the same can be done using a digitally graded quasi-alloy (Sundaram *et al.*, 1991). In this harmonic-oscillator potential, the energy spectrum consists of equally spaced levels, $E_n = \hbar\omega_0(n + 1/2)$. The oscillator frequency is related to the shape of the parabola by

$$\omega_0 = \sqrt{\frac{8\Delta}{m^*W^2}} \qquad (101)$$

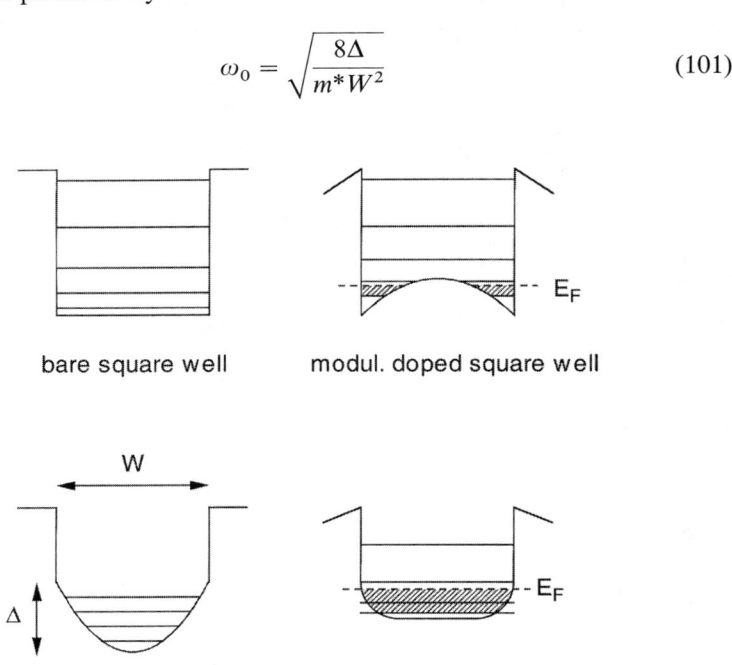

FIG. 37. Schematic of a rectangular (top) and parabolic (bottom) quantum well, before (left) and after (right) introducing electrons through modulation doping. The width W and depth Δ of the parabolic QW are indicated as well as the subbands and the Fermi energy.

where Δ is the depth of the parabola and W is its width (at the top edge). The ratio Δ/W^2 is proportional to the curvature of the parabola. Thus one can expect that a parabolic QW will exhibit resonant intersubband absorption at $\omega = \omega_0$. There is also a remarkable facet to this. Suppose that electrons are introduced in the parabolic QW through modulation doping. As a consequence, the electrons transferred to the well give rise to an additional Hartree potential, which is, according to Poisson's equation, also parabolic and exactly compensates the bare potential over a certain width near the minimum of the parabola, the width depending on the electron density. In this way, a high-mobility quasi-three-dimensional electron system can be tailored (Shayegan et al., 1988), which was the main motivation at the outset of this work. The remarkable phenomenon now is that, no matter how many electrons are located in the parabolic QW and change the self-consistent potential drastically, the resonant absorption always occurs at the bare oscillator frequency ω_0 (Karrai et al., 1989a, 1989b). This is a consequence of the generalized Kohn's theorem (Kohn, 1961; Brey et al., 1989), which states that in a parabolically confined system, the low-frequency excitations are independent of electron-electron interactions, since they only couple to the center-of-mass coordinates of the system. The resonance frequency can also be viewed as the plasma frequency of an 3D electron gas with 3D density $n = 8\varepsilon\varepsilon_0\Delta/e^2W^2$. Figure 37 schematically shows a rectangular and a parabolic quantum well, both before and after introducing carriers through modulation doping.

Such systems were realized based on GaAs–AlGaAs with Δ of the order of 100 meV and W around 1000 Å, leading to a resonance frequency of $\omega_0 \approx 10$ meV. The electron density can be varied with a gate voltage from the low to the mid 10^{11} cm^{-2} range. Situations with several occupied subbands were also investigated. Figure 38 shows absorption spectra (Wixforth et al., 1994) of a 2000-Å-wide parabolic QW for three different densities (from 1.6 to 2.5 × 10^{11} cm^{-2}). Here a grating coupler was used to couple to the intersubband–plasma resonance (see also Wendler et al., 1997). Deviations from a perfectly harmonic potential usually manifest themselves in the occurrence of additional absorption features (Wixforth et al., 1991).

Other experiments have been performed with a magnetic field applied parallel to the layers (Voigt geometry) or tilted from the surface normal. A detailed account can be found in the review by Wixforth et al. (1994).

A nice extension of this concept is realized by embedding a strongly coupled superlattice into a wide parabolic quantum well (Jo et al., 1990), which leads to interesting optical properties (Brey et al., 1990; Streibl et al., 1996).

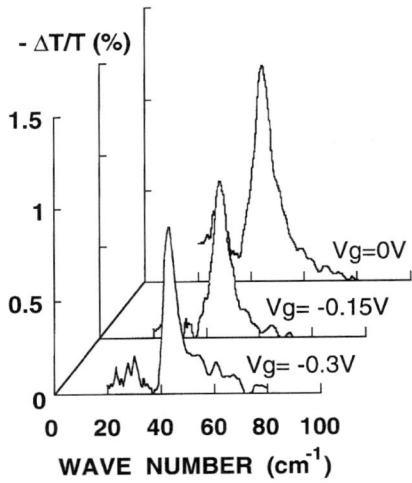

FIG. 38. Absorption spectra of a 200-nm-wide parabolic quantum well for three different gate voltages, as indicated, corresponding to electron concentrations from 1.6 to 2.5 × 10^{11} cm^{-2}. The resonance remains at the same frequency (from Wixforth et al., 1994).

3. IMPURITIES

Although impurities are not really a topic of this chapter, we mention a few things that are related to intersubband transitions.

In a bulk semiconductor the energy levels of shallow donors can be well described in the usual scheme for hydrogen atoms with the quantum numbers N, l (angular momentum), m (magnetic quantum number); that is, the 1s, 2s, 2p ($m = \pm 1, 0$), 3s, 3p ($m = \pm 1, 0$), 3d ($m = \pm 2, \pm 1, 0$), etc., states. In two dimensions, the classification is different, and, for example, the 2p state is only twofold degenerate. The exact two-dimensional limit corresponds to a situation with a single subband, and is only of limited relevance for realistic QWs. In a quasi-2D system with several subbands it is still possible and useful to employ the 3D classification, which, however, becomes modified by the QW potential. The breaking of the translation symmetry in the z direction removes the degeneracy of some states (Greene and Bajaj, 1985), which become pinned to higher QW subbands. The subband index n associated with a certain hydrogenic level can be determined by the relation $n = l - |m| + 1$ (Cheng and McCombe, 1990). Correspondingly, the most important levels near the first subband are 1s, $2p_{\pm 1}$, and near the second subband $2p_0$ and $3d_{\pm 1}$ (for illustration, see Fig. 39).

1 THE BASIC PHYSICS OF INTERSUBBAND TRANSITIONS   87

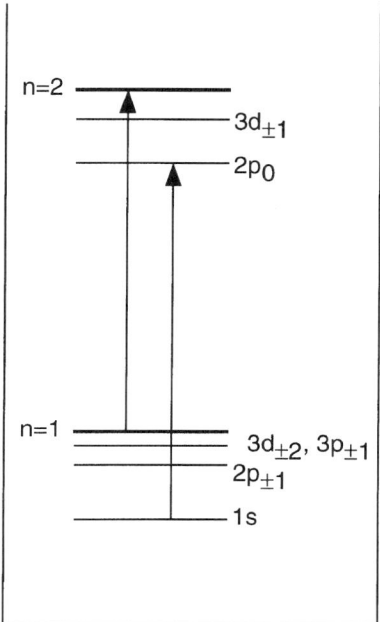

FIG. 39. Schematic of the most important shallow-donor states in a quantum well. The threefold degeneracy of the $2p$ state is lifted and the $2p_0$ state moves up to the second subband, where it becomes the ground state of a new hydrogenic series. Indicated are the intersubband transition and the $1s-2p_0$ donor transition. For details, see text.

The $2p_0$ (or $2p_z$) state actually becomes the ground state of the hydrogenic series associated with the $n = 2$ subband. It turns out that optical transitions are allowed for the $1s-2p_z$ transitions when the light is polarized perpendicular to the layers. So this transition follows the same selection rule as the usual intersubband transition and also occurs at nearly the same energy, thus it can be regarded as an "impurity-shifted intersubband transition." A difference comes about only through the different (and position dependent) binding energies (Lane and Greene, 1986; Helm et al., 1992) of the donors with respect to the $n = 1$ and $n = 2$ subbands. This is the reason why the $1s-2p_z$ transition has been observed only in wide quantum wells, where the donor binding energy is comparable to E_{21} (Helm et al., 1991, 1992), or in superlattices, where it is energetically well separated from the critical points and can actually be used to measure the miniband width of the first and second minibands (Helm et al., 1993). For illustration, we refer back to Fig. 22 in Section VI. For observation of the impurity transition, the doping of the QWs must also be low enough so that the impurity states ("impurity band") have not completely merged with the conduction band.

4. PHOTON DRAG EFFECT

Finally, we mention an effect, where the momentum transfer rather than the energy transfer from the photon to the electron system is relevant. This is the so-called photon-drag effect (Luryi, 1987; Grinberg and Luryi, 1988; Stockman *et al.*, 1990), which is well known in connection with intervalence band transitions in bulk semiconductors such as Ge, but has also been observed with intersubband transitions (Wieck *et al.*, 1990).

Although the photon momentum q is very small as compared to a typical electron momentum (such as the Fermi momentum k_F), it is not completely negligible. Figure 40 shows the intersubband absorption process taking into account the finite momentum transfer. The energy balance for the absorption process reads

$$\hbar\omega - \hbar\omega_{21} = \frac{\hbar^2}{2m^*}(\mathbf{k}_\perp + \mathbf{q}_\perp)^2 - \frac{\hbar^2 \mathbf{k}_\perp^2}{2m^*} \cong \frac{\hbar^2}{m^*}(\mathbf{k}_\perp \cdot \mathbf{q}_\perp) \qquad (102)$$

for $\mathbf{q}_\perp \ll \mathbf{k}_\perp$ ($\mathbf{k}_\perp$ and $\mathbf{k}_\perp + \mathbf{q}_\perp$ are the initial and final in-plane wavevectors, respectively). This can be regarded as a Doppler shift of the resonance

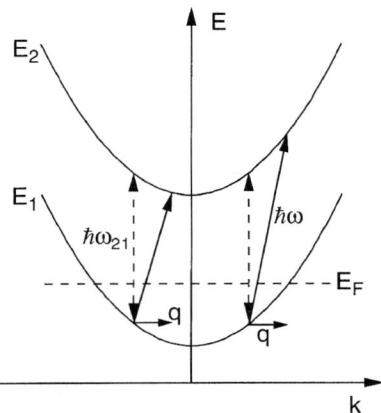

FIG. 40. Illustration of the photon-drag effect in intersubband absorption. The size of the photon momentum q is drawn vastly exaggerated for clarity.

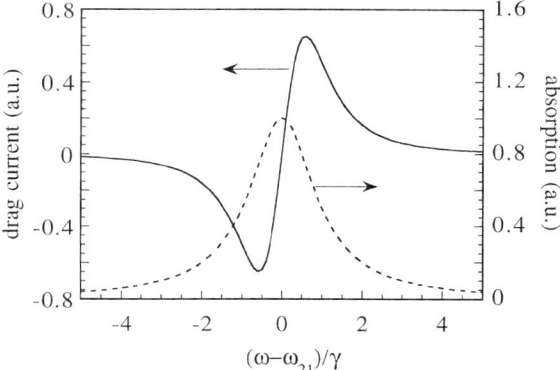

FIG. 41. Dependence of the photon-drag current on the relative detuning, $(\omega - \omega_{21})/\gamma$, calculated for a Lorentzian broadened intersubband transition (solid curve). The absorption line shape (dashed curve) is also shown for comparison.

frequency. The net drag current can be written as

$$\mathbf{j} = -\frac{e\hbar}{m^*}[\tau_2(\mathbf{k}_\perp + \mathbf{q}_\perp) - \tau_1 \mathbf{k}_\perp] \cdot \dot{n} \tag{103}$$

where τ_1 and τ_2 are the momentum relaxation times in the first and second subbands, respectively, and $\dot{n}$ is the number of excitations per unit time and area (Sigg, 1992). The two terms correspond to a drag current of an electron in the second subband and a quasi-hole left behind in the first subband, respectively. Obviously, the current will be large, when the two relaxation times are sufficiently different. This is indeed the case in a modulation doped QW with $\hbar\omega_{21} > \hbar\omega_{op}$, the optical phonon energy. Then τ_2 is determined by optical-phonon scattering and is of the order of 0.5 ps, whereas τ_1 is limited by remote impurity scattering and can be 10 times larger. The quasi-holes are then responsible for the drag current, which gives rise to a dispersion-like line shape as a function of detuning, $(\omega - \omega_{21})/\gamma$, where γ is the HWHM of the absorption line (see Fig. 41). Also shown is the absorption coefficient for comparison. This behavior has been observed experimentally (Wieck et al., 1990). Due to the intrinsic speed of the photon drag effect, which is limited only by the momentum relaxation time, it can be used for extremely fast infrared detectors (Sigg et al., 1995).

XII. Concluding Remarks and Outlook

We have attempted to give an introductory overview over the basic physics of intersubband transitions in quantum wells and its present comprehension. Herein a wealth of different aspects of solid state physics and optics have been shown to play a significant role. The thorough understanding achieved to date has enabled researchers to develop useful applications such as infrared detectors and lasers, which are discussed in other chapters of this volume. At present, much research effort is being devoted to the investigation of coherent and quantum optical effects related to intersubband transitions. Among these are microcavities (Duboz, 1996; Berger et al., 1997; Liu, 1997; Faist et al., 1996a; for a survey, see Burstein and Weisbuch, 1995; Rarity and Weisbuch, 1996), Fano-resonances and electromagnetically induced transparency (Imamoglu and Ram, 1994; Faist et al., 1996b; Schmidt and Imamoglu, 1996; Schmidt et al., 1997), and dressed states (Sadeghi et al., 1995). Many of these new developments rely on the similarity of intersubband transitions with atomic transitions. Although the relaxation and dephasing rates in semiconductors are many orders of magnitudes larger than in atoms, the steady improvement of the material quality may render possible the observation of presently unforeseen phenomena.

Acknowledgments

Most of all, I am grateful to my former graduate student and collaborator Thomas Fromherz, whose help has been invaluable, even after he left academic research. It is my pleasure to acknowledge uncounted hours of discussion with him. Without him, the section concerning the valence band would have never come into existence. I am also grateful to Helga M. Böhm for discussions about and help with many-body theory, to M. Zaluzny for advice about intersubband electromagnetics, and to Rui Yang for discussions on normal-incidence absorption. Special thanks also to S. J. Allen and M. S. Sherwin for their kind hospitality during a one-month stay at the University of California, Santa Barbara, where this chapter was started, and to K. Unterrainer for discussions about many theoretical and experimental aspects. I also want to express my gratitude to H. C. Liu, R. J. Warburton, A. Wixforth, and G. Strasser for providing me with figures from their publications or unpublished data. Finally, I would like to thank my former graduate students W. Hilber, P. Kruck, and M. Seto for their contributions to our joint research, and to G. Bauer for his continuous support and encouragement.

References

Abramovich, Y., Poplawski, J., Ehrenfreund, E., Gershoni, D., Brar, B., and Kroemer, H. (1994). *Phys. Rev. B* **50**, 8922.
Ahn, D., and Chuang, S. L. (1986). *Phys. Rev. B* **34**, 9034.
Ahn, D., and Chuang, S. L. (1987). *Phys. Rev. B* **35**, 4149.
Allen, S. J., Tsui, D. C., and Vinter, B. (1976). *Solid State Commun.* **20**, 425.
Altarelli, M. (1986). In *Heterojunctions and Semiconductor Superlattices*, eds. Allan, G., Bastard, G., Boccara, N., Lannoo, M., and Voos, M. (Springer, Berlin), p. 12.
Altarelli, M., Ekenberg, U., and Fasolino, A. (1985). *Phys. Rev. B* **32**, 5138.
Altschul, V. A., Fraenkel, A., and Finkman, E. (1992). *J. Appl. Phys.* **71**, 4382.
Andersson, J. Y., and Lundqvist, L. (1991). *Appl. Phys. Lett.* **59**, 857.
Andersson, J. Y., and Lundqvist, L. (1992). *J. Appl. Phys.* **71**, 3600.
Andersson, J. Y., Lundqvist, L., and Paska, Z. F. (1991). *Appl. Phys. Lett.* **58**, 2264.
Ando, T. (1977a). *Solid State Commun.* **21**, 133.
Ando, T. (1977b). *Z. Phys. B* **26**, 263.
Ando, T., Eda, T., and Nakayama, M. (1977). *Solid State Commun.* **23**, 751.
Ando, T. (1978). *J. Phys. Soc. Jpn.* **44**, 475.
Ando, T., Fowler, A. B., and Stern, F. (1982). *Rev. Mod. Phys.* **54**, 437.
Asai, H., and Kawamura, Y. (1990). *Appl. Phys. Lett.* **56**, 1149.
Bandara, K. S. M... V., Coon, D. D., Byungsung, O., Lin, Y. F., and Francombe, M. H. (1988). *Appl. Phys. Lett.* **53**, 1931; erratum: (1989). *Appl. Phys. Lett.* **55**, 206.
Bastard, G. (1981). *Phys. Rev. B* **24**, 5693.
Bastard, G. (1982). *Phys. Rev. B* **25**, 7584.
Bastard, G. (1988). *Wave Mechanics Applied to Semiconductor Heterostructures* (Les editions de Physique, Les Ulis, France).
Bastard, G., Mendez, E. E., Chang, L. L., and Esaki, L. (1983). *Phys. Rev. B* **28**, 3241.
Batke, E. (1991). *Festkörperprobleme* (*Adv. Solid State Phys.*) **31**, 297.
Batke, E., Weimann, G., and Schlapp, W. (1989). *Phys. Rev. B* **39**, 11171.
Batke, E., Weimann, G., and Schlapp, W. (1991). *Phys. Rev. B* **43**, 6812.
Bäuerle, R. J., Elsaesser, T., Kaiser, W., Lobentanzer, H., Stolz, W., and Ploog, K. (1988). *Phys. Rev. B* **38**, 4307.
Beadie, G., Rabinovich, W. S., Katzer, D. S., and Goldenberg, M. (1997). *Phys. Rev. B* **55**, 9731.
Beinvogl, W., and Koch, F. (1977). *Solid State Commun.* **24**, 687.
Beinvogl, W., and Koch, F. (1978). *Phys. Rev. Lett.* **40**, 1736.
Berger, V., Vermeire, G., Demeester, P., and Weisbuch, C. (1995). *Appl. Phys. Lett.* **66**, 218.
Berger, V., Duboz, J.-Y., Ducloux, E., Lafon, F., Pavel, I., Boucaud, P., Gauthier-Lafaye, O., Julien, F. H., Tchelnokov, A., and Planel, R. (1997). *Mat. Res. Soc. Symp. Proc.* **450**, 135.
Berreman, D. W. (1963). *Phys. Rev.* **130**, 2193.
Bir, G. L., and Pikus, G. E. (1974). *Symmetry and Strain-Induced Effects in Semiconductors* (Wiley, New York).
Bloss, W. (1989). *J. Appl. Phys.* **66**, 3639.
Boucaud, P., Gao, L., Moussa, Z., Visocekas, F., Julien, F. H., Lourtioz, J.-M., Sagnes, I., Campidelli, Y., and Badoz, P.-A. (1995). *Appl. Phys. Lett.* **67**, 2948.
Boucaud, P., Julien, F. H., Prazeres, R., Ortega, J.-M., Berger, V., Nagle, J., and Leburton, J.-P. (1996). *Electron. Lett.* **32**, 2357.
Boykin, T. B., and Chui, H. C. (1997). *Phys. Rev. B* **55**, 7091.
Braun, M., and Rössler, U. (1985). *J. Phys. C* **18**, 3365.
Brey, L., Johnson, N. F., and Halperin, B. I. (1989). *Phys. Rev. B* **40**, 647.
Brey, L., Johnson, N. F., and Dempsey, J. (1990). *Phys. Rev. B* **42**, 2886.

Brown, E. R., and Eglash, S. J. (1990). *Phys. Rev. B* **41**, 7559.
Brown, E. R., Eglash, S. J., and McIntosh, K. A. (1992). *Phys. Rev. B* **46**, 7244.
Burstein, E., and Weisbuch, C. (eds.). (1995). *Confined Electrons and Photons* (Plenum Press, New York).
Burt, M. G. (1992). *J. Phys. Condens. Matter* **4**, 6651.
Byungsung, O., Choe, J. W., Francombe, M. H., Bandara, K. M. S. V., Coon, D. D., Li, Y. F., and Takei, W. J. (1990). *Appl. Phys. Lett.* **57**, 503.
Campman, K. L., Schmidt, H., Imamoglu, A., and Gossard, A. C. (1996). *Appl. Phys. Lett.* **69**, 2554.
Capasso, F., Sirtori, C., Faist, J., Sivco, D. L., Chu, S.-N. G., and Cho, A. Y. (1992). *Nature* **358**, 565.
Capasso, F., Sirtori, C., and Cho, A. Y. (1994). *IEEE J. Quantum Electron.* **30**, 1313.
Chang, Y.-C., and James, R. B. (1989). *Phys. Rev. B* **39**, 12672.
Chen, W. P., Chen, Y. J., and Burstein, E. (1976). *Surf. Sci.* **58**, 263.
Cheng, J. P., and McCombe, B. D. (1990). *Phys. Rev. B* **42**, 7626.
Chuang, S. L. (1995). *Physics of Optoelectronic Devices* (Wiley, New York).
Chuang, S. L., Luo, M. S.-C., Schmitt-Rink, S., and Pinczuk, A. (1992). *Phys. Rev. B* **46**, 1897.
Chui, H. C., Martinet, E. L., Fejer, M. M., and Harris, J. S., Jr. (1994). *Appl. Phys. Lett.* **64**, 736.
Chun, S. K., and Wang, K. L. (1992). *Phys. Rev. B* **46**, 7682.
Chun, S. K., Pan, D. S., and Wang, K. L. (1993). *Phys. Rev. B* **47**, 15638.
Cohen-Tannoudji, C., Dupont-Roc, J., and Grynberg, G. (1989). In *Photons and Atoms* (Wiley, New York), p. 325–326.
Cole, B. E., Chamberlain, J. M., Henini, M., Nakov, V., and Gobsch, G. (1996). *J. Appl. Phys.* **80**, 6058.
Cole, T., and McCombe, B. D. (1984). *Phys. Rev. B* **29**, 3180.
Corbin, E., Wong, K. B., and Jaros, M. (1994). *Phys. Rev. B* **50**, 2339.
Craig, K., Felix, C. L., Heyman, J. N., Markelz, A. G., Sherwin, M. S., Campman, K. L., Hopkins, P. F., and Gossard, A. C. (1994). *Semicond. Sci. Technol.* **9**, 627.
Craig, K., Galdrikian, B., Heyman, J. N., Markelz, A. G., Williams, J. B., Sherwin, M. S., Campman, K., Hopkins, P. F., and Gossard, A. C. (1996). *Phys. Rev. Lett.* **76**, 2382.
Cui, D., Chen, Z., Pan, S., Lu, H., and Yang, G. (1993). *Phys. Rev. B* **47**, 6755.
Dahl, D. A., and Sham, L. J. (1977). *Phys. Rev. B* **16**, 651.
Davé, D. P., and Taylor, H. F. (1994). *Phys. Lett. A* **184**, 301.
Dingle, R., Wiegmann, W., and Henry, C. H. (1974). *Phys. Rev. Lett.* **33**, 827.
Duboz, J. Y. (1996). *J. Appl. Phys.* **80**, 5432.
Dupont, E., Corkum, P. B., Liu, H. C., Buchanan, M., and Wasilewski, Z. R. (1995). *Phys. Rev. Lett.* **74**, 3596.
Dür, M., Goodnick, S. M., and Lugli, P. (1996). *Phys. Rev. B* **54**, 17794.
Ekenberg, U. (1987). *Phys. Rev. B* **36**, 6152.
Ekenberg, U. (1989). *Phys. Rev. B* **40**, 7714.
Elsaesser, T., Bäuerle, R. J., Kaiser, W., Lobentanzer, H., Stolz, W., and Ploog, K. (1989). *Appl. Phys. Lett.* **54**, 256.
Ensslin, K., Heitmann, D., and Ploog, K. (1989). *Phys. Rev. B* **39**, 10879.
Eppenga, R., Schuurmans, M. F. H., and Colak, S. (1987). *Phys. Rev. B* **36**, 1554.
Esaki, L., and Sakaki, H. (1977). *IBM Techn. Disclosure Bull.* **20**, 2456.
Faist, J., Capasso, F., Hutchinson, A. L., Pfeiffer, L. N., and West, K. W. (1993a). *Phys. Rev. Lett.* **71**, 3573.
Faist, J., Capasso, F., Sirtori, C., Sivco, D. L., Hutchinson, A. L., Chu, S.-N. G., and Cho, A. Y. (1993b). *Appl. Phys. Lett.* **63**, 1354.

Faist, J., Capasso, F., Sivco, D. L., Sirtori, C., Hutchinson, A. L., and Cho, A. Y. (1994a). *Science* **264**, 553.
Faist, J., Sirtori, C., Capasso, F., Pfeiffer, L., and West, K. W. (1994b). *Appl. Phys. Lett.* **64**, 872.
Faist, J., Gmachl, C., Striccoli, M., Sirtori, C., Capasso, F., Sivco, D. L., and Cho, A. Y. (1996a). *Appl. Phys. Lett.* **69**, 2456.
Faist, J., Sirtori, C., Capasso, F., Chu, S.-N. G., Pfeiffer, L. N., and West, K. W. (1996b). *Opt. Lett.* **21**, 985.
Flatté, M. E., Young, P. M., Peng, L. H., and Ehrenreich, H. (1996). *Phys. Rev. B* **53**, 1963.
Fowler, A. B., Fang, F. F., Howard, W. E., and Stiles, P. J. (1966). *Phys. Rev. Lett.* **16**, 901.
Fromherz, T., Koppensteiner, E., Helm, M., Bauer, G., Nützel, J. F., and Abstreiter, G. (1994). *Phys. Rev. B* **50**, 15073.
Fromherz, T., Kruck, P., Helm, M., Bauer, G., Nützel, J. F., and Abstreiter, G. (1996). *Appl. Phys. Lett.* **68**, 3611.
Gauer, C., Wixforth, A., Kotthaus, J. P., Abstreiter, G., Weimann, G., and Schlapp, W. (1995). *Europhys. Lett.* **30**, 111.
Greene, R. L., and Bajaj, K. K. (1985). *Phys. Rev. B* **31**, 4006.
Grinberg, A. A., and S. Luryi, S. (1988). *Phys. Rev. B* **38**, 87.
Goossen, K. W., and Lyon, S. A. (1985). *Appl. Phys. Lett.* **47**, 1257.
Goossen, K. W., Lyon, S. A., and Alavi, K. (1988). *Appl. Phys. Lett.* **53**, 1027.
Gunapala, S. D., Levine, B. F., and Chand, N. (1991). *J. Appl. Phys.* **70**, 305.
Gunnarson, O., and Lundqvist, B. I. (1976). *Phys. Rev. B* **13**, 4274.
Harbecke, B. (1986). *Appl. Phys. B* **39**, 165.
Harbecke, B., Heinz, B., and Grosse, P. (1985). *Appl. Phys. A* **38**, 263.
Hartig, M., Haacke, S., Deveaud, B., and Rota, L. (1996). *Phys. Rev. B* **54**, 14269.
Hartig, M., Haacke, S., Selbmann, P. E., Deveaud, B., Taylor, R. A., and Rota, L. (1998). *Phys. Rev. Lett.* **80**, 1940.
Harwit, A., and Harris, J. S., Jr. (1987). *Appl. Phys. Lett.* **50**, 685.
Hasnain, G., Levine, B. F., Bethea, C. G., Logan, R. A., Walker, J., and Malik, R. J. (1989). *Appl. Phys. Lett.* **54**, 2515.
Haug, H., and Koch, S. W. (1993). *Quantum Theory of the Optical and Electronic Properties of Semiconductors* (World Scientific, Singapore).
Hedin, L., and Lundqvist, B. I. (1971). *J. Phys. C* **4**, 2064.
Heitmann, D., Kotthaus, J. P., and Mohr, E. G. (1982). *Solid State Commun.* **44**, 715.
Heitmann, D., and Mackens, U. (1986). *Phys. Rev. B* **33**, 8269.
Helm, M. (1995). *Semicond. Sci. Technol.* **10**, 557.
Helm, M., Peeters, F. M., DeRosa, F., Colas, E., Harbison, J. P., and Florez, L. T. (1991). *Phys. Rev. B* **43**, 13983.
Helm, M., Peeters, F. M., DeRosa, F., Colas, E., Harbison, J. P., and Florez, L. T. (1992). *Surface Sci.* **263**, 518.
Helm, M., Hilber, W., Fromherz, T., Peeters, F. M., Alavi, K., and Pathak, R. N. (1993). *Phys. Rev. B* **48**, 1601.
Helm, M., Kruck, P., Fromherz, T., Weichselbaum, A., Seto, M., Bauer, G., Moussa, Z., Boucaud, P., Julien, F. H., Loutzioz, J.-M., Nützel, J. F., and Abstreiter, G. (1977). *Thin Solid Films* **294**, 330.
Herman, M. A., Bimberg, D., and Christen, J. (1991). *J. Appl. Phys.* **70**, R1.
Hertle, H., Schuberth, G., Gornik, E., Abstreiter, G., and Schäffler, F. (1991). *Appl. Phys. Lett.* **59**, 2977.
Heyman, J. N., Craig, K., Galdrikian, B., Sherwin, M. S., Campman, K., Hopkins, P. F., Fafard, S., and Gossard, A. C. (1994). *Phys. Rev. Lett.* **72**, 2183.

Heyman, J. N., Unterrainer, K., Craig, K., Galdrikian, B., Sherwin, M. S., Campman, K., Hopkins, P. F., and Gossard, A. C. (1995). *Phys. Rev. Lett.* **74**, 2683.
Heyman, J. N., Unterrainer, K., Craig, K., Williams, J., Sherwin, M. S., Campman, K., Hopkins, P. F., Gossard, A. C., Murdin, B. N., and Langerak, C. J. G. M. (1966). *Appl. Phys. Lett.* **68**, 3019.
Hirayama, Y., Smet, J. H., Peng, L. H., Fonstad, C. G., and Ippen, E. P. (1993). *Appl. Phys. Lett.* **63**, 1663.
Höpfel, R. A., and Gornik, E. (1986). In *Heterojunctions and Semiconductor Superlattices*, eds. Allan, G. *et al.* (Springer, Berlin), p. 84.
Huang, D., Gumbs, G., and Manasreh, M. O. (1995). *Phys. Rev. B* **52**, 14126.
Hunsche, S., Leo, K., Kurz, H., and Köhler, K. (1994). *Phys. Rev. B* **50**, 5791.
Imamoglu, A., and Ram, R. J. (1994). *Opt. Lett.* **19**, 1744.
Jiang, M. Y. (1992). *Solid State Commun.* **84**, 81.
Jo, J., Santos, M., Shayegan, M., Suen, Y. W., Engel, L. W., and Lanzilotto, A. M. (1990). *Appl. Phys. Lett.* **57**, 2130.
Jogai, B. (1991). *J. Vac. Sci. Technol. B* **9**, 2473.
Julien, F. H., and P. Boucaud, P. (1997). In *Optical Spectroscopy of Low-Dimensional Semiconductors*, ed. Abstreiter, G. (Kluwer Academic Publishers, Dordrecht), p. 41.
Julien, F. H., Lourtioz, J.-M., Herschkorn, N., Delacourt, D., Pocholle, J. P., Papuchon, M., Planel, R., and Le Roux, G. (1988). *Appl. Phys. Lett.* **53**, 116; Erratum ibid. (1993) **62**, 2289.
Kaindl, R., Lutgen, S., Woerner, M., Elsaesser, T., Nottelmann, B., Axt, V. M., Kuhn, T., Hase, A., and Künzel, H. (1998). *Phys. Rev. Lett.* **80**, 3575.
Kamgar, A., Kneschaurek, P., Dorda, G., and Koch, J. F. (1974). *Phys. Rev. Lett.* **32**, 1251.
Kane, E. O. (1969). In *Tunneling Phenomena in Solids*, ed. Burstein, E., and Lundqvist, S. (Plenum, New York), p. 1.
Kane, M. J., Emeny, M. T., Apsley, N., Whitehouse, C. R., and Lee, D. (1988). *Semicond. Sci. Technol.* **3**, 722.
Karrai, K., Drew, H. D., Lee, H. W., and Shayegan, M. (1989a). *Phys. Rev. B* **39**, 1426.
Karrai, K., Ying, X., Drew, H. D., and Shayegan, M. (1989b). *Phys. Rev. B* **40**, 12020.
Karunasiri, G., Park, J. S., Chen, J., Shih, R., Scheihing, J. F., and Dodd, M. A. (1995). *Appl. Phys. Lett.* **67**, 2600.
Katz, J., Zhang, Y., and Wang, W. I. (1992). *Appl. Phys. Lett.* **61**, 1697.
Keilmann, F. (1994). *Solid State Commun.* **92**, 223.
Khurgin, J. (1993). *Appl. Phys. Lett.* **62**, 1390.
Kim, B. W., and Majerfeld, A. (1995). *J. Appl. Phys.* **77**, 4552.
Kim, K. T., Lee, S. S., and Chuang, S. L. (1990). *J. Appl. Phys.* **69**, 6617.
Kneschaurek, P., Kamgar, A., and Koch, J. F. (1976). *Phys. Rev. B* **14**, 1610.
Kohn, W. (1961). *Phys. Rev.* **123**, 1242.
Kohn, W., and Sham, L. J. (1965). *Phys. Rev.* **140**, A1133.
Lane, P., and Greene, R. L. (1986). *Phys. Rev. B* **33**, 5871.
Lee, C., and Wang, K. L. (1992). *Appl. Phys. Lett.* **60**, 2264.
Lee, C., and Wang, K. L. (1994). *Appl. Phys. Lett.* **64**, 1256.
Lee, S.-C., Galbraith, I., and Pidgeon, C. R. (1995). *Phys. Rev. B* **52**, 1874.
Lee, S.-C., and Galbraith, I. (1997). *Phys. Rev. B* **55**, R16025.
Lenz, G., and Salzman, J. (1990). *Appl. Phys. Lett.* **56**, 871.
Levenson, J. A., Dolique, G., Oudar, J. L., and Abram, I. (1990). *Phys. Rev. B* **41**, 3688.
Levine, B. F., Malik, R. J., Walker, J., Choi, K. K., Bethea, C. G., Kleinman, D. A., and Vandenberg, J. M. (1987). *Appl. Phys. Lett.* **50**, 273.
Levine, B. F., Gunapala, S. D., Kuo, J. M., Pei, S. S., and Hui, S. (1991). *Appl. Phys. Lett.* **59**, 1864.

Levine, B. F. (1993). *J. Appl. Phys.* **74**, R1.
Lew Yan Voon, L. C., Willatzen, M., and Ram-Mohan, L. R. (1995). *J. Appl. Phys.* **78**, 295.
Lew Yan Voon, L. C., Willatzen, M., Cardona, M., and Ram-Mohan, L. R. (1996). *J. Appl. Phys.* **80**, 600 (Comment); Reply: Peng, L. H., and Fonstad, C. G. (1996). *J. Appl. Phys.* **80**, 603.
Li, H. S., Karunasiri, R. P. G., Chen, Y. W., and Wang, K. L. (1993). *J. Vac. Sci. Technol. B* **11**, 922.
Li, S. S., and Su, Y. K. (1998). *Intersubband Transitions in Quantum Wells: Physics and Devices* (Kluwer Academic Publishers, Dordrecht).
Li, W. J., McCombe, B. D., Chambers, F. A., Devane, G. P., Ralston, J., and Wicks, G. (1990a). *Surf. Sci.* **228**, 164.
Li, W. J., McCombe, B. D., Chambers, F. A., Devane, G. P., Ralston, J., and Wicks, G. (1990b). *Phys. Rev. B* **42**, 11953.
Li, W. J., and McCombe, B. D. (1992). *J. Appl. Phys.* **71**, 1038.
Liu, A. (1994). *Phys. Rev. B* **50**, 8569.
Liu, A. (1997). *Phys. Rev. B* **55**, 7101.
Liu, H. C., Levine, B. F., and Andersson, J. Y. (eds.). (1994). *Quantum Well Intersubband Transition Physics and Devices* (Kluwer Academic Publishers, Dordrecht).
Liu, H. C. (1993). *J. Appl. Phys.* **73**, 3062.
Liu, H. C., Buchanan, M., and Wasilewski, Z. R. (1998). *Appl. Phys. Lett.* **72**, 1682.
Loehr, J. P., and Manasreh, M. O. (1993). In *Semiconductor Quantum Wells and Superlattices for Long-Wavelength Infrared Detectors*, ed. Manasreh, M. O. (Artech House, Boston).
Luo, K., Zheng, H., Lu, Z., Xu, J., Xu, Z., Zhang, T., Li, C., Yang, X., and Tian, J. (1997). *Appl. Phys. Lett.* **70**, 1155.
Luo, M. S.-C., Chuang, S.-L., Schmitt-Rink, S., and Pinczuk, A. (1993). *Phys. Rev. B* **48**, 11086.
Luryi, S. (1987). *Phys. Rev. Lett.* **58**, 2263.
Lutgen, S., Kaindl, R. A., Woerner, M., Elsaesser, T., Hase, A., Künzel, H., Meglio, D., and Lugli, P. (1996a). *Phys. Rev. Lett.* **77**, 3657.
Lutgen, S., Kaindl, R. A., Woerner, M., Elsaesser, T., Hase, A., and Künzel, H. (1996b). *Phys. Rev. B* **54**, 17343.
Luttinger, J. M., and Kohn, W. (1955). *Phys. Rev.* **97**, 869.
Maan, J. C. (1984). In *Two-Dimensional Systems, Heterostructures, and Superlattices*, eds. Bauer, G., Kuchar, F., and Heinrich, H. (Springer, Berlin), p. 183.
Man, P., and Pan, D. S. (1992). *Appl. Phys. Lett.* **61**, 2799.
Manasreh, M. O., Szmulowicz, F., Vaughan, T., Evans, K. R., Stutz, C. E., and Fischer, D. W. (1991). *Phys. Rev. B* **43**, 9996.
McCombe, B. D., Holm, R. T., and Schafer, D. E. (1979). *Solid State Commun.* **32**, 603.
Meney, A. T., Gonul, B., and O'Reilly, E. P. (1994). *Phys. Rev. B* **50**, 10893.
Merlin, R. (1987). *Solid State Commun.* **64**, 99.
Mii, Y. J., Wang, K. L., Karunasiri, R. P. G., and Yuh, P. F. (1990a). *Appl. Phys. Lett.* **56**, 1046.
Mii, Y. J., Karunasiri, R. P. G., Wang, K. L., Chen, M., and Yuh, P. F. (1990b). *Appl. Phys. Lett.* **56**, 1986.
Milonni, P. W., and Eberly, J. H. (1988). *Lasers* (Wiley, New York).
Murdin, B. N., Knippels, G. M. H., van der Meer, A. F. G., Pidgeon, C. R., Langerak, C. J. G. M., Helm, M., Heiss, W., Unterrainer, K., Gornik, E., Geerinck, K. K., Hovenier, N. J., and Wenckebach, W. Th. (1994). *Semicond. Sci. Technol.* **9**, 1554.
Murdin, B. N., Heiss, W., Langerak, C. J. G. M., Lee, S.-C., Galbraith, I., Strasser, G., Gornik, E., Helm, M., and Pidgeon, C. R. (1997). *Phys. Rev. B* **55**, 5171.
Nakayama, M. (1975). *J. Phys. Soc. Jpn.* **39**, 265.
Nakayama, M. (1977). *Solid State Commun.* **21**, 587.

Nee, S.-M., Claessen, U., and Koch, F. (1984). *Phys. Rev. B* **29**, 3449.
Nelson, D. F., Miller, R. C., and Kleinman, D. A. (1987). *Phys. Rev. B* **35**, 7770.
Newson, D. J., and Kurobe, A. (1988). *Semicond. Sci. Technol.* **3**, 786.
Nikonov, D. E., Imamoglu, A., Butov, L. V., and Schmidt, H. (1997). *Phys. Rev. Lett.* **79**, 4633.
Oberli, D. Y., Wake, D. R., Klein, M. V., Klem, J., Henderson, T., and Morkoc, H. (1987). *Phys. Rev. Lett.* **59**, 696.
Oelting, S., Merkt, U., and Kotthaus, J. P. (1986). *Surface Sci.* **170**, 402.
Olszakier, M., Ehrenfreund, E., Cohen, E., Bajaj, J., and Sullivan, G. J. (1989). *Phys. Rev. Lett.* **62**, 2997.
Park, J. S., Karunasiri, R. P. G., and Wang, K. L. (1992). *Appl. Phys. Lett.* **61**, 681.
Peeters, F. M., Matulis, A., Helm, M., Fromherz, T., and Hilber, W. (1993). *Phys. Rev. B* **48**, 12008.
Peng, L. H., and Fonstad, C. G. (1993). *Appl. Phys. Lett.* **62**, 3342.
Peng, L. H., and Fonstad, C. G. (1995). *J. Appl. Phys.* **77**, 747.
Peng, L. H., Smet, J. H., Broekaert, T. P. E., and Fonstad, C. G. (1992). *Appl. Phys. Lett.* **61**, 2078.
Peng, L. H., Smet, J. H., Brokaert, T. P. E., and Fonstad, C. G. (1993). *Appl. Phys. Lett.* **62**, 2413.
People, R., and Sputz, S. K. (1990). *Phys. Rev. B* **41**, 8431.
People, R., Bean, J. C., Bethea, C. G., Sputz, S. K., and Peticolas, L. J. (1992a). *Appl. Phys. Lett.* **61**, 1122.
People, R., Bean, J. C., Sputz, S. K., Bethea, C. G., and Peticolas, L. J. (1992b). *Thin Solid Films* **222**, 120.
Perera, A. G. U., Choe, J.-W., and Francombe, M. H. (1997). *Thin Films* **23**, 217.
Persson, A., and Cohen, R. M. (1988). *Phys. Rev. B* **38**, 5568.
Pikus, G. E., and Bir, G. L. (1960). *Sov. Phys. Solid State* **1**, 1502.
Pillath, J., Batke, E., Weimann, G., and Schlapp, W. (1989). *Phys. Rev. B* **40**, 5879.
Pinczuk, A., and Worlock, J. M. (1982). *Surf. Sci.* **113**, 69.
Pinczuk, A., and Abstreiter, G. (1989). In *Light Scattering in Solids V*, eds. Cardona, M., and Güntherodt, G. (Springer, Berlin), p. 153.
Pinczuk, A., Schmitt-Rink, S., Danan, G., Valladares, J. P., Pfeiffer, L. N., and West, K. W. (1989). *Phys. Rev. Lett.* **63**, 1633.
Ralston, J. D., Gallagher, D. F. G., Bittner, P., Fleissner, J., Dischler, B., and Koidl, P. (1992). *J. Appl. Phys.* **71**, 3562.
Ramsteiner, M., Ralston, J. D., Koidl, P., Dischler, B., Biebl, H., Wagner, J., and Ennen, H. (1990). *J. Appl. Phys.* **67**, 3900.
Rarity, J., and Weisbuch, C. (eds.). (1996). *Microcavities and Photonic Bandgaps: Physics and Applications* (Kluwer, Dordrecht).
Reisinger, H., and Koch, F. (1981). *Solid State Commun.* **37**, 429.
Rikken, G. L. J. A., Sigg, H., Langerak, C. J. G. M., Myron, H. W., Perenboom, J. A. A. J., and Weimann, G. (1986). *Phys. Rev. B* **34**, 5590.
Rössler, U. (1984). *Solid State Commun.* **49**, 943.
Rosencher, E., and Bois, Ph. (1991). *Phys. Rev. B* **44**, 11415.
Rosencher, E., Vinter, B., and Levine, B. F. (eds.). (1992). *Intersubband Transitions in Quantum Wells* (Plenum Press, New York).
Rosencher, E., Fiore, A., Vinter, B., Berger, V., Bois, Ph., and Nagle, J. (1996). *Science* **271**, 168.
Sa'ar, A. (1993). *J. Appl. Phys.* **74**, 5263.
Sadeghi, S. M., Young, J. F., and Meyer, J. (1995). *Phys. Rev. B* **51**, 13349.
Sakaki, H., Noda, T., Hirakawa, H., Tanaka, M., and Matsusue, T. (1987). *Appl. Phys. Lett.* **51**, 1934.

Schlesinger, Z., Hwang, J. C. M., and Allen, S. J., Jr. (1983). *Phys. Rev. Lett.* **50**, 2098.
Schmidt, H., and Imamoglu, A. (1996). *Opt. Commun.* **131**, 333.
Schmidt, H., Campman, K. L., Gossard, A. C., and Imamoglu, A. (1997). *Appl. Phys. Lett.* **70**, 3455.
Schulman, J. N., and Chang, Y.-C. (1985). *Phys. Rev. B* **31**, 2056.
Seilmeier, A., Hübner, H. J., Abstreiter, G., Weimann, G., and W. Schlapp, W. (1987). *Phys. Rev. Lett.* **59**, 1345.
Shaw, M. J., and Jaros, M. (1994). *Phys. Rev. B* **50**, 7768.
Shayasteh, S. F., Dumelow, T., Parker, T. J., Mirjalili, G., Vorobjev, L. E., Donetsky, D. V., and Kastalsky, A. (1996). *Semicond. Sci. Technol.* **11**, 323.
Shayegan, M., Sajoto, T., Santos, M., and Silvestre, C. (1988). *Appl. Phys. Lett.* **53**, 791.
Sherwin, M. S., Craig, K., Galdrikian, B., Heyman, J., Markelz, A., Campman, K., Fafard, S., Hopkins, P. F., and A. C. Gossard (1995). *Physica D* **83**, 229.
Shik, A. (1988). *Sov. Phys. Semicond.* **22**, 1165.
Shik, A. (1992). In *Intersubband Transitions in Quantum Wells*, eds. Rosencher, E., Vinter, B., and Levine, B. F. (Plenum Press, New York), p. 319.
Sigg, H. (1992). In *Intersubband Transitions in Quantum Wells*, eds. Rosencher, E., Vinter, B., and Levine, B. (Plenum Press, New York), p. 83.
Sigg, H., Kwakernaak, M., Margotte, B., Erni, D., van Son, P., and Köhler, K. (1995). *Appl. Phys. Lett.* **67**, 2827.
Sirtori, C., Capasso, F., Faist, J., Sivco, D. L., Chu, S.-N. G., and Cho, A. Y. (1992). *Appl. Phys. Lett.* **61**, 898.
Sirtori, C., Capasso, F., Faist, J., and Scandolo, S. (1994). *Phys. Rev. B* **50**, 8663.
Smet, J. H., Peng, L. H., Hirayama, Y., and Fonstad, C. G. (1994). *Appl. Phys. Lett.* **64**, 986.
Smith, J. S., Chiu, L. C., Margalit, S., Yariv, A., and Cho, A. Y. (1983). *J. Vac. Sci. Tech. B* **1**, 376.
Stern, F., and Howard, W. E. (1967). *Phys. Rev.* **163**, 816.
Stockman, M. I., Pandey, L. N., and George, T. F. (1990). *Phys. Rev. Lett.* **65**, 3433.
Stoklitsky, S. A., Holtz, P. O., Monemar, B., Zhao, Q. X., and Lundström, T. (1994). *Appl. Phys. Lett.* **65**, 1706.
Stoklitsky, S. A., Zhao, Q. X., Holtz, P. O., Monemar, B., and Lundström, T. (1995). *J. Appl. Phys.* **77**, 5256.
Streibl, M., Warburton, R. J., Wixforth, A., Campman, K. L., and Gossard, A. C. (1996). In *Proc. 23rd Int. Conf. on the Physics of Semiconductors*, eds. Scheffler, M., and Zimmermann, R. (World Scientific, Singapore), p. 1739.
Sundaram, M., Chalmers, S. A., Hopkins, P. F., and Gossard, A. C. (1991). *Science* **254**, 1326.
Szmulowicz, F., and Brown, G. J. (1995). *Phys. Rev. B* **51**, 13203; *Appl. Phys. Lett.* **66**, 1659.
Szmulowicz, F., and Manasreh, M. O. (1992). *J. Vac. Sci. Technol. B* **10**, 1341.
Szmulowicz, F., Manasreh, M. O., Stutz, C. E., and Vaughan, T. (1994). *Phys. Rev. B* **50**, 11618.
Tatham, M. C., Ryan, J. F., and Foxon, C. T. (1989). *Phys. Rev. Lett.* **63**, 1637.
Terzis, A. F., Liu, X. C., Petrou, A., McCombe, B. D., Dutta, M., Shen, H., Smith, D. D., Cole, M. W., Taysing-Lara, M., and Newman, P. G. (1990). *J. Appl. Phys.* **67**, 2501.
Tsang, L., and Chuang, S. L. (1995). *IEEE J. Quant. Electron.* **31**, 20.
Ullrich, C. A., and Vignale, G. (1998). *Phys. Rev. B* **58**, 15756.
Vinter, B. (1976). *Phys. Rev. B* **13**, 4447.
Vinter, B. (1977). *Phys. Rev. B* **15**, 3947.
Vodopyanov, K. L., Chazapis, V., and Phillips, C. C. (1996). *Appl. Phys. Lett.* **69**, 3405.
Vodopyanov, K. L., Chazapis, V., Phillips, C. C., Sung, B., and Harris, J. S. (1997). *Semicond. Sci. Technol.* **12**, 708.
von Allmen, P. (1992). *Phys. Rev. B* **46**, 13351.

von Allmen, P., Berz, M., Petrocelli, G., Reinhart, F.-K., and Harbeke, G. (1988). *Semicond. Sci. Technol.* **3**, 1211.
Wang, K. L., and Karunasiri, R. P. G. (1993). In *Semiconductor Quantum Wells and Superlattices for Long-Wavelength Infrared Detectors*, ed. Manasreh, M. O. (Artech House, Boston), p. 139.
Wang, K. L., Lee, C., and Chun, S. K. (1994a). In *Quantum Well Intersubband Transition Physics and Devices*, eds. Liu, H. C., Levine, B. F., and Andersson, J. Y. (Kluwer Academic Publishers, Dordrecht), p. 221.
Wang, Y. H., Li, Sheng, S., Ho, Pin, and Manasreh, M. O. (1993). *J. Appl. Phys.* **74**, 1382.
Wang, Y. H., Li, Sheng, S., Chu, J., and Ho, Pin. (1994b). *Appl. Phys. Lett.* **64**, 727.
Warburton, R. J., Gauer, C., Wixforth, A., Kotthaus, J. P., Brar, E., and Kroemer, H. (1996). *Phys. Rev. B* **53**, 7903.
Warburton, R. J., Weilhammer, K., Kotthaus, J. P., Thomas, M., and Kroemer, H. (1998). *Phys. Rev. Lett.* **80**, 2185.
Wendler, L., Kraft, T., Hartung, M., Berger, A., Wixforth, A., Sundaram, M., English, J. H., and Gossard, A. C. (1997). *Phys. Rev. B* **55**, 2303.
West, L. C., and Eglash, S. J. (1985). *Appl. Phys. Lett.* **46**, 1156.
West, L. C., and Roberts, C. W. (1994). In *Quantum Well Intersubband Transition Physics and Devices*, eds. Liu, H. C., Levine, B. F., and Andersson, J. Y. (Kluwer, Dordrecht), p. 501.
Wieck, A. D., Batke, E., Heitmann, D., and Kotthaus, J. P. (1984). *Phys. Rev. B* **30**, 4653.
Wieck, A. D., Maan, J. C., Merkt, U., Kotthaus, J. P., Ploog, K., and Weimann, G. (1987). *Phys. Rev. B* **35**, 4145.
Wieck, A. D., Bollweg, K., Merkt, U., Weimann, G., and Schlapp, W. (1988). *Phys. Rev. B* **38**, 10158.
Wieck, A. D., Thiele, F., Merkt, U., Ploog, K., Weimann, G., and Schlapp, W. (1989). *Phys. Rev. B* **39**, 10879.
Wieck, A. D., Sigg, H., and Ploog, K. (1990). *Phys. Rev. Lett.* **64**, 463.
Wiesinger, K., Reisinger, H., and Koch, F. (1982). *Surf. Sci.* **113**, 102.
Winkler, R., and Rössler, U. (1993). *Phys. Rev. B* **48**, 8918.
Wixforth, A., Kaloudis, M., Rocke, C., Ensslin, K., Sundaram, M., English, J. H., and Gossard, A. C. (1994). *Semicond. Sci. Technol.* **9**, 215 (and references therein).
Wixforth, A., Sundaram, M., Ensslin, K., English, J. H., and Gossard, A. C. (1991). *Phys. Rev. B* **43**, 10,000.
Xie, H., Piao, J., Katz, J., and Wang, W. I. (1991a). *J. Appl. Phys.* **70**, 3152.
Xie, H., Katz, J., and Wang, W. I. (1991b). *Appl. Phys. Lett.* **59**, 3601.
Xie, H., Katz, J., and Wang, W. I. (1992a). *J. Appl. Phys.* **72**, 3681.
Xie, H., Katz, J., and Wang, W. I. (1992b). *Appl. Phys. Lett.* **61**, 2694.
Xie, H., Katz, J., Wang, W. I., and Chang, Y. C. (1992c). *J. Appl. Phys.* **71**, 2844.
Xie, H., Wang, W. I., Meyer, J. R., Hoffman, C. A., and Bartoli, F. J. (1993). *J. Appl. Phys.* **74**, 1195.
Xie, H., Wang, W. I., and Meyer, J. R. (1994). *J. Appl. Phys.* **76**, 92.
Xu, B., and Hu, Q. (1997). *Appl. Phys. Lett.* **70**, 2511.
Xu, Wenlan, Fu, Y., and Willander, M. (1993). *Phys. Rev. B* **48**, 11477.
Xu, Wenlan, Willander, M., and Shen, S. C. (1994). *Phys. Rev. B* **49**, 13760.
Yang, C.-L., Pan, D.-S., and Somoano, R. (1989). *J. Appl. Phys.* **65**, 3253.
Yang, D. D., Julien, F. H., Lourtioz, J.-M., Boucaud, P., and Planel, R. (1990). *IEEE Photon. Technol. Lett.* **2**, 398.
Yang, R. Q. (1995a). *Appl. Phys. Lett.* **66**, 959.
Yang, R. Q. (1995b). *Phys. Rev. B* **52**, 11958.
Yang, R. Q., Xu, J. M., and Sweeny, M. (1994). *Phys. Rev. B* **50**, 7474.

Yi, K. S., and Quinn, J. J. (1983). *Phys. Rev. B* **27**, 2396.
Yoo, K. H., Ram-Mohan, L. R., and Nelson, D. F. (1989). *Phys. Rev. B* **39**, 12808.
Yu, L. S., Li, S. S., Wang, Y. H., and Kao, Y. C. (1992). *J. Appl. Phys.* **72**, 2105.
Yuh, P. F., and Wang, K. L. (1988). *Phys. Rev. B* **38**, 8377.
Yuh, P. F., and Wang, K. L. (1989). *J. Appl. Phys.* **65**, 4377.
Yuh, P. F., Kuo, T. C., and Wang, K. L. (1990). *J. Appl. Phys.* **67**, 3199.
Zaluzny, M. (1989). *Phys. Rev. B* **40**, 8495.
Zaluzny, M. (1991). *Phys. Rev. B* **43**, 4511.
Zaluzny, M. (1992a). *Solid State Commun.* **82**, 565.
Zaluzny, M. (1992b). *Appl. Phys. Lett.* **60**, 1486.
Zaluzny, M. (1996). *Solid State Commun.* **97**, 809.
Zaluzny, M., and Nalewajko, C. (1997). *J. Appl. Phys.* **81**, 3323.
Zaluzny, M., and Nalewajko, C. (1999). *Phys. Rev. B* **59**, 13043.
Zanier, S., Berroir, J. M., Guldner, Y., Vieren, J. P., Sagnes, I., Glowacki, F., Campidelli, Y., and Badoz, P. A. (1995). *Phys. Rev. B* **51**, 14311.
Zawadzki, W. (1983). *J. Phys. C* **16**, 229.
Zawadzki, W. (1987). *Semicond. Sci. Technol.* **2**, 550.
Zhang, Y., Baruch, N., and Wang, W. I. (1993). *Appl. Phys. Lett.* **63**, 1068.
Zhang, Y., Baruch, N., and Wang, W. I. (1994). *J. Appl. Phys.* **75**, 3690.
Zheng, L., Schaich, W. L., and MacDonald, A. H. (1990). *Phys. Rev. B* **41**, 8493.

CHAPTER 2

Quantum Interference Effects in Intersubband Transitions

Jerome Faist

INSTITUTE OF PHYSICS
UNIVERSITY OF NEUCHÂTEL
NEUCHÂTEL, SWITZERLAND

Carlo Sirtori

THOMSON-CSF
LABORATOIRE CENTRAL DE RECHERCHES
ORSAY, FRANCE

Federico Capasso, Loren N. Pfeiffer, Ken W. West, Deborah L. Sivco, and Alfred Y. Cho

BELL LABORATORIES, LUCENT TECHNOLOGIES
MURRAY HILL, NEW JERSEY

I. INTRODUCTION	101
II. EXPERIMENTAL SETUP	104
III. BOUND STATE ABOVE A QUANTUM WELL THROUGH ELECTRON WAVE INTERFERENCE	104
IV. QUANTUM INTERFERENCE IN A COUPLED-WELL SYSTEM	108
V. FANO INTERFERENCE IN INTERSUBBAND ABSORPTION	112
VI. CONTROL OF QUANTUM INTERFERENCE BY TUNNELING TO A CONTINUUM . . .	116
1. Absorption Experiments	118
2. Emission Experiments	122
REFERENCES	127

I. Introduction

Quantum interference is at the heart of quantum mechanics. Paths interfere when they connect identical initial and final states and the sign of the interference is determined by the phase difference accumulated between the paths. Phase-breaking processes such as collisions and coupling of the system under consideration to the continuum of degrees of freedom of the surrounding environment, for example, via dissipation processes, tend to destroy the coherence between the paths and thus the interference. An interesting situation arises when this coherence can be manipulated externally. Recent examples in atomic physics are based on the coherent control of interfering atoms coupled by laser beams. They include interference between multiple absorption pathways in two-photon absorption (Stewart and Diebold, 1986) and in molecular dissociation (Class-Manjean et al., 1988); absorption cancellation in dressed three-level atomic systems, which may lead to laser without inversion (Zibrov et al., 1995; Padmadandu et al., 1996), and the production of a large index of refraction with vanishing absorption by atoms prepared in a coherent superposition of an excited state doublet (Scully, 1991).

The development of molecular beam epitaxy (MBE) (Cho, 1991) and band-structure engineering (Capasso, 1991) have made possible the design and realization of quantum semiconductor structures with new and unusual optical properties (Capasso et al., 1997). A key factor in this success has been the synthesis of potentials of practically any shape. This has enabled researchers to control the boundary conditions of the envelope functions, their phase, and their relative overlap to an unprecedented degree.

The importance of coherence has recently been demonstrated in time-resolved experiments such as the generation of terahertz radiation from charge oscillations in coupled wells (Roskos et al., 1993) or in transport experiments such as the "resistance resonance" (Palevski et al., 1990). Interference phenomena between one- and two-photon absorption in quantum well infrared photodetectors have also been demonstrated (Dupont et al., 1995).

In this chapter, we explore quantum interference phenomena associated with intersubband optical absorption in high-quality heterostructures. Because of the deltalike joint density of states, intersubband transitions—that is, those connecting quantized electronic states *within the same band*—are especially well suited to explore the new coherence effects in semiconductors inspired by atomic physics. We summarize the systems studied in Fig. 1. First, we explore how one is able to build a bound state above the classical continuum using coherent Bragg reflection from coupled quantum wells placed on each sides of a single quantum well (Fig. 1a). We then show how, by coupling two quantum wells through a thin barrier and controlling the coupling by the application of an external electric field, cancellation of the absorption from the ground state can be achieved (Fig. 1b).

Of particular interest is coherent control through tunneling into an energy continuum. The simplest case is the one discussed in the landmark paper of Fano (Fano, 1961). In the Fano effect, two optical absorption paths from the ground state, a direct one and an indirect one mediated by a resonance,

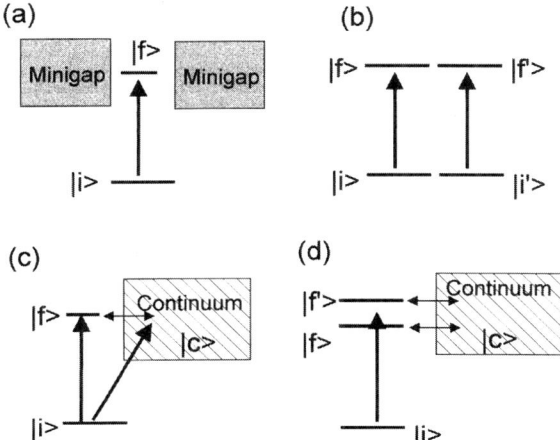

FIG. 1. Summary of the different systems studied in this chapter: (a) a bound state is created in the continuum by electron interference in the barriers, (b) interference between two quantum wells modulates the optical absorption, (c) Fano interference, and (d) interference in absorption between two states coupled to a single continuum.

to the same energy continuum interfere to produce an asymmetric absorption line with a zero near the maximum. The 180° phase shift in the absorption matrix element occurs as the wavelength is scanned across the resonance (Fig. 1c). Similar interference effects are produced when both paths are mediated by two different resonances. Here the two resonances, broadened by tunneling into a continuum, arise from the mixing of the states of two quantum wells (see Fig. 1d). Reversing the direction of tunneling from this doublet of resonances reverses the sign of quantum interference (constructive or destructive) in optical absorption between the ground state and the two resonances.

The practical interest of systems exhibiting Fano resonance arose because, Akripin and Heller (1983) showed that in these situations there is lack of reciprocity between absorption and emission line shapes; that is, the emission remains finite even at the photon energy at which the absorption vanishes. They proposed the use of this effect as the basis to achieve a laser without population inversion (LWI). A few years later, Harris showed that a system such as shown in Fig. 1d), where two distincts states may tunnel to a single continuum, would provide a new and much more efficient way of obtaining a LWI (Harris, 1989). The main reason is that the two resonances will concentrate their oscillator strength in a narrow energy band determined by the escape rate Γ and the splitting between the two states. An implementation of this proposal using intersubband transitions in coupled quantum wells was discussed in a theoretical paper by Imamoglu and Ram (1994).

II. Experimental Setup

Because of the well-known selection rule, intersubband absorption occurs only for light polarized with a component of the electric field perpendicular to the plane of the quantum well layers. To maximize it we used a waveguide geometry. The samples, grown on semi-insulate substrates (therefore transparent to the infrared radiation), were cleaved in stripes and the cleaved edges were polished at 45° to provide a multipass waveguide. The absorption spectra were measured with a Nicolet 800 Fourier transform infrared (FTIR) spectrometer. The absorbance for polarization parallel to the plane of the layers ($-\log_{10} T_{\parallel}$) is subtracted from the absorbance normal to the plane of the layer ($-\log_{10} T_{\perp}$), to remove instrumental, substrate, and free-carrier absorption contributions.

III. Bound State above a Quantum Well through Electron Wave Interference

Consider a thin quantum well with thick barriers, designed to have a single bound state of energy E_1. At energies greater than the barrier height, there are only scattering states. At certain energies, corresponding to a semi-integer number of de Broglie wavelengths across the well, one finds enhanced transmission resonances (Bastard, 1990). Although at these energies the electron amplitude in the well layer is enhanced, the corresponding states remain extended. Figure 2a represents the energy diagram of a series of such wells consisting of 3.2-nm layers of GaInAs separated by 15-nm-thick AlInAs barriers. Our reference sample contains 20 of these wells n-type doped with silicon (3×10^{11} cm^{-2}). Note that at energy $> \Delta E_c$, bands called minibands are formed from the coupling between wells. On the other hand, the confined states of the wells (E_1) do not form a miniband because of localization effects associated with unavoidable scattering and imperfections, given their negligible tunneling amplitude.

To construct bound states (that is, states with square-integrable wave functions) at energies greater than the barrier height, we replace the barriers with AlInAs–GaInAs superlattice barriers of equal height (Capasso et al., 1992; Sirtori et al., 1992). The superlattice is designed so that Bragg

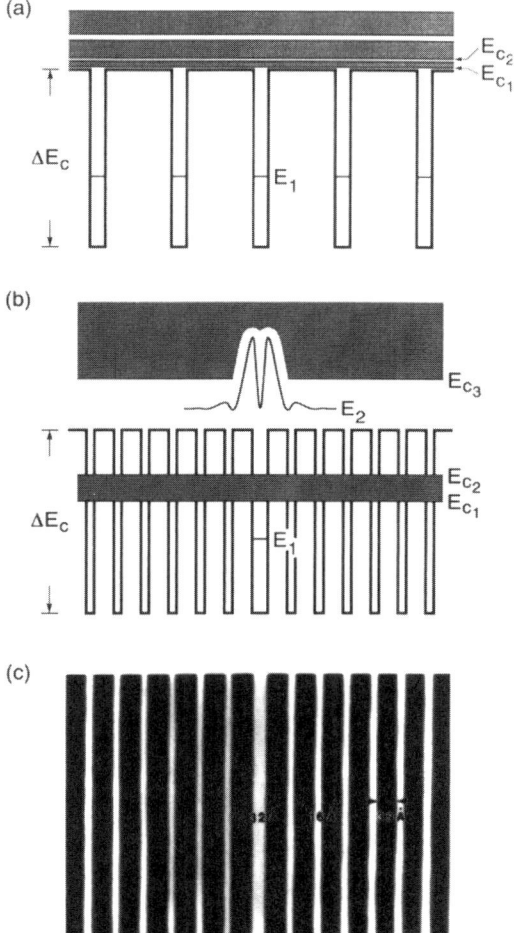

FIG. 2. Conduction band diagrams of AlInAs–GaInAs heterostructures. All energies are measured from the bottom of the wells; E_{c_1} and E_{c_2} indicate conduction miniband edges. (a) GaInAs quantum wells (3.2 nm) with AlInAs barriers (reference sample). Shown are the ground state of the well ($E_1 = 240$ meV) and the minibands in the continuum. (b) GaInAs quantum well (3.2 nm) with Bragg reflector barriers produced by the AlInAs–GaInAs superlattice. Shown is probability density $|\psi|^2$ of the bound state ($E_2 = 560$ meV) localized above the well in the superlattice minigap (266 meV). (c) Transmission electron micrograph of a cleaved cross section of the structure with superlattice Bragg reflectors, showing the atomic abruptness of the interfaces.

reflections spatially localize the state corresponding to the first continuum resonance of the well. The energy of this resonance E_2 is found from

$$k_w L_W = \pi \qquad (1)$$

where L_w is the well thickness and $k_w = \sqrt{2m_w^*(E)E/\hbar^2}$ is the electron wavevector in the well material (GaInAs), where E is the electron energy and $m_w^*(E)$ is the energy-dependent electron effective mass (Bastard, 1990). From Eq. (1) we obtain $E_2 = 560$ meV for $L_w = 3.2$ nm. All the waves partially reflected by the superlattice interfaces constructively interfere to form a bound state at the energy E_2 within the superlattice minigap (Fig. 2b). To achieve this, the thickness of the superlattice wells (d_w) and barriers (d_B) are chosen to be a quarter of the corresponding de Broglie wavelength at the energy of the first transmission resonance E_2, that is,

$$k_w d_w = \frac{\pi}{2} \qquad (2)$$

$$k_B d_B = \frac{\pi}{2} \qquad (3)$$

where $k_B = \sqrt{(2m_B^*(E_2)(E_2 - \Delta E_c)/\hbar^2)}$ is the wavevector in the barrier material at the energy E_2; $m_B^*(E_2)$ is the effective mass at the energy E_2 in the barrier material. From Eqs. (2) and (3), one finds $d_w = L_w/2 = 1.6$ nm and $d_B = 3.9$ nm. The effect of cladding the wells with Bragg mirrors can be viewed as the equivalent of using high-reflectivity (ideally unity) quarter-wave dielectric stacks to produce a Fabry–Pérot optical cavity. The energy level E_2 can be also understood as a deep level in the superlattice bandgap, arising from the introduction of an "artificial defect" (the central well in Fig. 2b) in a periodic superlattice structure.

Our structure with Bragg reflectors consists of 22 six-period undoped AlInAs–GaInAs superlattices, designed as already described, separated by 20 3.2-nm GaInAs wells doped n-type with silicon to give an electron sheet density of 3×10^{11} cm^{-2} necessary for infrared absorption between subbands. Figure 2c is a transmission electron micrograph of a cleaved cross section, showing a GaInAs well sandwiched between the two superlattice barriers acting as Bragg reflectors.

Figure 3 shows the absorption spectra at 10 K of the multiple quantum well structures with and without Bragg reflectors. In the spectrum of the reference structure, the low-energy cutoff ($\hbar\omega \approx \Delta E_c - E_1 = 0.3$ eV, where ω is an angular frequency) corresponds to the onset of the classical continuum

2 QUANTUM INTERFERENCE EFFECTS IN INTERSUBBAND TRANSITIONS   107

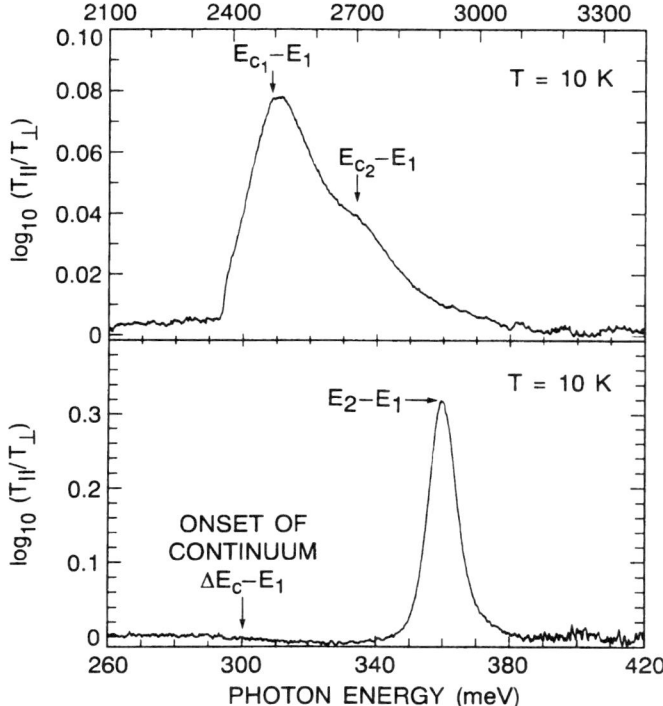

FIG. 3. Absorption spectra at 10 K for the reference structure of Fig. 2a (top) and the structure with superlattice Bragg reflectors of Fig. 2b (bottom). The absorbance for polarization parallel to the plane of the layers ($-\log_{10} T_{\parallel}$) is subtracted from the absorbance normal to the plane of the layers ($-\log_{10} T_{\perp}$), to remove the contribution of free-carrier absorption in the buffer layers. The resulting quantity is proportional to the intersubband absorption coefficient. The transition to the confined state above the well (E_2 in Fig 2b) corresponds to the peak at 360 meV, in excellent agreement with the calculated value ($E_2 - E_1$).

(Fig. 2a). The energy of the absorption peak is close to the energy difference between the edge of the lower miniband and the bound state of the well $E_{c_1} - E_1$, very near the onset of the continuum, as expected from the calculated spectrum and band diagram. The shoulder at $E_{c_2} - E_1$ represents the onset of absorption by the second miniband, in good agreement with the theoretical value. Absorption to higher bands is not observed because the matrix elements are too small. In the multiwell structure with Bragg reflectors we instead observe an isolated narrow peak corresponding to a transition to a state at an energy above the barrier height ΔE_c. It is centered at an energy of 360 meV and has a full width at half maximum of 9.7 meV;

its width at 300 K is 17.5 meV. The position of this peak agrees well with the calculations. The width of the absorption peak at $\hbar\omega = 360$ meV is virtually equal to that of bound state–bound state transitions measured by us in conventional 55-Å GaInAs quantum wells with 300-Å AlInAs barriers doped to a comparable level. These wells were designed to have two bound states below the barrier height. The use of a $\lambda/4$ stacks to enhance barrier reflectivity has been used extensively in quantum cascade lasers. Readers are referred to the article by Faist et al. in this volume.

IV. Quantum Interference in a Coupled-Well System

We investigate here a coupled-quantum-well system that exhibits quantum interference in absorption between electronic bound states (Faist *et al.*, 1993). This phenomenon manifests itself in an electric field–induced suppression of intersubband absorption, accompanied by a negligible Stark shift of the corresponding transition. As such, the large modulation of the dipole matrix element in our effect *does not originate from a reduced overlap between the wave functions*, as in the case of the quantum confined Stark effect (Chen and Andersson, 1992), but has its origin in the quantum interference between the two coupled wells.

The sample, grown by MBE on a semi-insulating GaAs substrate, comprises 50 modulation-doped coupled quantum wells. Each period consists of two GaAs wells, respectively 6.2 and 7.2 nm thick, separated by a 2.0-nm $Al_{0.33}Ga_{0.67}As$ barrier. A 145-nm $Al_{0.33}Ga_{0.67}As$ spacer layer separates the coupled-well periods. To supply the electron charge in the wells, a δ-doped Si layer ($1 \times 10^{12}/cm^2$) is inserted in the spacer layers to ensure a symmetric charge transfer.

Figure 4a shows the conduction band diagram of a period of the sample at zero bias. Indicated are the energy levels and the modulus squared of the wave functions. The energy levels and wave functions are computed by solving Schrödinger's and Poisson's equations in the envelope-function formalism (Sirtori *et al.*, 1994; Bastard, 1990).

To achieve a better insight into the behavior of the coupled-well system as a function of the applied electric field, let us first consider the two quantum wells, denoted here as well a and well b, coupled by the barrier using a tight-binding approach. In such a model, the calculated wave functions ψ_i ($i = 1, 2, 3, 4$) of this system are expanded in terms of the eigenfunctions $\varphi_{1,2}^{a,b}$ of the first two bound states 1 and 2 of the two isolated wells. In the tight-binding approximation, the transition matrix element $z_{1i} = \langle \psi_1 | z | \psi_i \rangle$ ($i = 3, 4$) between the first and the third or fourth state of

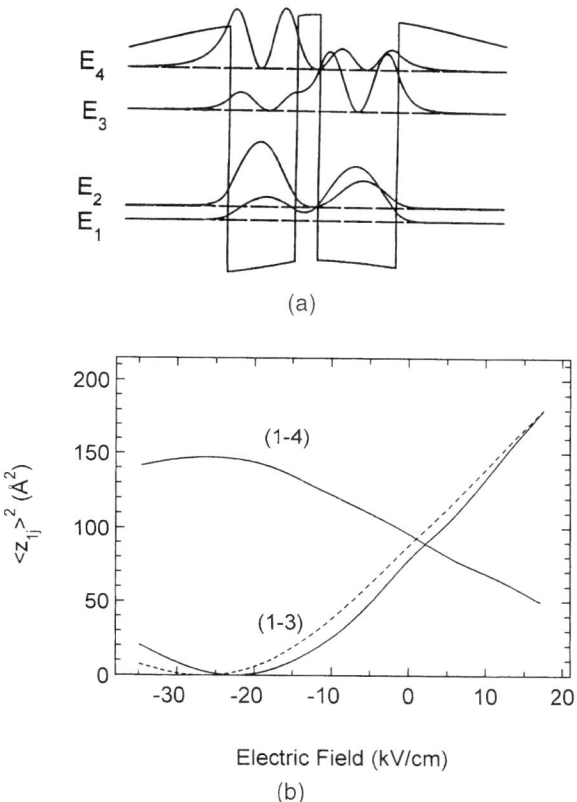

FIG. 4. (a) The energy band diagram of a single period of the GaAs–Al$_{0.33}$Ga$_{0.67}$As modulation-doped coupled quantum well exhibiting interference in absorption. Shown are the positions of the calculated energy subbands and the corresponding modulus squared of the wave functions. (b) Square of the intersubband transition matrix elements $(z_{13})^2$ and $(z_{14})^2$ calculated as a function of the electric field. Dotted line: tight binding model for $(z_{13})^2$.

the coupled-well system can now be written as the sum of the contribution from the two wells a and b

$$z_{1i} = \langle \psi_1 | \varphi_1^a \rangle \langle \psi_i | \varphi_2^a \rangle z_{12}^a + \langle \psi_1 | \varphi_1^b \rangle \langle \psi_i | \varphi_2^b \rangle z_{12}^b \qquad (4)$$

where z_{12}^a and z_{12}^b are the transition matrix elements computed for the isolated wells. As ψ_1 is the ground state of the system, $\langle \psi_1 | \varphi_1^a \rangle$ and $\langle \psi_1 | \varphi_1^b \rangle$ have the same sign. On the contrary, since the second excited state ψ_3

crosses zero twice and is constructed with the antisymmetric wave functions $\varphi_2^{a,b}$, $\langle\psi_3|\varphi_2^a\rangle$ and $\langle\psi_3|\varphi_2^a\rangle$ have opposite signs. Therefore, if we consider a transition between the first and third state of the coupled-well system, the two terms of Eq. (4) have opposite signs. One thus expects large values of z_{13} for large absolute values of the electric field where both wave functions are localized either in well a or well b (first or last term of Eq. (4) dominates) and a null for some intermediate value of the electric field. At this field the absorption will be suppressed. This behavior is clearly apparent in Fig. 4b, where we display $(z_{13})^2$ as given by Eq. (4) along with the exact calculation, using the parameters of the sample. As expected, z_{13} decreases with the applied field and has a null for an electric field of $-22\,\text{kV/cm}$. It is important to note that the suppression of the transition matrix element ($z_{13} = 0$) does not arise from the parity selection rule because for all electric fields, the wave functions ψ_1 and ψ_3 do not have a well-defined parity.

The interference phenomenon occurs for the (1–3) transition because the two terms in Eq. (4) have opposite signs. On the contrary, as $\langle\psi_4|\varphi_2^a\rangle$ and $\langle\psi_4|\varphi_2^b\rangle$ have the same sign, the two terms of Eq. (4) for the (1–4) transition add constructively. For large values of the electric field, the ground state wave function ψ_1 will be localized in one well and ψ_4 in the other. Because the two wave functions ψ_1 and ψ_4 now have a small overlap, the resulting dipole matrix element z_{14} will be small. This behavior appears clearly in Fig. 4 where $(z_{14})^2$ is displayed as a function of the applied field; z_{14} is maximum at about $-22\,\text{kV/cm}$ and decreases for positive fields or larger negative fields.

For the experiments, the samples were processed into square mesas (800 μm side) and ohmic contacts were provided to the n^+ contact layers. They were then processed into a two-pass waveguide. In Fig. 5 the low-temperature absorption spectra are shown for different applied biases. Positive bias refers to the band diagram configuration in which the thick well is lowered below the thin well. The height of the absorption peak corresponding to the 1–3 transition (136 meV) decreases regularly when the applied bias U_b is changed from 30 to $-20\,\text{V}$, has a minimum at this value, and rises again if the bias is decreased further to $-30\,\text{V}$. At $-20\,\text{V}$, the integrated absorption of the peak has decreased to 1.6% of its value at 30 V. Meanwhile, as shown also in Fig. 6a, the (1–3) transition energy has changed by less than 1 meV in the same bias range. Thus the modulation of the absorption is not due to an electric field induced shift of the transition like in the Stark effect. As expected from the computations, the (1–4) transition has the opposite behavior, its oscillator strength is maximum for $U_b = -20\,\text{V}$ and not observable anymore at 30 V.

The characteristic anticrossing of the first two levels as a function of the applied bias is clearly apparent from Fig. 6a, where the measured transition energies E_{13} and E_{23} are reported along with the theoretical predictions

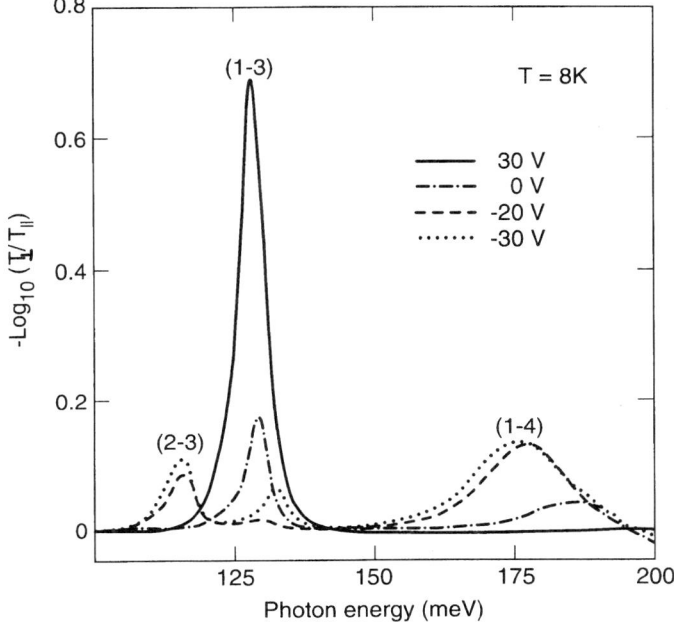

FIG. 5. Intersubband absorption spectra in a two-pass waveguide at $T = 8$ K for different applied biases, as indicated. A strong modulation of the absorption coefficient of transitions (1–3) and (1–4) is observed.

using the full self-consistent model. As shown in Fig. 6, the experimental points are in excellent agreement with the theory. In Fig. 6b, $(z_{13})^2$ is derived from the measured integrated absorption and the nominal electron sheet density $n_s = 5 \times 10^{11}$ cm^{-2}. The agreement with the theoretical prediction is excellent; $(z_{13})^2$ decreases first as the field is increased, has a minimum for $U_b = -20$ V, and then rises again.

As expected from an interference phenomenon, the amplitude of the wave functions in each well enter in Eq. (4), and these wave functions must remain coherent across the coupled-well system. A sufficient amount of disorder will localize the wave functions in either well and completely destroy the interference effect. Indeed, this localization effect has been observed (Vodjdani et al., 1991). In this work, the absorption of a coupled-well system with a thicker tunnel barrier (4 nm) is reported as a function of the applied field. The wave functions are localized in the two wells and no modulation of the transition element z_{13} is observed (Vodjdani et al., 1991). The modulation of the absorption is entirely due to a transfer of electrons between the

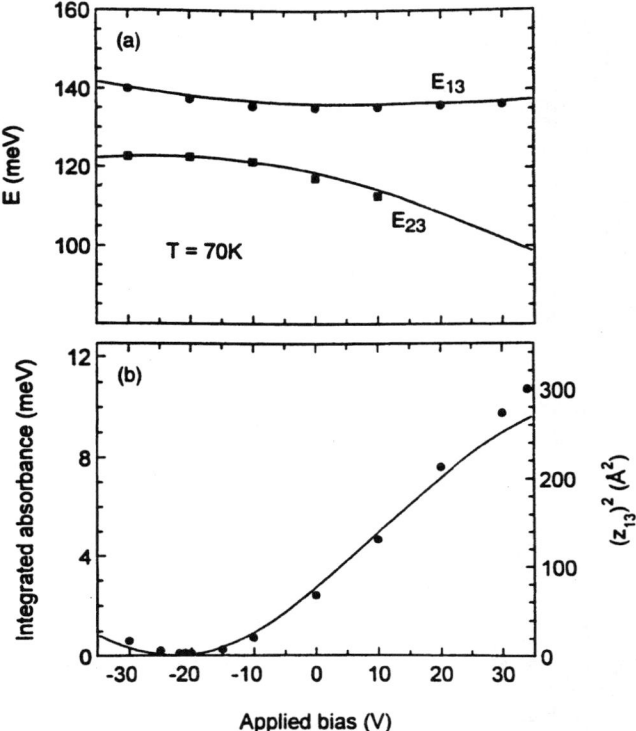

FIG. 6. (a) Measured and calculated (solid lines) transition energies E_{13} and E_{23}. To track E_{23} for a larger range of applied fields, the temperature was raised to 70 K to induce a thermal population of the second level. (b) Square of the transition matrix element $(z_{13})^2$ (right axis), as derived experimentally from the integrated absorbance of the (1–3) absorption peak (left axis) at $T = 8$ K. The solid line is the calculated value.

ground states of two quantum wells with different transition energy to the first excited state.

V. Fano Interference in Intersubband Absorption

With the excited state $|\varphi\rangle$ of a quantum system is coupled to a continuum $|\psi\rangle$ at the same energy, this state broadens due to the finite lifetime τ introduced by the coupling to the continuum. The absorption spectrum from a ground state $|1\rangle$ to this excited state will be Lorenzian with a half width at half maximum $\Gamma = \hbar/2\tau$. However, a peculiar situation arises when the transition matrix element $\langle 1|z|\psi\rangle$ from the ground state to the continuum is nonvanishing: the absorption line shape changes dramatically; it becomes asymmetric and displays a zero within a few Γ from the absorption peak. This phenomenon is called Fano interference (Fano, 1961) and was observed in many atomic, molecular, or solid-state systems (Nunes *et al.*, 1993; Oberli *et al.*, 1994; Maske *et al.*, 1991; Jin *et al.*, 1995; Siegner *et al.*, 1995).

The structures for the observation of Fano interference in intersubband transitions are grown by molecular beam epitaxy on semi-insulating GaAs substrates and consist of 10 periods. A 400-nm GaAs contact layer, *n*-type doped to $\sim 10^{18}\,\mathrm{cm}^{-3}$ started and finished the growth. As shown in Fig. 7, each period consists of a GaAs coupled well confined by a high, 40-nm-thick $Al_{0.33}Ga_{0.67}As$ barrier on the right and a thick low $Al_{0.165}Ga_{0.835}As$ barrier on the left (Faist *et al.*, 1996). The thickness $L = 200\,\mathrm{nm}$ of the $Al_{0.165}Ga_{0.835}As$ barrier is chosen such that it is much longer than the electron's coherence length $\lambda_c \sim 20-50\,\mathrm{nm}$ and therefore the states of this

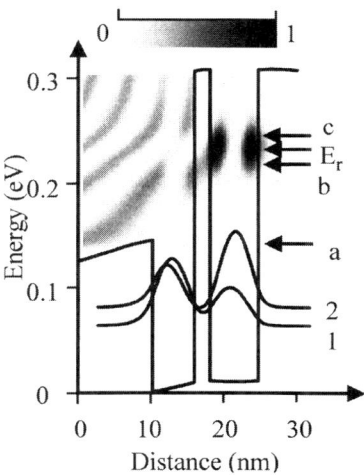

FIG. 7. Conduction band diagram of a portion of the structure for the observation of Fano interference. The layer thicknesses and doping used in this self-consistent calculation correspond to sample A. The moduli squared of the wave functions of the $n = 1$ and $n = 2$ states are displayed. The modulus squared of the wave function in the continuum $|\Psi\rangle$ is represented as a gray-scale density plot. Points *a*, *b*, and *c* represent the final state energies corresponding to the onset of the continuum, the zero, and the maximum of the absorption spectrum (see Fig. 8).

region behave as a continuum. Two structures with different quantum well thickness were grown. Sample A had the strongest coupling due to the relatively thin 2.0-nm barrier coupling the 5.2-nm left well to the 6.4-nm right well. A Si δ-doping sheet in the $Al_{0.165}Ga_{0.855}As$, separated from the quantum wells by a 25-nm spacer layer provides the $2.5 \times 10^{11} \, cm^{-2}$ electron sheet density in the coupled-well region. Sample B had a weaker coupling due to the thicker 2.5-nm barrier coupling the 5.5-nm left well coupled to the 6.5-nm well. The electron sheet density was $n_s = 5 \times 10^{10} \, cm^{-2}$ and the spacer layer was 50 nm.

In these samples, the individual ground states of the two wells are coupled by the thin intermediate barrier forming a doublet with a splitting of about 20 meV. The same barrier also couples the excited state of the right well $|\varphi\rangle$ with energy E_r to the energetically degenerate continuum which broadens the state $|\varphi\rangle$ by $\Gamma \cong 12$ meV for sample A and $\Gamma \cong 6$ meV for sample B. In both cases, the condition $\Gamma \gg \Gamma_d$ is satisfied, where $\Gamma_d \sim 1\text{-}2$ meV is the broadening (homogeneous and inhomogeneous) of $|\varphi\rangle$ associated with interface roughness and optical phonon scattering. Since the ground state wave function spans both wells it has a strong dipole matrix element to state $|\varphi\rangle$ and the continuum state $|\psi\rangle$ above the left well. Thus the structure fulfills the requirements for the observation of Fano interference. This effect is caused by the interference between two distinct paths to the continuum. The matrix elements from the ground state directly to the continuum and from the ground state to the continuum resonance destructively interfere at a well-defined photon energy corresponding to a 180° phase difference between the paths. This gives rise to the absorption minimum near E_r. These features are also clearly apparent in the plot of the modulus squared of the eigenfunction $|\Psi\rangle$ of the whole system; displayed as a gray-scale plot in Fig. 7. As predicted the calculated maximum and zero of the absorption (points c and b in Fig. 8) lie above and below E_r, respectively. This occurs because $\langle 1|z|\varphi\rangle$ and $\langle 1|z|\psi\rangle$ interfere with *opposite phase difference* on the two sides of the resonance. The expected energy-dependent phase shift experienced by the wave function $|\Psi\rangle$ is evident in Fig. 7 as an abrupt shift of the position of the minimum of the modulus squared of the wave functions as the energy crosses the resonant energy E_r.

In our structure, Γ is determined by the intermediate barrier thickness. Assuming a high potential barrier, we see that Fano's q parameter describing the asymmetry of the line shape (Fano, 1961) is

$$q^2 \approx \frac{\Gamma_{1-c} f_{1-b}}{\Gamma f_{1-c}} \tag{5}$$

where $2\Gamma_{1-c}$ is the energy range over which the absorption to the continuum is spread ($2\Gamma_{1-c} \sim 80$ meV, typically) and f_{1-c} and f_{1-b} are the oscillator strengths from the ground state to the continuum and to the state $|\varphi\rangle$, respectively. The values of f_{1-b} and f_{1-c} depend mainly on the respective weight of the ground state wave function in the two wells and we estimate $q = 1.2$ for sample A and $q = 3.3$ for sample B.

To prevent any thermal population of the $n = 2$ state, the absorption measurements were performed at low temperature ($T = 5$ K). A modulation technique was used to enhance the sensitivity in these samples having a relatively small number of periods and low doping. To this end, the samples were processed into square mesas with top and bottom contacts as described in the preceding paragraph. At an applied bias $U \sim -5$ V (lowering the left-hand side barrier in Fig. 7), the injected current, blocked by the 50-nm-thick $Al_{0.33}Ga_{0.67}As$ barriers, is negligible ($<1 \mu A$). At this bias, the electronic charge in the coupled well region is transferred to the $Al_xGa_{1-x}As$ barrier region by tunneling. As a result, the differential transmission $\Delta T/T = (T(U = -5V) - T(U = 0))/T(U = 0)$ is a measure of the intersubband absorption from the active region at $U = 0$ V. The absorption for both samples is reported in Fig. 8 along with the calculated spectra. The Fano interference is contained in the calculated spectra automatically since the absorption is computed using the wave function $|\Psi\rangle$, which is an eigenfunction of the whole system, coupled well plus continuum. Thus the agreement between the measured and calculated spectra is the proof that these samples exhibit Fano interference. Note, however, that the spectra cannot be fitted in a satisfactory way with the approximate form $+(q^2 - 1 + 2q\varepsilon)/(1 + \varepsilon^2)$ (Fano, 1961) where $\varepsilon = (E - E_0)/\Gamma$ is the reduced energy. This is so because the assumptions used to derive this expression (invariance of both the matrix element and the coupling strength as function of energy) do not hold in our case. Both spectra also show the qualitative features of the Fano line shapes with a zero close to the asymmetric absorption peak. The shift (~ 100 meV) between the absorption peak and the onset of the continuum is another feature that is specific of these samples exhibiting Fano resonance. As expected, the main peak is broader for sample A (50 meV) than for sample B (30 meV) due its stronger coupling to the continuum. It also displays a lower ratio r between the amplitude of the two peaks on both side of the zero ($r = 2$ for sample A versus $r = 24$ for sample B) in agreement with the lower value of q for the sample A.

Sample B was designed so that the Fermi energy at $U = 0$ V lies well below the band edge at the location of the delta-doped Si sheet. Therefore, applying a positive bias to the structure will not transfer more electrons but will strongly modify the shape of the potential by transforming the continuum into a triangular well of variable width (see the insets in Fig. 9). The absorption spectra depend strongly on the effective width L of this triangular well (taken at the resonant energy E_r). The absorptions for a few representative applied biases are reported in Fig. 9 and show the expected

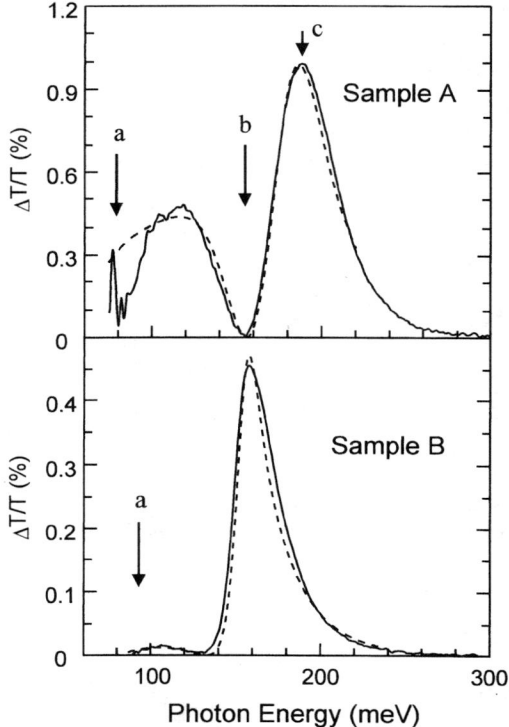

FIG. 8. (Upper) Full line, measured absorption spectrum for sample A having a strong coupling and asymmetry. Points *a*, *b*, and *c* refer to the onset of the continuum, the zero, and the maximum of the absorption (see Fig. 7). The dashed line is the calculated spectrum. (Lower) Measured absorption spectrum for sample B. Note the shift between the absorption peak and the onset of the continuum, which is a feature specific of these samples exhibiting Fano interference.

transition from a bound-to-continuum to a bound-to-bound system. Similar spectral features have been observed in quantum well infrared photodetectors under appropriate bias conditions (Lenchyshyn *et al.*, 1996).

VI. Control of Quantum Interference by Tunneling to a Continuum

In this section, we demonstrate a system where quantum interference arises between absorption paths to two states coupled to a common

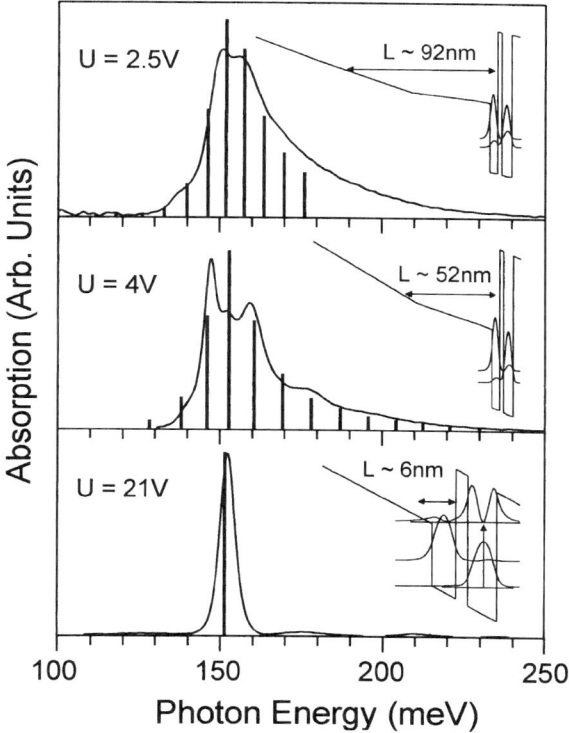

FIG. 9. Absorption spectra of sample B for different applied voltages, as indicated. The insets show the shape of the potential, along with the effective length L of the triangular well formed by the applied field. The position of the bars correspond to the calculated transition energies; their height is proportional to the oscillator strength of each transition.

continuum by tunneling, but where at the same time the optical matrix from the ground state to this continuum is negligible (Harris, 1989). The key features of our structures are a deep and a shallow well coupled by tunneling to a single continuum of energies (Fig. 10). The ground state of the shallow well and the first excited state of the deep well mix (i.e., quantum mechanically anticross) to create two excited energy levels ($|2\rangle$ and $|3\rangle$ in Fig. 10). Because these states are strongly coupled to a continuum, they should be viewed as continuum resonances characterized by a width. They can be represented quantum mechanically by a coherent superposition of the ground state of the shallow well and the first excited state of the deep well. The boundary condition on one side of the structure is set by a very thick

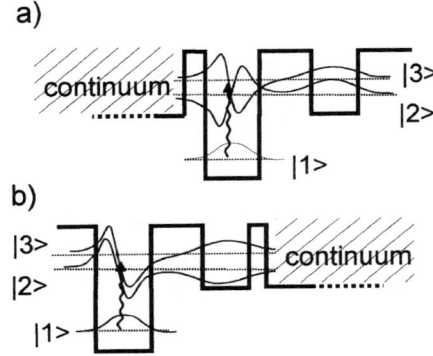

FIG. 10. Schematic band diagram of the quantum well structures used to demonstrate quantum interference in optical absorption from the ground state $|1\rangle$ to the two continuum resonant states $|2\rangle$ and $|3\rangle$, which are predominantly broadened by tunneling. The solid curves are schematics of the electronic wave functions at the ground state and resonance energies. The wavy arrows represent the optical transition. (a) When the thin tunneling barrier is *adjacent to the deep well*, the wave function of the final state of the transition experiences a 180° phase shift as the energy of the incident photon is scanned between the two resonances. This causes a *change of sign* of the transition matrix element and therefore a zero in the absorption coefficient (*destructive* interference) at an intermediate energy. (b) When the thin barrier is adjacent to the shallow well there is no 180° phase shift, leading to *constructive interference*.

(ideally infinitely thick) barrier to suppress any tunneling. This is a key feature that enables a description of the wave functions with real numbers. As shown in Fig. 10, there are two ways of constructing such a quantum system, with either the shallow or the deep well adjacent to the ultrathin barrier.

1. ABSORPTION EXPERIMENTS

From a quantum mechanical point of view, the probability amplitude for absorption of a photon of a given energy can be thought as the superposition of two paths, one via the lower and one via the upper resonant states. We show that the sign of the interference is opposite in the two structures, leading to either constructive or destructive interference. Consider the structure of Fig. 10a. We have chosen the arbitrary phase factor of the wave functions such that the latter, inside the infinitely thick barrier, is real and positive for all energies. In our experiments, the two resonances are probed by optical absorption from the ground state: as the photon energy is changed, the absorption matrix element changes sign as one moves from the lower to the upper resonances because the sign of the wave function in the deep well changes. This change of sign implies that at a certain energy the modulus of the matrix element vanishes (i.e., one has destructive interference). In the other structure (Fig. 10b), on the other hand, the matrix element does not change sign because the wave function in the deep well does not experience a phase shift; this leads to constructive interference.

For an analytical description of these phenomena, consider the normalized absorption cross section σ_{abs} in a quantum system in which two upper states are coupled to a common continuum by tunneling. In the limit of a vanishing dipole matrix element from the ground state to the continuum, σ_{abs} reads (Harris, 1989):

$$\sigma_{abs} = \frac{(\eta_3 + \eta_2 q)^2}{\eta_2^2 \eta_3^2 + (\eta_2 + \eta_3)^2} \tag{6}$$

In this expression $\eta_3 = 2(E_{31} - h\nu)/\Gamma_3$ and $\eta_2 = 2(E_{21} - h\nu)/\Gamma_2$ are the normalized detunings, $h\nu$ is the photon energy, E_{31} and E_{21} are the transition energies between the $n = 1$ ground state and the two resonances, and Γ_3 and Γ_2 are the tunneling rates from the latter to the continuum. It is assumed that broadening associated with scattering is negligible compared to Γ_3 and Γ_2. In the same spirit as that of Fano's original work (Fano, 1961), the parameter

$$q = \frac{z_{13}}{z_{12}} \sqrt{\frac{\Gamma_2}{\Gamma_3}} \tag{7}$$

is the weighted ratio of the matrix element to levels 2 and 3 and controls the magnitude and sign of the interference effect. Note that in Eqs. (6) and (7), the matrix elements are real numbers with their signs. As in Fano's work, the sign of the matrix element to the continuum, even though not explicit in Eqs. (6) and (7), guarantees that the sign of q does not depend on an arbitrary phase factor of the wave functions.

The gray-scale density plot of the wave functions for both structures, displayed in Fig. 11, provides additional physical understanding of the origin of this asymmetry. If the tunnel barrier is placed on the side of the GaAs deep well (Fig. 11a), the resonance present in the AlGaAs shallow well provides the phase shift for the wave function, which causes the cancellation of the absorption between the two resonances (solid line in the upper graph in Fig. 12). On the contrary, by placing the tunnel barrier on the side of the

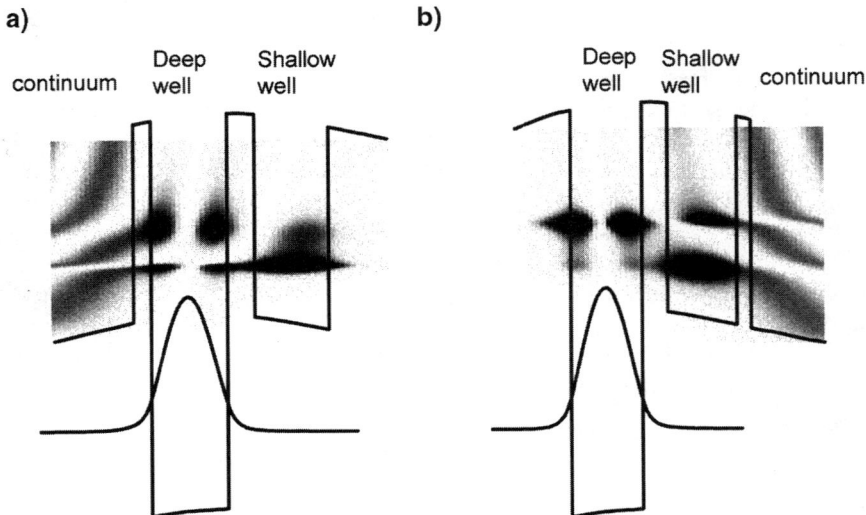

FIG. 11. Calculated conduction band diagram of a portion of the two grown semiconductor structures consisting of a deep, 7.3-nm-thick GaAs well coupled to a shallow, 6.8-nm-thick $Al_{0.165}Ga_{0.835}As$ well by a 2.5-nm-thick $Al_{0.33}Ga_{0.67}As$ tunnel barrier. The two structures differ only for the position of the 1.5-nm-thick $Al_{0.33}Ga_{0.67}As$ tunnel barrier that couples the double well to the continuum. The $Al_{0.33}Ga_{0.67}As$ barrier on the opposite side of the double well is 6.0 nm thick. The growth direction is from right to left. The gray-scale density plot of the calculated modulus squared of the wave function in the continuum helps us visualize the effect of the phase shift experienced by the wave function in the deep well of structure (a) as the energy crosses the resonance of the shallow well. This phase shift is responsible for destructive interference in absorption. Its absence in structure (b) leads instead to constructive interference. The modulus squared of the wave function of the $n = 1$ states is also displayed, calculated self-consistently.

shallow well the phase of the wave function in the deep well remains the same (Fig. 11b) and the two absorption resonances interfere constructively (dashed lines in Fig. 12, upper graph).

Our structures are grown by molecular beam epitaxy on semi-insulating GaAs substrates. After a 1-μm-thick undoped GaAs buffer layer, a 150-nm-thick $Al_{0.165}Ga_{0.835}As$ spacer layer region was grown with a sheet ($1 \times 10^{12}\,cm^{-2}$) of Si n-type doping (δ-doping) inserted 100 nm from the interface with the GaAs buffer layer. Subsequently, the double-well structure was grown. The thickness of the layers is given in the caption of Fig. 11. On top of the double-well structure (on its left side in Figs. 11a and 11b), a 231-nm-thick $Al_{0.165}Ga_{0.835}As$ thick spacer layer was grown with two n-type δ-doped layers ($1 \times 10^{12}\,cm^{-2}$), one inserted 22 nm and the other 187 nm from the left edge of the deep well. A 10-nm-thick undoped GaAs

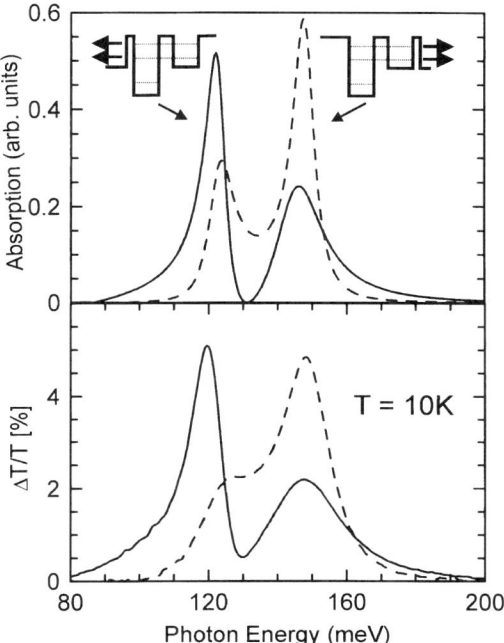

FIG. 12. (Upper) Computed absorption for the two structures with the two different location of the tunnel barrier coupling the step-coupled well to the continuum, as shown by the schematic band diagram of the insets. The location of the tunnel barrier forces destructive (solid line) or constructive (dashed line) interference between the absorption amplitudes to the two continuum resonances. (Lower) Absorption measurement for the corresponding structures at $T = 10$ K. The departure from the calculated line shapes is attributed to additional broadening mechanisms such as interface roughness scattering, which are not included in the model. The dependence of the line shape on the location of the tunnel barrier is a demonstration of the interference effect.

region capped the structure. The spacer layer thickness is adjusted to preserve the same distance between the δ-doping layers and the step-coupled-well system, and therefore the electrostatic potentials are identical in both structures. The δ-doping provides a two-dimensional electron gas in the deep well with a calculated sheet electron density of $n_s = 4 \times 10^{11}\,\text{cm}^{-2}$.

For the absorption measurements, we processed our samples in multipass (six) 45° wedge waveguide. The absorption was measured with a FTIR spectrometer using a step-scan modulation technique in which the electron gas in the coupled-step well is depopulated by a Ti/Au Shottky gate. The latter is evaporated on the surface of the sample and indium balls alloyed into the layer contact the two-dimensional electron gas. In Fig. 12, the absorption measurements at $T = 10\,\text{K}$ for both structures are compared to the calculated values. As predicted, the absorption strength at photon energies between the two resonances is strongly suppressed or enhanced by the interference effect depending on the location of the barrier. This proves that tunneling through the ultrathin (1.5-nm-thick) barrier controls the sign of interference effect when the broadening of the states is dominated by tunneling. However, the finite broadening introduced by interface disorder prevents full quantum interference; this is the main reason for the departure of the data from the calculated profiles and specifically the reason why the absorption does not vanish completely in the sample with destructive interference. Indeed, absorption linewidth measurements on samples with the same coupled-well structure but with negligible tunneling to the continuum exhibited a half width at half maximum of $\Gamma = 2.5\,\text{meV}$. This was achieved by inserting an identical double quantum well between two 60-nm-thick $Al_{0.33}Ga_{0.67}As$ barriers. This value is a measure of the non-tunneling contribution to the broadening of the optical transitions; it is lower than the calculated tunneling broadening $\Gamma \sim 8\,\text{meV}$ through the 1.5-nm barrier, but not completely negligible.

Destructive interference in intersubband absorption in a double-well structure coupled by tunneling to a continuum was also inferred from a fit of the absorption line shape to a model that included the collision broadening in a phenomenological manner (Smidt *et al.*, 1997). The present experiment gives more direct evidence of tunneling-induced quantum interference by showing that tunneling can be used to control the sign of the latter.

2. Emission Experiments

The observation of quantum interference by tunneling into a continuum is relevant for the design of semiconductor lasers without population inversion. The latter has so far been observed only in gases. Essential for LWI is nonreciprocity between emission and absorption. A possible semiconductor LWI scheme would use the present quantum well structure (Fig. 11a) for the active regions. The latter would be arranged in a quantum cascade laser geometry (i.e., alternated with electron injectors). An electron would be injected from the thick barrier side at an energy between the two resonances where the absorption cross section is minimum to ensure strong nonreciprocity between intersubband absorption and emission.

Although the realization of such a laser would be scientifically significant, its technological value would be limited due to the short lifetime of the excited state that is required to achieve strong interference (Kurghin and Rosencher, 1996). Another limitation occurs because the bound states in multiquantum well heterostructures are actually subbands with free electron-like dispersion in the plane, care must be taken when translating the atomic concepts of LWI into the semiconductor world. For example, a laser without population inversion based on intersubband transitions cannot operate at near zero temperature (i.e., both subbands are characterized by a quasi-Fermi distributions and $E_f > kT$) since the emission is forbidden by the Pauli exclusion principle in this case. On the other hand, lasing without intersubband population inversion may also occur without the presence of interference effects due to subband nonparabolicities (Faist et al., 1996b).

Our structures were grown by molecular beam epitaxy on a semi-insulating GaAs substrate. One of the structures (91796.1) consists of 35 active regions, alternated with thick (150-nm) $Al_{0.15}Ga_{0.85}As$ layers (see Fig. 13). Each active region is basically a repetition of the structure of Fig. 11a, the only difference being a thin (6-nm) barrier to inject electrons into the doublet of excited states. Structures with constructive and destructive interference were grown, and the existence of the interference effect was inferred from an analysis of the absorption in both cases, as was done in the preceding section (Faist et al., 1997b).

The absorption for the structure of Fig. 11a was also measured as a function of the applied electric field in a sample processed into a 800-μm-square mesa. The location of the two absorption peaks is plotted as a function of the applied bias in Fig. 14a). The excellent agreement between these data and our self-consistent model (solid lines) provides direct evidence of the anticrossing of the two resonances over a wide bias range. It

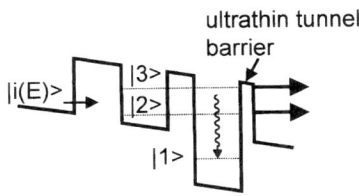

FIG. 13. Schematic band diagram of one period of the active region of sample 91796.1 under applied bias. The states of the shallow and deep wells are coupled by a thin tunnel barrier to form a doublet that interacts with the energy continuum via an ultrathin (1.5-nm-thick) tunnel barrier, creating destructive interference in absorption between the ground state $|1\rangle$ and excited states $|2\rangle$ and $|3\rangle$. Electron injection takes place through the left 6.0-nm-thick barrier and gives rise to electroluminescence from the strongly broadened excited states, as schematically shown by the wavy arrow.

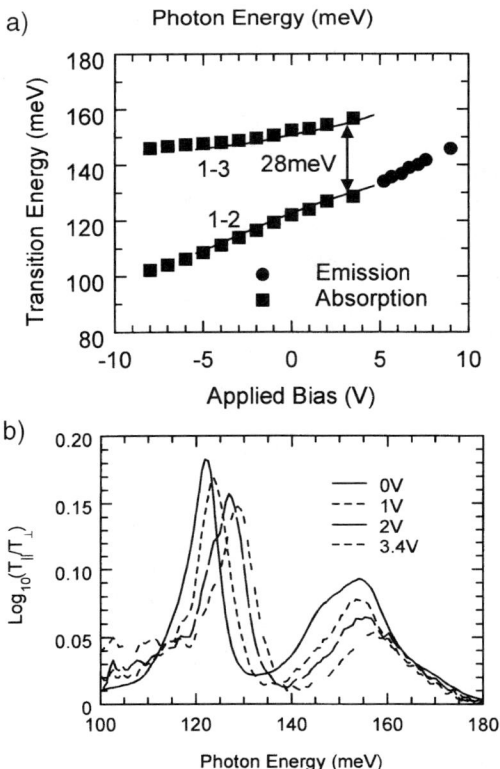

FIG. 14. (a) Intersubband transition energy as function of applied field for absorption (squares) and emission (dots) in the structure of Fig. 13. The solid lines are the theoretical predictions based on a self-consistent model. (b) Absorption spectra for positive applied bias between $U = 0$ and 3.5 V measured in the same sample using a four-pass waveguide.

was shown (Smidt et al., 1997) experimentally, and expected theoretically, that the interference effect is maximum for biases where the energy splitting due to the anticrossing is minimum ($U = 3$ V in our structures). Indeed, as shown in Fig. 14b, our absorption measurements up to $U = 3.4$ V do not show any decrease in the interference effect. In our structures, it is impossible to measure absorption and emission at the same bias, the quantum wells being depleted already at $U \sim 4$ V. Based on simple physical arguments, however, it is reasonable to expect that the interference effect observed up to $U = 3.4$ V will persist at the biases at which the electroluminescence is measured (5–9 V). The electroluminescence is collected through a polished

45° wedge and analyzed in our FTIR spectrometer with a step-scan and lock-in detection technique. The current in our 200-μm-diameter etched mesa diode was pulsed with a 50% duty cycle.

In Fig. 15a, electroluminescence spectra are shown for injection currents from 25 to 500 mA for the sample of Fig. 13. In contrast to earlier experiments in which the intersubband electroluminescence occurs through transitions between *bound* states, the emission spectra displayed in Fig. 15 and the absorption spectra displayed in Fig. 14 have very different line shapes, with the strong asymmetry of the absorption line not apparent in the emission spectrum. As compared to an atomic system, however, the line

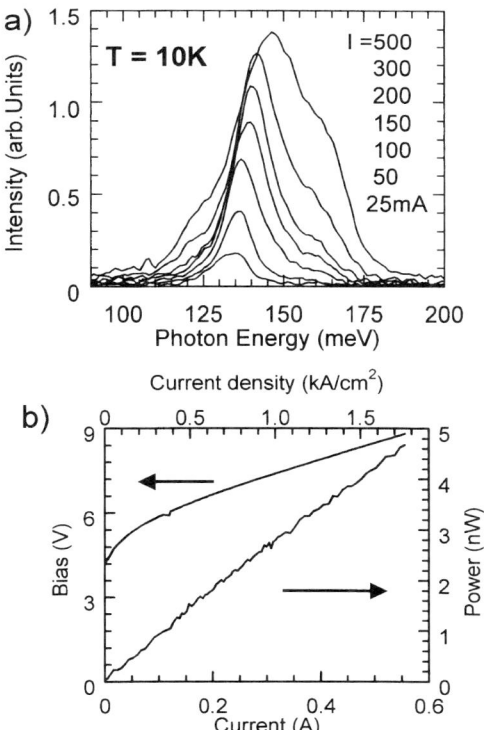

FIG. 15. (a) Luminescence spectra for various injection currents, as indicated for the structure of Fig. 1. The spectra broaden significantly as the current is increased. (b) Measured luminescence power as a function of injected current and current–voltage characteristics for the same sample. The measured electroluminescence efficiency is about 50 times lower than that of a conventional intersubband light emitting diode at the same wavelength due to the very short lifetime of the upper state.

shape is more complicated because it is inherently affected by two additional factors: the Stark effect (already discussed) in Fig. 14 and the distribution of electrons in the continuum of injector states $|i(E)\rangle$. Although the field dependence of the location of the absorption peaks is well predicted by theory (as shown in Fig. 14), the emission has clearly a different behavior with an apparent stronger shift, which we attribute to a change in the electron distribution in the injector. It is apparent from Fig. 15 that not only does the emission peak shift but the emission line shape is also strongly dependent on the injection current; namely, the full width at half maximum of the emission spectrum increases from 14 to 34 meV when the current is increased from 50 to 500 mA. In Fig. 15b, the electroluminescence power and bias voltage are displayed as a function of bias current. The linearity of the power–current curve demonstrates that the broadening of the spectrum, observed in the current range 50–500 mA cannot be explained by a decrease of the electron lifetime in the same interval. The change in emission line shape simply reflects the change in the electron distribution function of the initial injector states $|i(E)\rangle$ as the bias and current are increased. The appearance of a high-energy shoulder and the fact that the width of the low-energy side of the peak is constant suggests that we are approaching the regime of broadband incoherent pumping of both states ($|3\rangle$ and $|2\rangle$) at our highest injection currents ($I = 500$ mA). The physical origin for the modification of the injected electron distribution when the applied electric field is increased up to its maximum value of about 30 kV/cm is twofold. The field, combined with the high injection current, will directly heat the electron distribution and also create a triangular well at the injection barrier whose lowest state energy will increase with field.

Due to the complicating factors mentioned earlier, it is difficult to prove the existence of the nonreciprocity between absorption and emission from these measurements. However, the fact that the separation between the main peak and the high-energy shoulder observed in emission (18 meV), which is smaller than the splitting observed in absorption (28 meV), and the absence of a significant dip between these two peaks both suggest a nonreciprocity between absorption and emission line shapes.

We have shown that the pumping mechanism (i.e., the electron distribution in the injection region) is a key parameter that must be taken into account to understand the relationship between the emission and absorption profiles. Moreover, our measurements indicate that for typical situations encountered in semiconductor structures, this distribution has a width that is smaller or comparable to the broadening due to the tunnel effect into the continuum, complicating the interpretation of luminescence profiles but offering the potential for lower lasing thresholds.

Acknowledgments

We are grateful to A. L. Hutchinson for valuable help in the processing of the samples.

References

Akripin, V. G., and Y. I. Heller. (1983). *Phys. Lett.* **98A**, 12.
Bastard, G. (1990). *Wave Mechanics Applied to Semiconductor Heterostructures* (les editions de Physiques, Paris).
Capasso, F. (1991). *MRS Bull.* **16**, 23.
Capasso, F., C. Sirtori, J. Faist, D. L. Sivco, S. N. G. Chu, and A. Y. Cho. (1992). *Nature* **358**, 565.
Capasso, F., C. Sirtori, and J. Faist. (1997). *J. Math. Phys.* **37**, 4775.
Chen, W., and T. G. Andersson. (1992). *Semicond. Sci. Technol.* **7**, 828.
Cho, A. Y. (1991). *J. Cryst. Growth* **111**, 1.
Class-Manjean, M., H. Frohlich, and J. A. Beskick. (1988). *Phys. Rev. Lett.* **61**, 157.
Dupont, E., P. B. Corkum, H. C. Liu, M. Buchanan, and Z. R. Wasilewski. (1995). *Phys. Rev. Lett.* **74**, 3596.
Faist, J., F. Capasso, A. L. Hutchinson, L. Pfeiffer, and K. W. West. (1993). *Phys. Rev. Lett.* **71**, 3573.
Faist, J., C. Sirtori, F. Capasso, A. L. Hutchinson, L. Pfeiffer, and K. W. West. (1996a). *Opt. Lett.* **21**, 985.
Faist, J., F. Capasso, C. Sirtori, D. L. Sivco, A. L. Hutchinson, and A. Y. Cho. (1996b). *Phys. Rev. Lett.* **76**, 411.
Faist, J., F. Capasso, C. Sirtori, K. W. West, and L. N. Pfeiffer. (1997a). *Nature* **390**, 589.
Faist, J., F. Capasso, C. Sirtori, A. L. Hutchinson, K. W. West, and L. N. Pfeiffer. (1997b). *Appl. Phys. Lett.* **71**, 3477.
Fano, U. (1961). *Phys. Rev.* **124**, 1866.
Harris, S. E. (1989). *Phys. Rev. Lett.* **62**, 1033.
Imamoglu, A., and R. J. Ram. (1994). *Opt. Lett.* **19**, 1744.
Jin, K. J., S. H. Pan, and G. Z. Yang. (1995). *Phys. Rev. B* **51**, 9764.
Kurghin, J., and E. Rosencher. (1996). *IEEE J. Quantum Electron.* **32**, 1882.
Lam, J. F., S. R. Forrest, and G. L. Tangonan. (1991). *Phys. Rev. Lett.* **66**, 1614.
Lenchyshyn, L. C., H. C. Liu, M. Buchanan, and Z. R. Wasilewski. (1996). *J. Appl. Phys.* **79**, 3307.
Mashke, K., P. Thomas, and E. O. Göbel. (1991). *Phys. Rev. Lett.* **67**, 2646.
Nunes, L. A. O., L. Ioriatti, L. T. Florez, and J. P. Harbinson. (1993). *Phys. Rev. B* **47**, 13011.
Oberli, D. Y., G. Böhm, G. Weimann. (1994). *Phys. Rev. B* **49**, 5757.
Padmadandu, G. G., G. R. Welch, I. N. Shubin, E. S. Fry, D. E. Nikonov, M. D. Lukin, and M. O. Scully. (1996). *Phys. Rev. Lett.* **76**, 2053.
Palevski, A., F. Beltram, F. Capasso, L. Pfeiffer, and K. W. West. (1990). *Phys. Rev. Lett.* **65**, 1929.
Roskos, H. G., M. C. Nuss, J. Shah, K. Leo, D. A. B. Miller, A. M. Fox, S. Schmitt-Rink, and K. Köhler. (1993). *Phys. Rev. Lett.* **68**, 2216.
Scully, M. O. (1991). *Phys. Rev. Lett.* **67**, 1855.

Siegner, U., M. A. Mycek, S. Glutsch, and D. S. Chemla. (1995). *Phys. Rev. B* **51**, 4953.
Sirtori, C., F. Capasso, J. Faist, D. L. Sivco, S. N. G. Chu, and A. Y. Cho. (1992). *Appl. Phys. Lett.* **61**, 898.
Sirtori, C., F. Capasso, J. Faist, and S. Scandolo. (1994). *Phys. Rev. B* **50**, 8663.
Smidt, H., K. L. Campman, A. C. Gossard, and A. Imamoglu. (1997). *Appl. Phys. Lett.* **70**, 3455.
Stewart, R. B., and G. J. Diebold. (1986). *Phys. Rev. A* **34**, 2547.
Vodjdani, N., B. Vinter, V. Berger, E. Bockenhoff, and E. Costard. (1991). *Appl. Phys. Lett.* **59**, 555.
Zibrov, S., M. D. Lukin, D. E. Nikonov, L. Hollberg, M. O. Scully, V. L. Velichansky, and H. G. Robinson (1995). *Phys. Rev. Lett.* **75**, 1499.

CHAPTER 3

Quantum Well Infrared Photodetector Physics and Novel Devices

H. C. Liu

INSTITUTE FOR MICROSTRUCTURAL SCIENCES
NATIONAL RESEARCH COUNCIL
OTTAWA, ONTARIO, CANADA

I.	INTRODUCTION	129
	1. Background	129
	2. Simple Description of Intersubband Transition	132
	3. Wavelengths Covered by GaAs-Based Structures	135
II.	DETECTOR PHYSICS	137
	1. Dark Current	137
	2. Photocurrent	150
	3. Detector Performance	160
	4. Design of an Optimized Detector	167
III.	NOVEL STRUCTURES AND DEVICES	169
	1. Multicolor and Multispectral Detectors	169
	2. High-Frequency Detectors	176
	3. Integration with LEDs	183
IV.	CONCLUSIONS	188
	1. Concluding Remarks	188
	2. List of Symbols	190
	REFERENCES	193

I. Introduction

1. BACKGROUND

The earliest studies of optical intersubband transition (ISBT) in semiconductors were on two-dimensional (2D) electron systems in metal-oxide-semiconductor inversion layers (see Ando *et al.*, 1982, and references therein). Suggestions to use quantum wells for infrared (IR) devices were first documented by Chang *et al.* (1977) and Esaki and Sakaki (1977). The first experiment on making use of quantum wells for IR detection was reported by Chui *et al.* (1983) and Smith *et al.* (1983). Proposals of specific embodiments of photodetectors and related theoretical considerations were made by Coon and Karunasiri (1984), Coon *et al.* (1985, 1986), and Goossen and Lyon (1985, 1988). The first experiment on ISBT in quantum wells was carried out by West and Eglash (1985). Subsequently, strong intersubband absorption and the Stark shift were observed by Harwit and Harris (1987). The first clear demonstration of quantum well infrared photodetectors (QWIPs) was made by Levine *et al.* (1987); since then, tremendous progress has been made by the Bell Laboratories group (see Levine, 1993, and references therein). Today, large focal plane arrays with excellent uniformity are being fabricated, as discussed in detail in Chapter 4 by Gunapala and Bandara. Several review articles (Levine, 1993; Whitney *et al.*, 1993; Liu, 1996) relevant to QWIPs have been published. Collections of papers related to ISBT in quantum wells can be found in Rosencher *et al.* (1992a), Liu *et al.* (1994), and Li and Su (1998). A book on QWIPs was written by Choi (1997). A literature search of number of papers related to ISBT in quantum wells is graphed in Fig. 1. It seems that the peak of research activities was reached in the mid-1990s.

A natural question arises concerning the motivation of studying QWIPs. Afterall, technologies based on HgCdTe and InSb are well developed for IR detection and imaging in the wavelength region of about 2–20 μm. What advantages do QWIPs provide? The first and the most important advantage

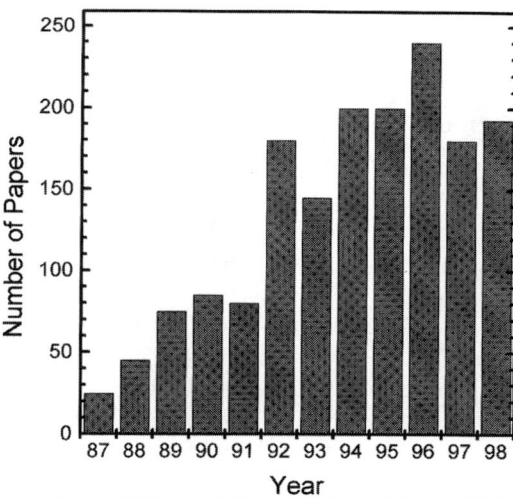

FIG. 1. Number of papers related to intersubband transition in semiconductors vs year.

relates to the availability of a mature material and processing technology for QWIPs based on GaAs. Figure 2 shows the good uniformity of device dark current–voltage characteristics of 40 individual devices. QWIPs based on Si and InP have also been investigated, but they are far less mature. With the mature technology, it is anticipated that the cost of an imaging device based on QWIP would be substantially less than that based on HgCdTe or InSb, and that a large volume production capability can be easily established. The second advantage is the flexibility of the QWIP approach for enhanced device functionality and for its ease to integrate with other devices. Some of these aspects are discussed in Section III.

The emphasis of this chapter is on (1) the physics of QWIPs and (2) related novel structures and devices. Efforts have been made to treat the material using physical pictures and intuition with a minimum of mathematics. In the rest of this section of the chapter, we briefly introduce the ISBT in quantum wells (QWs) and the wavelengths covered by GaAs-based

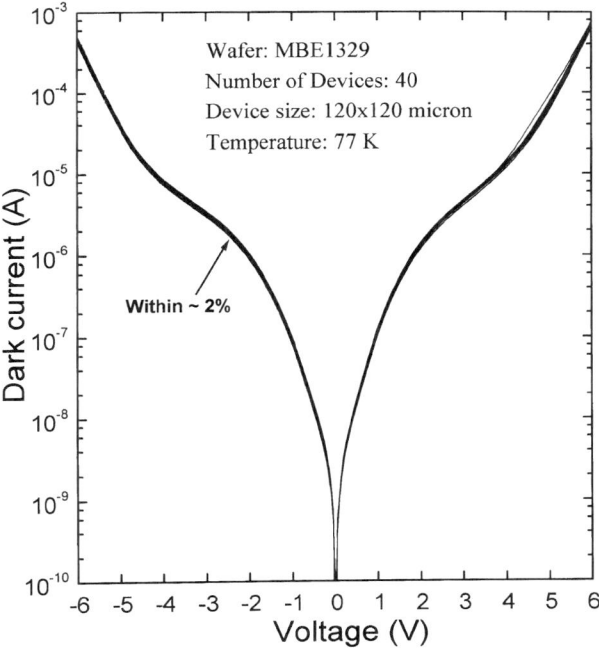

FIG. 2. Uniformity test results. Dark current–voltage characteristics of 40 nominally identical devices from an area of about 5×5 mm^2. The QWIP structure is a "standard" one with 32 wells, having a response in the 8- to 12-μm region.

structures. Section II discusses, in detail, the physics of QWIPs, which includes dark current, photocurrent, and detector performance and design. Section III discusses novel structures and devices, which includes multicolor and multispectral detectors, high frequency detectors, and QWIPs integrated with light emitting diodes (LEDs).

2. Simple Description of Intersubband Transition

The physics related to the optical ISBT in quantum wells is treated in great detail in Chapter 1 of this volume by Helm. Other materials of specific relevance to QWIPs can be found in Liu (1996) and Choi (1997).

In this chapter, we discuss ISBT only in the conduction band. Samples presented here are all made by GaAs-based molecular beam epitaxy (MBE). For this system, a polarization selection rule was realized in the early days by Coon and Karunasiri (1984): only the light polarized in the growth direction can cause ISBT. The selection rule is valid for quantum wells where the single isotropic effective mass approximation holds. Since the argument is based on the effective mass approximation, the selection rule is naturally not rigorous. As band mixing is the cause of the breakdown of the selection rule, physical intuition tells us that the accuracy of the selection rule should be related to the ratio of the energy scales involved. The closest band to the conduction band (at least for the GaAs case) is the valence band. The relevant energy ratio is then E_n/E_g, where E_n are the eigenenergies of the confined states in the conduction band quantum well and E_g is the bandgap. (Note that the reference point — zero energy — for E_n is chosen at the conduction bandedge of the well.) For common quantum wells used in QWIPs (see, e.g., Levine, 1993), the ratio is small, and therefore the selection rule is expected to be quite accurate, with a deviation of at most the 10% level, as shown experimentally by Liu et al. (1998c).

The band-edge profile of a GaAs–AlGaAs quantum well is depicted in Fig. 3. An optical ISBT in the conduction band (CB) is shown by an arrow. The vertical energy scale is drawn roughly in the correct proportion. It is seen that the closest bands, valence band (VB) and spin-orbit split-off (SO) band, are far away in comparison with the ISBT energy involved. It is therefore natural to expect that the effect of mixing VB or SO bands should be small, and hence the selection rule should hold quite accurately.

Because of the selection rule, a normal incidence geometry (i.e., light incident normal to the wafer and along the growth direction) is not suited. A commonly employed 45° edge facet geometry is shown in Fig. 4, as first used by Levine et al. (1987). This geometry "throws away" one half of the

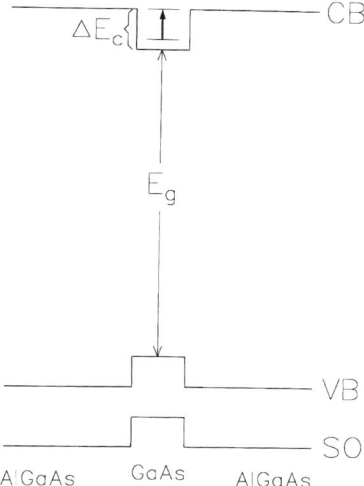

FIG. 3. Schematic bandedge profile of a GaAs–AlGaAs quantum well. Conduction band (CB), valence band (VB), and spin-orbit split-off (SO) band are shown. The bandgap energy (E_g) is about 10 times the conduction band offset (ΔE_c).

light, but is simple and convenient to use, and is commonly used to obtain a detector performance benchmark. The majority of applications of QWIPs requires large 2D arrays where the facet geometry is not suited. Various diffraction gratings are used for these practical devices, discussed in Chapter 4 by Gunapala and Bandara.

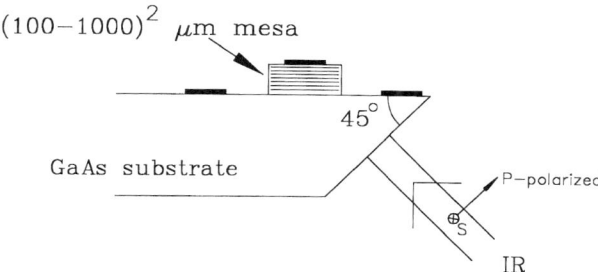

FIG. 4. The 45° edge facet light coupling geometry. The figure is not drawn to scale. The substrate thickness is usually in the range of 400–700 μm. For testing individual detector performance, mesa devices of areas from about 100^2 to 1000^2 μm^2 are used. The IR light is shone normal to the facet surface. The P and S polarizations are defined with respect to the light incidence on the quantum wells.

To obtain a simple picture of ISBT, much of the physics can be illustrated using the simplest model of an infinitely high barrier square quantum well, as in West and Eglash (1985). In this case, the eigenstate wavefunction and energy are trivial:

$$\psi_n(\vec{k}_{xy}) = \sqrt{\frac{2}{L_w A}} \sin\left(\frac{\pi n z}{L_w}\right) \exp(i\vec{k}_{xy} \cdot \vec{x}) \qquad (1)$$

$$E_n(\vec{k}_{xy}) = \frac{\hbar^2}{2m} \frac{\pi^2 n^2}{L_w^2} + k_{xy}^2 \qquad (2)$$

where L_w is the well width, A is the normalization area in the x–y plane, n is a positive integer, $\vec{k}_{xy}$ is the in-plane wave vector, and m is the effective mass in the well. We have chosen the well to be situated from $z = 0$ to $z = L_w$. Equation (2) explains the reason for the term "subband" instead of a single state. For a given quantized state, one can put many electrons occupying different in-plane momenta. For the ground state in equilibrium, the occupation of electrons leads to a Fermi energy determined by $n_{2D} = (m/\pi\hbar^2)E_f$, where n_{2D} is the 2D quantum well carrier density and $m/\pi\hbar^2$ is the 2D density of states.

The dipole matrix moment between any two states (e.g., n and n') with opposite parity is

$$e\langle z \rangle = eL_w \frac{8}{\pi^2} \frac{nn'}{(n^2 - n'^2)^2} \qquad (3)$$

Note that the in-plane momentum remains the same between initial and final states.

The oscillator strength is

$$f \equiv \frac{2m\omega}{\hbar} \langle z \rangle^2 = \frac{64}{\pi^2} \frac{n^2 n'^2}{(n^2 - n'^2)^3} \qquad (4)$$

The absorption probability for an IR beam polarized in the plane of incidence and propagating at an angle of θ with respect to the growth axis is

$$\eta = \frac{e^2 h}{4\varepsilon_0 n_r mc} \frac{\sin^2\theta}{\cos\theta} n_{2D} f \frac{1}{\pi} \frac{\Delta E}{(E_{n,n'} - \hbar\omega)^2 + (\Delta E)^2} \qquad (5)$$

where ε_0 is the vacuum permitivity, n_r is the index of refraction, c is the speed of light, $E_{n,n'} \equiv E_n - E_{n'}$, and ΔE is the broadening half width. The lineshape associated with broadening is modeled by a Lorenzian. At the peak ($E_{n,n'} = \hbar\omega$) the absorption is inversely proportional to ΔE. For a given n_{2D}, the absorption is inversely proportional to m; that is, the smaller the effective mass the larger the absorption. Note that for a given E_f, however, the absorption is independent of m since $n_{2D} = (m/\pi\hbar^2)E_f$. The derivation of Eq. (5) is straightforward using a dipole interaction Hamiltonian and the Fermi's golden rule (see Chapter 1 in this volume).

The factor $\sin^2\theta$ in Eq. (5) comes from the polarization selection rule discussed earlier. The factor $\cos\theta$ seems to give a unphysical result when $\theta \to 90°$, however, since the meaning of η is the absorption probability of light passing through the well, in this extreme the passing length becomes infinitely long resulting in an infinitely large absorption. If one had considered a quantity of absorption constant α defined by $\eta = \alpha \times$ (length), where (length) $= L_w/\cos\theta$ is the propagation length, this quantity would have been always finite. For a real quantum well with finite barrier height, it would be more physical to choose the length using the quantum well structure thickness (including barriers) instead of only L_w taken here.

Let us put some typical numbers into Eq. (5) to get a feel for how strong the ISBT absorption is. For a typical 8- to 12-μm peaked QWIP, the half width is about $\Delta E = 0.01$ eV. For ground state to first excited state transition, the oscillator strength is $f = 0.961$ (see Eq. (4) with $n' = 1$ and $n = 2$). For 77-K operation, the carrier density is set to about $n_{2D} = 5 \times 10^{11}$ cm^{-2} (see later for the reason). For a GaAs well, the reduced effective mass is $m^* = 0.067$ ($m = m^* \times m_e$, where m_e is the free-electron mass) and the refractive index is about $n_r = 3.3$. With these values and for a 45° angle ($\theta = 45°$), the absorption for a single quantum well is $\eta = 0.54\%$ (for polarized light). Let us also evaluate the absorption constant for the case of $\theta = 90°$, with other parameters the same as just described. Taking a quantum well structure thickness of 50 nm (including barriers), the peak absorption constant is $\alpha = 1520$ cm^{-1}.

3. Wavelengths Covered by GaAs-Based Structures

For a standard QWIP, the optimum well design is the one having the first excited state in resonance with the top of the barrier. This configuration gives *at the same time* both a large absorption (similar to the bound state to bound state transition discussed earlier) and a rapid escape for the excited electrons. The optimum design configuration was recognized and experimentally proven by Steele *et al.* (1991) and Liu (1993). To design a quantum well for a given QWIP wavelength, one must know how the barrier height (conduction band offset) relates to heterosystem parameters; that is, the Al fraction (x) in the GaAs–Al$_x$Ga$_{1-x}$As case. Surveying many samples by comparing the calculated transition energy with the experimental peak absorption, we find a range of values for the conduction band offset $\Delta E_c = (0.87 \pm 0.04) \times x$ eV, where x is the Al fraction. The calculation is a simple eigenenergy calculation of a square quantum well. Higher order effects, such as nonparabolicity and many-body effects, influence the precise values of the calculation (as discussed by Helm in this volume).

For thermal imaging, the spectrum region of 3–12 μm is most interesting. QWIPs based on GaAs can easily cover this region. Figure 5 shows spectral response curves of six individual QWIPs with InGaAs or GaAs wells and AlGaAs barriers. Using special designs, QWIPs covering wavelengths much shorter than 3 μm may be possible (see, for example, Liu et al., 1998b). The short wavelength is limited by the lack of available high barriers. For beyond 12 μm, QWIPs with cutoff wavelengths as long as 28 μm have been demonstrated by Perera et al. (1998). However, these longer wavelength QWIPs have small potential energy barriers, which require much lower than 77 K cooling for operation. The region of 30–40 meV (31–41 μm) is masked by the reststrahlen (transverse optical phonon absorption), making it

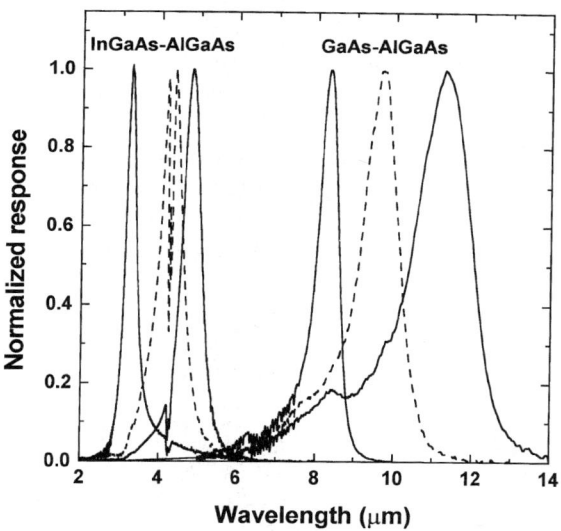

FIG. 5. Spectral response curves of six QWIPs covering the two atmospheric transmission windows of 3- to 5- and 8- to 12-μm wavelength regions. The dip in the second curve from the left at about 4.2 μm is due to the CO_2 absorption, and the noise from 5.5 to 7.5 μm is due to the water absorption.

difficult to design QWIPs here. For long wavelengths (longer than about 20 μm), free-carrier absorption also makes the design for highly efficient QWIPs difficult.

II. Detector Physics

1. DARK CURRENT

Before going into a detailed discussion of the physics of QWIP, we present the simplest picture of a QWIP made of n-type GaAs–AlGaAs in Fig. 6. The detector operation is based on photoemission of electrons from the quantum wells. The contacts on both sides are of the same n-type. The device is essentially a unipolar photoconductor. Usually many wells (10–100) are required for sufficient absorption.

A good understanding of the dark current is crucial for design and optimization of QWIPs because dark current contributes to the detector

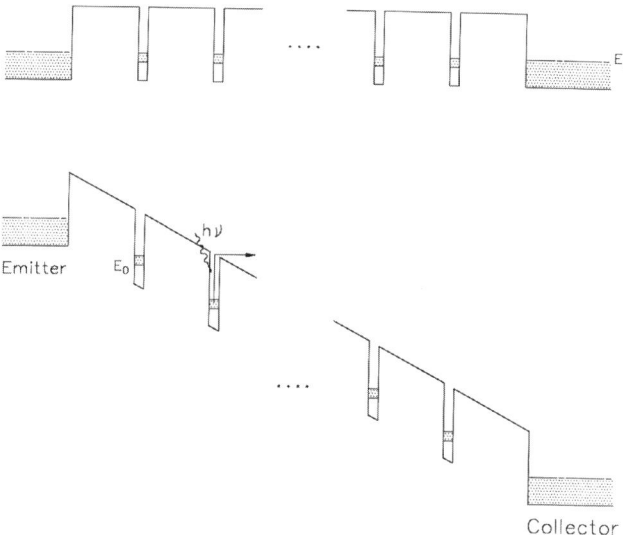

FIG. 6. Schematic conduction band-edge profile of a QWIP under zero (above) and finite (below) bias. The electron population in the n-type wells is provided by doping using silicon. The emitter and collector contact layers are doped with silicon. The ground state is labeled by E_0, and the Fermi energy in the well is E_f. Photons ($h\nu$) excite electrons from quantum wells, causing a photocurrent.

noise and dictates the operating temperature. We first present two simple physical models. As such, the results provide only an order-of-magnitude estimate, however, physical pictures are clearly seen. We then briefly discuss numerical models suited for guiding the fine-tuning of the device parameters and for interpreting higher order effects.

a. Simple Physical Models

For all the discussions in this subsection, there are several common assumptions or approximations made to define the physical regime. These are (a) the interwell tunneling contributes negligibly to the dark current, (b) the electron density in each well remains constant, (c) the heavily doped emitter serves as a perfectly injecting contact, and (d) the quantum well confines mainly one bound state including the case where the upper bound state is in resonance or very close to the top of the barrier. Assumption a is satisfied by requiring the barriers to be sufficiently thick. Assumption b is a good approximation but is not strictly valid especially at large bias voltages as shown experimentally by Liu *et al.* (1991b). Assumption c is expected to be valid for QWIPs with a large number of quantum wells, consistent with experimental results of Liu *et al.* (1997b). The effect of contact becomes important for QWIPs with a small number of quantum wells as simulated by Ershov *et al.* (1995). To produce good detectors, condition d is required, as stated in Subsection 3.

Once we have defined the physical regime, the dark current in a typical and standard QWIP is controlled by the flow of electrons above the barriers, and by the emission and capture of electrons in the wells. Figure 7 presents pictorially the electron distribution (top) and the processes controlling the dark current (bottom). The top part of the figure indicates that, at finite temperature, electrons are not only bound in the well but are also distributed outside of the well and on top of the barriers. The energy region for electrons contributing to the dark current is indicated by the large brace.

The lower part of the figure shows the dark current paths. In the barrier regions (on top of the barriers), the current flows in a three-dimensional (3D) fashion, and the current density is labeled as j_{3D}, which *equals* the dark current J_{dark}. In the vicinity of each well, the emission of electrons from the well contributes to the dark current (j_{em}). This current, which tends to lower the electron density in the well, *must* be balanced by the trapping or capture of electrons into the well under steady state ($j_{trap} = j_{em}$). Since the dark current is the same throughout the structure, j_{3D} and j_{em} (or j_{trap}) are related. If we define a trapping or capture probability p_c for an electron traversing a well with energy larger than the barrier height, we must have $j_{trap} = p_c j_{3D}$,

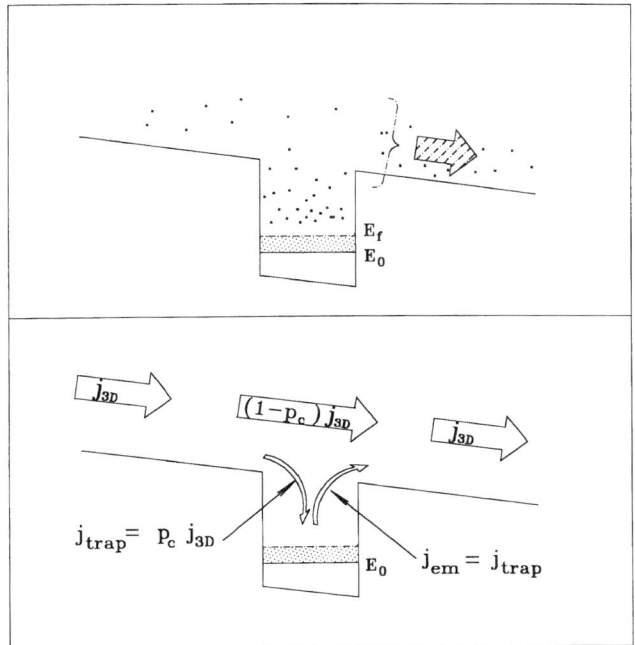

FIG. 7. Schematic representation of the electron distribution (top) and the processes controlling the dark current (bottom). Symbols are defined in the text.

and the sum of the captured and uncaptured fractions must equal the current in the barrier region: $j_{3D} = j_{trap} + (1 - p_c)j_{3D} = j_{em} + (1 - p_c)j_{3D}$. With this physical picture, one can model the dark current J_{dark} by calculating either j_{3D} directly or j_{em}, and in the latter case $J_{dark} = j_{em}/p_c$.

3D Carrier Drift Model. The first heuristic physical model calculates J_{dark} by directly estimating j_{3D}. A 3D electron density on top of the barriers n_{3D} is estimated and only drift contribution is taken into account (diffusion is neglected). The model was first presented in a very clear and concise paper by Kane et al. (1992). The dark current density is given by

$$J_{dark} = en_{3D}v(\mathscr{E}) \tag{6}$$

where $v(\mathscr{E})$ is the drift velocity as a function of electric field $\mathscr{E}$. The 3D density is calculated by treating the barriers as a bulk semiconductor. Superlattice band structure effects are neglected. This is justified because the barriers are thick (much thicker than the wells) and the resulting superlattice minibandgaps are less than the thermal energy $k_B T$ at device operating temperatures. The only 2D quantum well effect comes into the picture for the evaluation of the Fermi energy. QWIPs are normally degenerately doped in the wells. Assuming a completely ionization, the 2D doping density N_D equals the electron density within a given well. The Fermi energy is easily calculated by $N_D = (m/\pi\hbar^2)E_f$. Under all the preceding assumptions, a simple calculation yields

$$n_{3D} = 2(m_b k_B T / 2\pi\hbar^2)^{3/2} \exp(-E_{act}/k_B T) \qquad (7)$$

where m_b is the barrier effective mass; and E_{act} is the thermal activation energy, which equals the energy difference between the top of the barrier and the top of the Fermi sea in the well. We have assumed that $E_{act}/k_B T \gg 1$, appropriate for most practical cases.

The drift velocity takes the usual form

$$v(\mathscr{E}) = \frac{\mu \mathscr{E}}{[1 + (\mu \mathscr{E}/v_{sat})^2]^{1/2}} \qquad (8)$$

where μ is the low field mobility and v_{sat} is the saturated drift velocity.

A note on the validity of the model is in order. The key step is the evaluation of n_{3D}, by taking the equilibrium value at zero bias with the Fermi level determined by the well doping. It is therefore expected that the result is valid only for a low electric field. As an extension and perhaps an improvement, Man and Pan (1995) and Chu et al. (1987) proposed a model involving different carrier temperatures or Fermi levels for the 3D barrier electrons and the 2D well electrons. Man and Pan (1995) adopted an empirical expression relating the barrier hot electron temperature to the 2D well electron temperature and applied electric field; whereas Chu et al. (1987) calculated the 3D barrier electron Fermi level by balancing the tunneling escape rate and the capture rate due to electron–phonon scattering. In both cases, better fits to experiments were obtained.

The simple model compares well with experiments in the expected regime of a low applied electric field. We show a comparison with three samples designed to have high absorption for heterodyne detection, having high well doping and sacrificing dark current. The main difference between the samples is the doping density in the well. The three samples were grown by molecular beam epitaxy on semi-insulating GaAs substrates. The period of the 100-repeat multiple-quantum-well structure consists of a GaAs well and $Al_x Ga_{1-x}As$ barriers. The GaAs well center region is doped with Si to give

TABLE I

QUANTUM WELL PARAMETERS OF THE 100-WELL QWIPs

N_D (cm^{-2})	x	L_w (nm)	L_b (nm)	V_b (eV)	E_0 (eV)	E_{ex} (eV)
1×10^{12}	0.200	6.6	25.0	0.18	0.049	0.015
1.5×10^{12}	0.192	6.6	25.0	0.17	0.048	0.017
2×10^{12}	0.197	5.9	24.0	0.18	0.056	0.020

The symbols are x, Al fraction; L_w, well width; L_b, barrier width; N_D, doping density in the well; V_b, barrier height; E_0, ground state energy; and E_{ex}, exchange energy.

an equivalent 2D density of 1×10^{12}, 1.5×10^{12}, and 2×10^{12} cm^{-2}, respectively. The well width L_w is 6.6, 6.6, and 5.9 nm, the barrier width is 25.0, 25.0, and 24.0 nm, and the Al fraction x is 0.200, 0.192, and 0.197, respectively, for the three samples. The top and bottom GaAs contact layers are 400 and 800 nm thick, doped with Si to 2×10^{18} cm^{-3}. The quantum well parameters are listed in Table I.

The mobility and saturated velocity values used for fitting were $\mu = 1000$ cm^2/V s^{-1} and $v_{sat} = 10^7$ cm/s, respectively. The final parameter needed is the activation energy, which is $E_{act} = V_b - (E_0 - E_{ex}) + E_f$, where V_b is the barrier height and E_0 is the ground state eigenenergy. The energy shift in ground state due to exchange effect is taken into account using the expression of Bandara et al. (1988). With these parameters, we obtain an adequate fit to the experimental data in Fig. 8 in the low-field region (below a few kilovolts per centimeter), as expected for the model.

In fact, this simple 3D carrier drift model can be applied to more complicated structures. The assumption of having only one bound state can be relaxed. As an example, we apply this model to a set of three p-type QWIPs (Liu et al., 1998a) where several bound states (including both heavy and light hole states) are confined in the well. The main difference between the samples is the barrier height (or the x value). The period of the 100-repeat multiple-quantum-well structure consists of a GaAs well and 20 nm Al$_x$Ga$_{1-x}$As barriers. The well width L_w is 4.0, 4.1, and 4.1 nm, while the Al fraction x is 0.215, 0.245, and 0.29, respectively, for the three samples. The GaAs well center region is doped with Be to give an equivalent 2D density of 10^{12} cm^{-2}. The top and bottom GaAs contact layers are 200 and 600 nm thick, doped with Be to 8×10^{18} cm^{-3}. The quantum well parameters are listed in Table II. For completeness the calculated eigenenergies are given as well.

The three-dimensional density of free holes in the barrier region is estimated using the Fermi energy determined by the well doping. Both

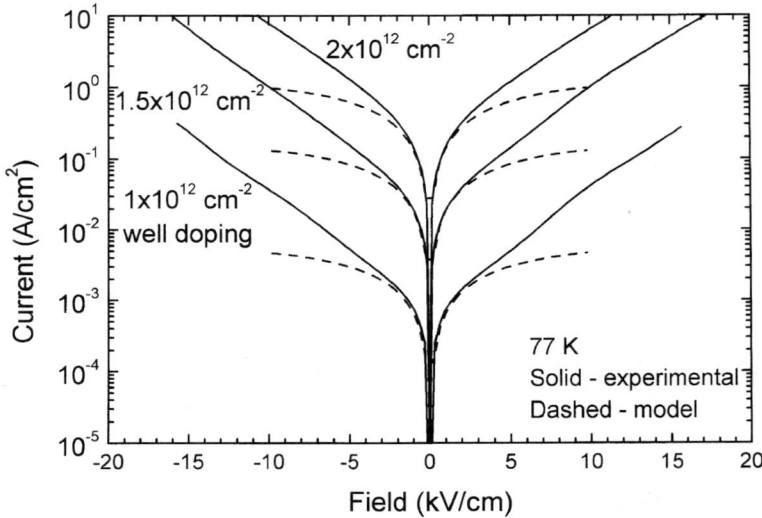

FIG. 8. Dark current characteristics at 77 K for a set of n-type QWIPs. Solid curves are experimental results, while dashed curves are calculated using the simple 3D drift model for a low field. The three samples differ mainly by the doping in the well.

heavy and light hole densities are included. The hole effective mass in the barrier was obtained by linearly interpolating the values for GaAs and AlAs. Since the hole mobility values are not known for our structures, we have used a value of $\mu_h = 100 \text{ cm}^2/\text{V s}^{-1}$ for holes in the barrier. The activation energy is $E_{act} = V_b - (HH1 - E_{ex}) + E_f$, where HH1 is the heavy hole ground state eigenenergy. Using Bandara et al. (1988), the calculated exchange

TABLE II

QUANTUM WELL PARAMETERS OF THE 100-WELL p-TYPE QWIPS

x	L_w (nm)	V_b (eV)	HH1 (eV)	HH2 (eV)	LH1 (eV)
0.215	4.0	0.114	0.029	0.101	0.057
0.245	4.1	0.130	0.029	0.106	0.059
0.290	4.1	0.154	0.031	0.114	0.064

The symbols are x, Al fraction; L_w, well width; V_b, barrier height; HH, heavy hole; LH, light hole; and the number after HH or LH is the eigenlevel index. All barriers are 20 nm thick, and all wells are doped with Be to 10^{12} cm^{-2}.

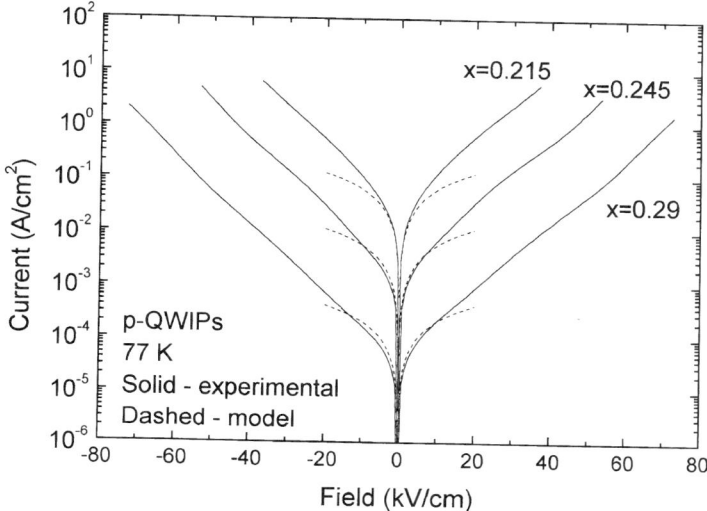

FIG. 9. Dark current characteristics at 77 K for a set of p-type QWIPs. Solid curves are experimental results, while dashed are calculated using the simple 3D drift model for low field. The three samples differ mainly by the barrier Al fraction (x), and therefore the barrier height.

energy E_{ex} is 26 meV. The eigenenergies and the Fermi energy are calculated using the 8×8 envelope-function approximation as in Szmulowicz and Brown (1995), which gives $E_f = 9$ meV, where E_f is referenced to the HH1 eigenlevel. Using these parameters, we obtain an adequate fit to the experimental data in Fig. 9 in the low-field region (below a few kilovolts per centimeter) as expected for the model.

Emission-Capture Model. The second model calculates j_{em} and then $J_{dark} = j_{em}/p_c$, as presented by Liu et al. (1993c) originally. This model does give adequate results for a large range of applied fields.

Before presenting the model, we first review and comment on several other published models in the literature. The formula given by Levine et al. (1990) (also used by Andrews and Miller, 1991) is identical to the one to be derived here. The physical arguments leading to the formula are, however, different. Levine et al. (1990) did not discuss the process of trapping or capture to balance the emission or escape. We feel that the derivation given here is more physically justified. We comment on Andrews and Miller (1991) who included an image charge effect using the usual formula for the image potential. This is incorrect for a quantum well with mainly an occupied ground state. The image charge effect is expected to be small for QWIPs. In a well-known paper, Kinch and Yariv (1989) formulated a model by estimating the 2D electrons distributed above the barriers. They also did not discuss the process of trapping or capture. In a more recent paper, Xu et al. (1995) used the same formula as Levine et al. (1990).

The model of Petrov and Shik (1991) estimates j_{em} and takes this as the total dark current. The estimate was done by integrating the product of the "velocity" v_z and the transmission coefficient, where the z coordinate is in the direction of the current flow. This approach models the electron escape only by direct tunneling and neglects the scattering assisted escape process. The latter process, which we think is the dominant one, involves the following. Electrons associated with the confined ground state in the well and distributed on the 2D in-plane dispersion curve undergo a scattering event to "get out" of the well and become a 3D mobile carrier in the barrier.

We now derive the dark current expression. We use the important theoretical result of Meshkov (1986), which can be stated as follows: including scattering processes, an electron tunneling rate in a one-dimensional potential is controlled by the *total* energy rather than the energy associated with the tunneling direction. For thick barriers, scattering assisted processes determine the tunneling transmission probability. We calculate j_{em} associated with the escape of electrons from the ground state subband (see the lower part of Fig. 7). We use the following expression:

$$j_{em} = e \int_{E_0}^{\infty} dn_{2D} D(E, \mathscr{E})/\tau_{scatt} \qquad (9)$$

$$dn_{2D} \equiv \frac{m}{\pi \hbar^2} \left[1 + \exp\left(\frac{E - E_f}{k_B T} \right) \right]^{-1} dE \qquad (10)$$

where E_0 is the ground state eigenenergy; E is the total energy of an electron, $D(E, \mathscr{E})$ is the transmission coefficient, which is taken to be unity for E higher than the barrier; τ_{scatt} is the scattering time to transfer electrons from the 2D ground state subband to a nonconfined 3D transport state on top of the barrier, and E_f is the Fermi energy. Strictly speaking, the scattering time τ_{scatt} is energy-dependent, but here as an approximation we will take τ_{scatt} out of the integral in Eq. (9). One can view this as taking an "averaged" τ_{scatt}. Since the two processes of carrier scattering out of the well and trapping into the well are inverse of each other, as another approximation, we take $\tau_{scatt} = \tau_{life}$, where τ_{life} is the lifetime associated with the carrier trapping process. These approximate steps are necessary to obtain the simple final result. We then have:

$$j_{em} = eN_{2D}/\tau_{life} \qquad (11)$$

$$N_{2D} \equiv \int_{E_0}^{\infty} D(E, \mathscr{E}) \, dn_{2D} \qquad (12)$$

The emission current j_{em} constitutes only part of the total dark current. Taking trapping and current injection into account (see the lower part of Fig. 7), the total dark current is

$$J_{dark} = j_{em}/p_c \qquad (13)$$

where the trapping or capture probability $p_c = \tau_{trans}/(\tau_{life} + \tau_{trans})$, and τ_{trans} is the transit time for an electron across *one* quantum region, partly including the surrounding barriers. (Note that τ_{trans} defined here is associated with only *one* period of the multiple quantum well device in contrast to Liu, 1992a, where the same symbol was used for the transit time across the entire detector.) In the limit of $p_c \ll 1$ (i.e., $\tau_{life} \gg \tau_{trans}$), as is true for actual devices at operating electric fields, the dark current becomes

$$\begin{aligned} J_{dark} &= \frac{eN_{2D}}{\tau_{life}} \times \frac{\tau_{life} + \tau_{trans}}{\tau_{trans}} \\ &\approx \frac{eN_{2D}}{\tau_{trans}} \\ &\equiv \frac{evN_{2D}}{\mathscr{L}} \end{aligned} \qquad (14)$$

where the electron "drift" velocity on top of the barriers is $v \equiv \mathscr{L}/\tau_{trans}$, and $\mathscr{L}$ is a relevant length scale.

There is an ambiguity concerning this relevant length scale $\mathscr{L}$ involved in both τ_{life} and τ_{trans}. The uncertainty is related to the quantum mechanical nature of the electron and to the question of exactly what length scale an electron is distributed over within its coherent lifetime. The definition of p_c and its physical meaning is, however, rigorous and clear since the same uncertainty exists in both τ_{life} and τ_{trans}. As an approximation, the length $\mathscr{L}$ is taken to be the period of the multiple quantum well structure $L_p = L_w + L_b$; that is, the sum of the well width L_w and the barrier width L_b. Physically, taking $\mathscr{L} = L_p$ should give a good approximation as long as the barrier width is large enough, but is not so large that L_b is longer than the electron coherence length. One should therefore expect a reduction in dark current when L_p is increased, but increasing L_b beyond a certain point will not reduce the dark current further.

Using Eqs. (10) and (14), and setting $\mathscr{L} = L_p$, the final expression for dark current is

$$J_{\text{dark}} = evN_{2D}/L_p$$
$$= ev \int_{E_0}^{\infty} \frac{m}{\pi\hbar^2 L_p} D(E, \mathscr{E}) \left[1 + \exp\left(\frac{E - E_f}{k_B T}\right)\right]^{-1} dE \quad (15)$$

Physically, we could arrive at the above expression only when the trapping process was taken into account. Early derivations were done by "converting" the 2D electron density N_{2D} into an "average" 3D density by multiplying by $1/L_p$ as in Levine et al. (1990) and Andrews and Miller (1991).

In our derivation, we have not explicitly included the effects of an upper bound state. Because the upper bound state (if it exists) is very close to the top of the barrier (by assumption) and the broadening is substantial, we believe that the contributions to the dark current from the upper subband is part of j_{3D} and has therefore been effectively taken into account already.

As before, the v vs $\mathscr{E}$ relationship is given by Eq. (8). The electric field dependence in Eq. (15) is given explicitly through Eq. (8) and implicitly through $D(E, \mathscr{E})$. Using the Wentzel–Kramers–Brillouin (WKB) approximation, $D(E, \mathscr{E})$ is given by

$$D(E, \mathscr{E}) = \exp\left[-2 \int_0^{z_c} dz \sqrt{2m_b(V - E - e\mathscr{E}z)}/\hbar\right] \quad (16)$$

for energy less than the barrier height $E < V$, where m_b is the barrier mass, and $V = V_b - e\mathscr{E}L_w/2$ includes a barrier lowering by applied bias field. This value for V takes into account the effective barrier lowering because E_0 (referenced to the center of the well) is approximately independent of $\mathscr{E}$, and $z_c = (V - E)/e\mathscr{E}$ defines the classical turning point. Figure 10 shows schematically the relevant quantities. The WKB approximation compares well with more exact calculations using the transfer matrix approach. Furthermore, we neglect nonparabolic effects, which would result in a small correction in the values of the calculated transmission coefficients.

Equation (15) can be simplified in the pure thermionic emission regime; that is, the tunneling contribution can be neglected. This is equivalent to setting $D(E, \mathscr{E}) = 0$ for E below the barrier. Furthermore, the condition $(E - E_f)/k_B T \gg 1$ for $E > V$ generally holds. Equation (15) then be-

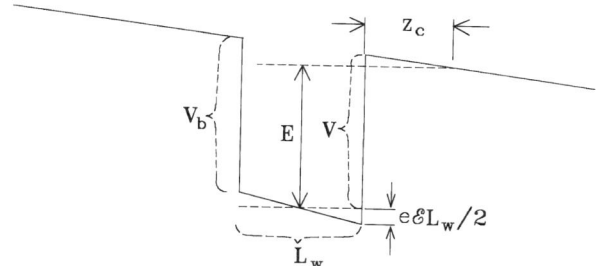

FIG. 10. Schematic illustration of relevant energies and the effective barrier lowering. The electron energy E is referenced to the center of the well.

comes

$$J_{\text{dark}} = ev\frac{m}{\pi\hbar^2 L_p} k_B T \exp(-E_{\text{act}}/k_B T) \qquad (17)$$

where the activation energy is $E_{\text{act}} = V - E_0 - E_f$. This expression closely resembles Eq. (6) together with Eq. (7). The main difference is that the activation energy here includes a barrier lowering (i.e., using V as opposed to V_b before). This leads to an improvement in the fit to the experiments.

The model discussed here has been compared with experiments extensively (see Levine et al., 1990; Liu, 1996). Given the simplicity of the model, the good agreement obtained exceeds the expectation. Figure 11 shows an example of model–experiment comparison. The samples have nominally the same parameters other than the number of wells. The well width is 6 nm, the barrier width is 25 nm, the barrier x value is 0.25, and the well doping is 9×10^{11} cm^{-2}. The measured curves in Fig. 11 display an asymmetry between positive and negative bias polarities, which was found to be due to the dopant segregation during growth (Liu et al., 1993b; Wasilewski et al., 1994).

A point worth noting about Eq. (17) is that it can be used for a much broader range of samples than Eq. (15). As an example, for a set of structures designed for high absorption heterodyne detection, the interwell tunneling contribution is large as soon as the applied field is larger than a few kilovolts per centimeter. Equation (15) cannot be used for these structures, however, to obtain a good estimate for the low field region, Eq. (17) can be used. The samples and parameters are the same as in Table I. Figure 12 shows the calculated and measured results. In comparison with Fig. 8, a slight improvement is seen: the agreement extends to a slightly larger bias field.

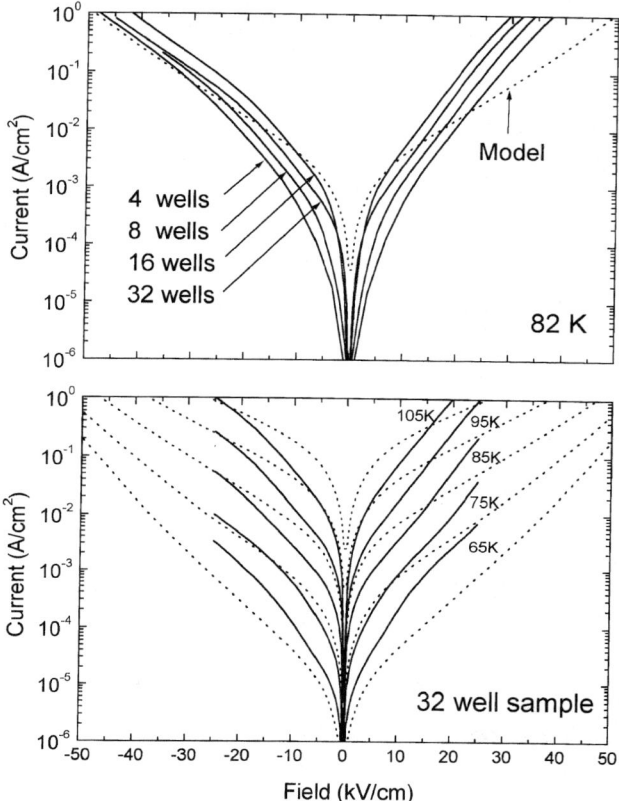

FIG. 11. Dark current of a set of nearly identical samples varying only the number of wells. The dashed curve is calculated using the emission-capture model.

b. *Self-Consistent and Numerical Models*

Self-Consistent Drift-Diffusion Model. The model developed by Ershov *et al.* (1995, 1996) calculates the QWIP characteristics by self-consistently solving three equations: (1) the Poisson equation, (2) the continuity equation for electrons in the barriers, and (3) the rate equation for electrons in the quantum wells. The inclusion of Poisson's equation is especially important for QWIPs with a small number of wells (<10) because the field can be substantially different (often higher) for the first few periods starting from the emitter in comparison with the rest of the wells (see Fig. 13). The continuity equation involves the current (expressed in the standard drift-

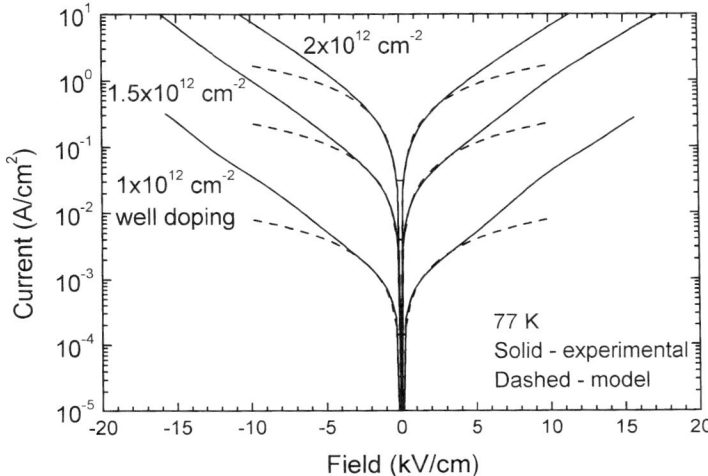

FIG. 12. Dark current of a set of samples with different well doping densities. The dashed curve is calculated using the emission-capture model but with only the thermionic emission contribution. The samples are the same as in Fig. 8.

diffusion form) and rates of thermal and optical generation and of recombination.

Using this numerical model, we were able to account for the observed unusual capacitance behavior (Ershov et al., 1997b) and explain the nonlinear photoconductivity at high excitation power using a CO_2 laser (Ershov

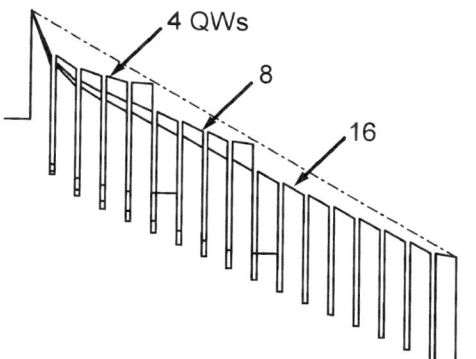

FIG. 13. Self-consistently calculated profiles of QWIPs with 4, 8, and 16 wells. The average applied field (10 kV/cm) is shown by a dash-dotted line (Courtesy of M. Ershov).

et al., 1997c). In addition, the model has the capability to predict transient and hence frequency characteristics (Ershov, 1996), as well as photoresponse under localized IR excitation (Ershov, 1998).

Self-Consistent Emission-Capture Model. Thibaudeau *et al.* (1996) presented a numerical model that extends the simple emission-capture model presented earlier. The model allows the electric field to be nonuniform, self-consistently determined by the Gauss law. The authors obtained better agreement with experiments than the simple model.

More recently, Ryzhii (1997) constructed an analytical model by solving Poisson's equation and an equation governing the electron balance in the quantum well. Interesting functional dependence of responsivity on the number of wells and the photon excitation power were found.

Numerical Monte Carlo Model. Publications by Ryzhii and Ryzhii (1998) and Ryzhii *et al.* (1998) presented Monte Carlo simulation results on QWIPs, in particular their ultrafast electron transport properties. Monte Carlo simulation sheds light on the hot electron distribution on top of the barriers, and should provide guidance to the optimization of QWIPs for ultra-high-speed operation.

c. *Remarks*

Although several models have been established, with varying degree of complexity, and good agreement between models and experiments has been obtained, to formulate a true first-principle QWIP model is a highly nontrivial task. This is so because the QWIP is a very complicated and "dirty" system. Given the wide barriers and narrow wells, the transport mechanism falls between ballistic and drift diffusion; and due to the high doping and high field, realistic calculations of scattering or trapping rates are extremely complicated and have not been performed so far. The situation becomes even more complicated to model (Petrov and Shik, 1998) for *p*-type structures (Brown and Szmulowicz, 1996).

2. PHOTOCURRENT

Photoconductivity phenomena in solids are well known, and many texts have been written on the subject, for example, Rose (1963), Kingston (1978), Long (1980), and Dereniak and Crowe (1984). The device operation of the photoconductive QWIPs is similar to that of extrinsic semiconductor detectors, discussed in detail by Sclar (1984). The distinct feature of QWIPs in contrast with the conventional intrinsic and extrinsic photoconductors is the discreteness; that is, incident photons are only absorbed in discrete quantum wells that are normally much narrower than the inactive barrier regions. In this section, we discuss the photocurrent caused by intersubband excitations in a QWIP and introduce the concept of photoconductive gain. Here we consider only the case of positive photoconductivity; that is, the effect of the incident IR light is to make the device resistance smaller. Negative photoconductivity is possible; for example, if one has a device with a negative differential resistance region (Choi et al., 1987; Liu et al., 1991a, 1993a).

a. Photoconductive Gain

Photoconductive gain is defined as the number of electrons flowing through the external circuit for each photon absorbed. A model (Liu, 1992a) specifically for photoconductive QWIPs has been constructed, which answers exactly what constitutes the mechanism of photoconductive gain and how it depends on device parameters such as the number of wells. The model also explains observations of large ($\gg 1$) photoconductive gains (Hasnain et al., 1990; Steele et al., 1992). In this subsection, we first present the physical picture, and then derive an expression for the gain. A comparison with experiments and an estimate of relevant time scales follow.

To visualize the physical process and the gain mechanism, a simple one-well structure is given in Fig. 14. The top part shows the dark current paths (the same as in the bottom part of Fig. 7). All these dark current paths remain *unchanged* when infrared light is shone on the detector. The additional processes as a result of the infrared are shown in the bottom part of Fig. 14. There is a *direct* photoemission of electrons from the well, and this, of course, contributes to the observed photocurrent in the collector. The photoconductive gain is a result of the extra current injection from the contact necessary to balance the loss of electrons from the well due to photoemission. The amount of the extra injection must be sufficiently large that its fraction trapped in the well equals the direct photoemission current. On the other hand, the fraction of the extra injected current that reaches the collector contact is in fact indistinguishable from the direct photoemitted current, and therefore contributes to the observed photocurrent. The total photocurrent consists of contributions from the direct photoemission and the extra current injection. Note that the physical mechanism given here is the same as for a conventional photoconductor (Rose, 1963), although this simple physical picture was presented only in the context of QWIP physics

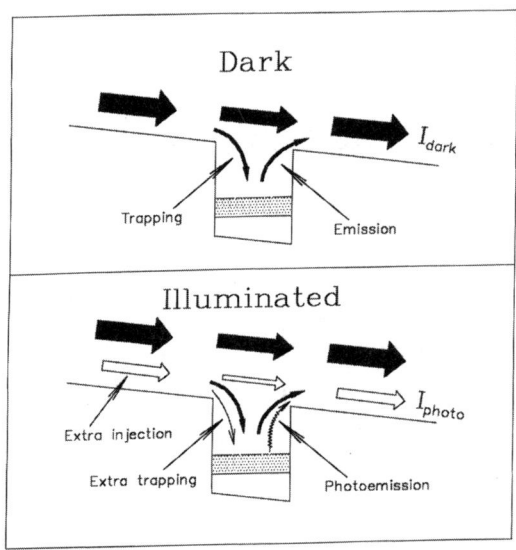

FIG. 14. The photoconductive gain mechanism. The top part shows the dark current paths, while the bottom indicates the direct photoemission and the extra current injection from the contact to balance the loss of electrons from the well. The dark current paths remain the same under illumination. The collected total photocurrent is the sum of the direct photoexcited and the extra injection contributions.

(Liu, 1992a). The common physical picture to explain larger than unity gain states that photoexcited electrons circulate around the circuit several times. This seems plausible, and appears in text books (see, e.g., p. 97 of Dereniak and Crowe, 1984), but is very misleading since a collector "absorbs" all electrons and the excess energy of the electrons gives rise to the ohmic heating.

An important fact, perhaps counterintuitive at first, is that the magnitude of photocurrent is independent of the number of wells if the absorption for each well is the same. To understand this pictorially, Fig. 15 shows the photocurrent paths for a two-well structure. If the first well is next to the emitter, one can see that for the second well, the processes of photoemission and refilling are identical to those in the first well. The only difference is that for the first well, the extra injection comes directly from the emitter, whereas for the second well, the injection is the resulting total photocurrent in the barrier separating the two wells. The same argument can be made for any subsequent wells. This means that the magnitude of the photocurrent is unaffected by adding more wells as long as the magnitude of absorption and hence photoemission from all the wells remains the same.

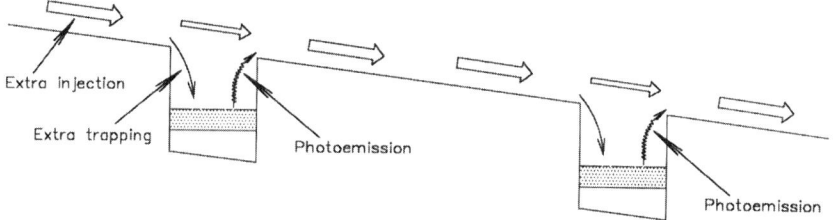

FIG. 15. A two-well case that illustrates the independence of photocurrent on number of wells. The left well is next to the emitter. The extra injection is the observed photocurrent, and the same injection balances the photoemission from both wells.

We present a derivation of the photoconductive gain expression explicitly for QWIPs. The model is constructed under the same assumptions a–d in Subsection 1a. A clarification of how the assumption of an injecting contact (assumption c in Subsection 1a) can be fulfilled is in order, because this is an important point both for the dark current model and for the concept of photoconductive gain. To have a good injecting contact, the barrier between the emitter contact and the multi-quantum-well (MQW) region must not be large. In most cases this barrier is the same as the barrier separating wells in the MQW region. If needed, an extra injection of electrons is achieved by increasing the electric field at the emitter–MQW junction. This is a self-consistent process: for example, if one adds an extra emission channel of electrons from the wells (e.g., by photoemission), the wells will tend to become slightly charged, which increases the electric field at the emitter–MQW junction and hence increases injection to balance the loss of electrons in the wells (Rosencher et al., 1992b). The mechanism is shown schematically in Fig. 16.

Having discussed the two pictures of one- and two-well cases (Figs. 14 and 15), a simple derivation can be constructed. We first calculate the emission current as a consequence of *direct* excitation of electrons into the continuum (the zigzag arrow shown in the bottom part of Fig. 14). The photoemission current directly ejected from *one* well is

$$i^{(1)}_{photo} = e\Phi \eta^{(1)} \frac{\tau_{relax}}{\tau_{relax} + \tau_{esc}} \equiv e\Phi \eta \frac{p_e}{N} \qquad (18)$$

where Φ is the incident photon number per unit time, the superscript (1) indicates quantities for one well, τ_{esc} is the escape time, τ_{relax} is the intersubband relaxation time, $\eta \equiv N\eta^{(1)}$ is the total absorption quantum

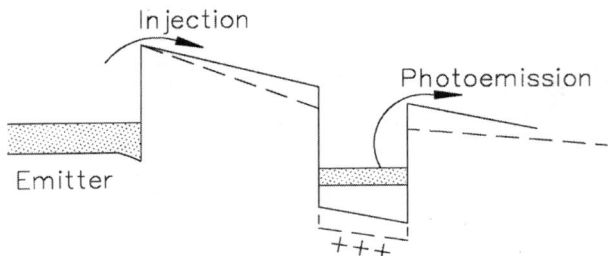

FIG. 16. The mechanism of injection. The band-edge profiles with and without the photoemission are shown in dashed and solid lines, respectively. Due to photoemission of electrons, the well becomes slightly charged, leading to an increase in the electric field and injection.

efficiency, N is the number of wells, and the escape probability for an excited electron from the well is given by

$$p_e \equiv \frac{\tau_{\text{relax}}}{\tau_{\text{relax}} + \tau_{\text{esc}}} \qquad (19)$$

We have assumed that the amount of absorption is the same for all the wells, that is, $\eta \equiv N\eta^{(1)}$. The photon flux could depend on the location of the well, but this is very specific to the detector light coupling geometry (e.g., 45° facet coupling) (Levine *et al.*, 1987) or grating coupling (Andersson and Lundqvist, 1991). The derivation of Eq. (18) is straightforward from a rate equation consideration: let n_{ex} be the number of the excited electrons, we have

$$\frac{dn_{\text{ex}}}{dt} = \Phi\eta^{(1)} - \frac{n_{\text{ex}}}{\tau_{\text{esc}}} - \frac{n_{\text{ex}}}{\tau_{\text{relax}}} \qquad (20)$$

Under steady state, $dn_{\text{ex}}/dt = 0$, we solve for n_{ex} from Eq. (20). Then the photoemitted current from one well is $en_{\text{ex}}/\tau_{\text{esc}}$, which gives Eq. (18).

As shown in Figs. 14 and 15, for each well, the injection current $i_{\text{photo}}^{(1)}/p_c$, which refills the well to balance the loss due to emission $i_{\text{photo}}^{(1)}$ *equals* the observed photocurrent. The photocurrent is then given by

$$I_{\text{photo}} = i_{\text{photo}}^{(1)}/p_c \qquad (21)$$

Using Eq. (18), we immediately get

$$I_{\text{photo}} = e\Phi\eta \frac{p_e}{Np_c} \equiv e\Phi\eta g_{\text{photo}} \qquad (22)$$

and

$$g_{\text{photo}} \equiv \frac{p_e}{Np_c} \tag{23}$$

is the photoconductive gain. As before in the discussion of dark current, the capture probability is given by

$$p_c = \frac{\tau_{\text{trans}}}{\tau_{\text{life}} + \tau_{\text{trans}}} \tag{24}$$

where τ_{trans} is the transit time for an electron across *one* quantum well region or the period of the structure. Note that Beck (1993), using the same physical model, obtained $g_{\text{photo}} = 1/(Np_c)$. The difference from Eq. (23) is purely due to the difference in the definition of quantum efficiency η. Beck's definition of quantum efficiency is not photon *absorption* quantum efficiency. His quantum efficiency is equivalent to $p_e\eta$ here, which results in the missing p_e factor in his photoconductive gain expression in comparison with Eq. (23). A similar photoconductive gain expression for QWIPs was given by Serzhenko and Shadrin (1991), valid for $p_c \ll 1$.

The photoconductive gain expression for QWIP should correspond to the expression given by the conventional theory of photoconductivity (Rose, 1963) $g_{\text{photo}} = \tau_{\text{life}}/\tau_{\text{trans,tot}}$, where $\tau_{\text{trans,tot}} = N\tau_{\text{trans}}$ is the total transit time across the detector active region. Under the approximation $p_e \approx 1$ and $p_c \approx \tau_{\text{trans}}/\tau_{\text{life}} \ll 1$, the gain expressions given by Eq. (23) and the conventional theory become the same:

$$g_{\text{photo}} \approx \frac{1}{Np_c} \approx \frac{\tau_{\text{life}}}{\tau_{\text{trans,tot}}} \tag{25}$$

For QWIPs, the lifetime τ_{life} is associated *only* with those processes that scatter an electron into the ground state subband in the well (trapping). For a simple square well, the condition $p_e \approx 1$ is met for a bound-to-continuum case (i.e., only one bound state is confined in the well); while for a bound-to-bound case (two bound states) this is no longer true (Xing et al., 1994). For structures where p_e and p_c can be designed independently (see, e.g., Schönbein et al., 1998) Eq. (25) cannot be used. If the absorption is proportional to N, as is a good approximation for a lot of practical cases, the photocurrent is independent of N since g_{photo} is inversely proportional to N. This was shown experimentally (see Fig. 17 and Steele et al., 1992). Photocurrent independence of N is equivalent to its independence of device length in the conventional theory. This independence does not mean that

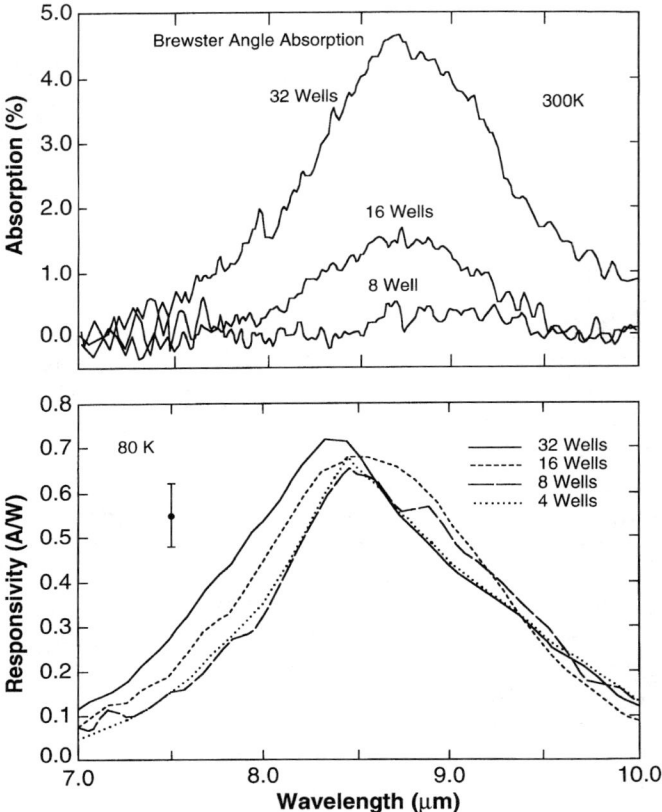

FIG. 17. Absorption (top) and responsivity (bottom) spectra of four samples with nominally identical parameters but the number of wells. This clearly evidences that the responsivity is independent of the number of wells although the absorption is increased proportionally. See Steele et al. (1992) for experimental details.

the detector performance is independent of the number of wells because of noise considerations (see Subsection 3).

Using Eq. (23) for different values of p_c and $p_e = 1$, the calculated photoconductive gain is plotted vs the number of wells in Fig. 18, together with some experimental data (at high fields) from Levine et al. (1990), Hasnain et al. (1990), Gunapala et al. (1991), Kane et al. (1992), and Steele et al. (1992). Our experiments (Steele et al., 1992) were performed on samples with comparable parameters (grown one after the other) except the number of

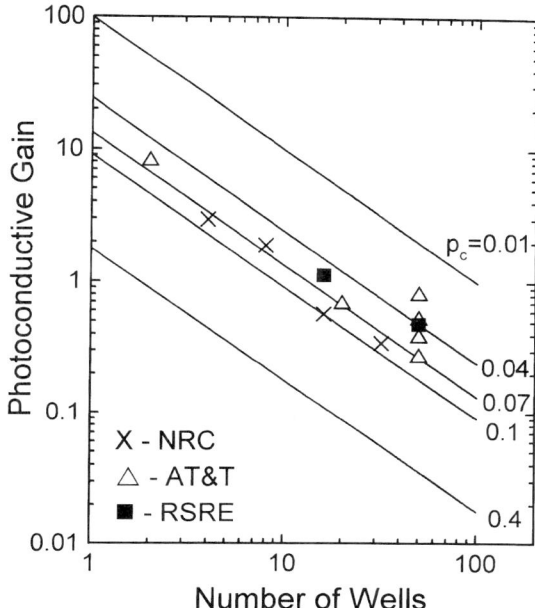

FIG. 18. Calculated photoconductive gain vs the number of wells for capture probability $p_c = 0.01$, 0.04, 0.07, 0.1, and 0.4, respectively. Experimental data are taken from Steele et al. (1992) (cross); Levine et al. (1990), Hasnain et al. (1990), Gunapala et al. (1991) (triangle); and Kane et al. (1992) (square).

wells. A capture probability of about 0.08 is inferred from our data. Most of the reported detector samples have 50 quantum wells. It is seen from Fig. 18 that a range of gain values from about 0.27 to 0.80 for 50 well samples have been observed, and hence the gain is quite sample dependent. The difference in τ_{trans} between samples is a possible reason for the spread of the observed gain values. The transit time τ_{trans} is mainly determined by the high field drift velocity of an excited electron in the barrier region and varies somewhat depending on the field strength and barrier material. The other possibility is the variation in τ_{life}. Processes that result in trapping are due to scattering by impurities and electrons in the well region, phonons, and interface roughness. Experiments of Gunapala et al. (1991) suggest that the impurity and electron–electron scattering may not be the dominant mechanism because the observed gain values did not decrease systematically with increasing well doping density. Phonon scattering would probably result in comparable values of capture probability for similar structures, and may not explain the strong sample dependence. Interfaces between AlGaAs and

GaAs could be very different from sample to sample and from one crystal growth facility to another.

We can make some estimates of the time scales involved. From the measured intersubband absorption line width (normally no narrower than about 5 meV), we know that τ_{life} is longer than 100 fs for a typical QWIP (i.e., 100 fs is a lower bound for τ_{life}). From time-resolved experiments, Tatham et al. (1989) projected an upper bound on the intersubband relaxation time of about 500 fs and Baier et al. (1996) deduced this time to be approximately 1 ps, both for a bound-to-bound state relaxation process. Using an excited state differential absorption technique, Faist et al. (1993) inferred a relaxation time of slightly less than 1 ps, again, for a bound-to-bound situation. Andersson (1995) performed calculations of τ_{life} for QWIPs and gave $\tau_{\text{life}} = 5.5$ ps for a typical structure. There is a difference between the bound-to-bound and continuum-to-bound relaxation processes. Since in the latter case the excited electron is extended in the barrier region, the relaxation should take more time. We therefore expect that τ_{life} for a typical QWIP should be approximately 5 ps.

The transit time can be estimated by $\tau_{\text{trans}} \approx L_p/v$, where L_p is the quantum well period and v is the drift velocity. For typical parameters of $v = 10^7$ cm/s and $L_p = 30$–50 nm, τ_{trans} is estimated to be in the range of 0.3–0.5 ps. One therefore expects a capture probability $[p_c = \tau_{\text{trans}}/(\tau_{\text{life}} + \tau_{\text{trans}}) \approx \tau_{\text{trans}}/\tau_{\text{life}}]$ to be in the range of 0.06–0.1 consistent with existing experiments (see Fig. 18).

To quantitatively calculate τ_{life} is a highly nontrivial problem (Goodnick and Lugli, 1988; Sotirelis et al., 1993). This problem is also of key importance to the operation of (interband) quantum well lasers (Blom et al., 1993b) as well as quantum cascade lasers (see Chapter 5 by Faist et al.). Large variations of carrier relaxation time from about 1 to 20 ps have been predicted, and the experimental evidence using many samples with varying well width supported the prediction (Blom et al., 1993b; Blom et al., 1993a; Muraki et al., 1996). There is a need for both theoretical and experimental investigations of intersubband relaxation and lifetime specifically for QWIPs; that is, heavily doped quantum wells with at most two bound states and high operating fields. A detailed first principle understanding may lead to a better QWIP design with enhanced lifetime and hence a better detector performance (see Subsection 3).

b. Detector Responsivity

In this subsection, we discuss factors that control the responsivity and estimate the values of relevant time scales. The magnitude of responsivity is controlled by both quantum efficiency and photoconductive gain. A high absorption does not necessarily result in a high detector responsivity. There must not only be high absorption but the photoexcited electrons must also escape the wells efficiently to give rise to a large photocurrent. Our experiments (Steele et al., 1991) and calculations (Liu, 1993) addressed the optimal design of quantum well for maximizing the responsivity. It was shown that the optimum occurs when the excited state is in close resonance with the top of the barrier. In general, the spectral current responsivity is given by

$$\mathcal{R}_i = I_{photo}/(hv\Phi) = \frac{e}{hv} \eta g_{photo} \qquad (26)$$

where v is the photon frequency. For QWIPs, g_{photo} is given by Eq. (23), which is

$$g_{photo} = \frac{\tau_{relax}}{\tau_{relax} + \tau_{esc}} \frac{\tau_{life} + \tau_{trans}}{\tau_{trans}} \frac{1}{N} \qquad (27)$$

where the first two fractions come from p_e and $1/p_c$, respectively. Let us discuss ways to maximize the responsivity. Under the assumption of $\eta \propto N$, there is nothing that can be done about the number of wells since $\mathcal{R}_i$ is independent of N. The escape probability must be made close to unity. This is done by ensuring $\tau_{esc} \ll \tau_{relax}$. For the bound-to-continuum case, the process of escape takes little time; that is, once an electron is excited, it is already in the continuum. In this case, $\tau_{esc} \approx 0$ and $p_e \approx 1$. For the bound-to-bound case, τ_{relax} is shorter than τ_{life}. From various experiments and calculations (see previous subsection), the value of τ_{relax} is about 1 ps (5 ps for τ_{life}). To ensure $p_e \sim 1$, we must have $\tau_{esc} \ll 1$ ps. This implies that if a bound-to-bound design is employed, one must have the excited state close to the top of the barrier so that the tunneling escape time is much less than 1 ps. For a typical 10-μm GaAs–AlGaAs QWIP under a typical field of 10 kV/cm, this dictates that the excited state should not be lower than about 10 meV below the top of the barrier. The upper state (E_1) tunneling escape time is estimated easily by considering an "attempt frequency" $v_1/2L_w$ and the transmission probability D, and is given by $\tau_{tunnel} \approx (2L_w/v_1)D^{-1}$, where $v_1 = \sqrt{2E_1/m}$. This estimate is semiclassical but does produce excellent results in comparison with rigorous calculations. The electric-field-dependent transmission probability is easily estimated using a WKB method.

3. Detector Performance

a. Detector Noise

In general, a photoconductor has several sources of noise: $1/f$ noise, Johnson noise, dark current noise, and photon noise (noise associated with the current induced by incident photons). The physical mechanism of $1/f$ noise is very complicated and is an on-going research topic. For GaAs QWIPs, experiments show that $1/f$ noise seldom limits the detector performance. We therefore neglect the contribution of $1/f$ noise. Johnson noise is inherent to all resistive devices and the noise mean square current is

$$i_{n,J}^2 = \frac{4k_B T}{R} \Delta f \qquad (28)$$

where Δf is the measurement bandwidth and R is the device differential resistance. Johnson noise is easily calculated once the device current–voltage (I–V) curve is known, and the contribution is usually small in a QWIP (Xing et al., 1994). Contributions from the dark current noise and the photon noise usually limit the detector ultimate performance in QWIPs. Here we concentrate on these two mechanisms.

In deriving noise expressions, I have found the simple physical picture given by Rose (1963, see pp. 97–99) to be extremely convenient. One identifies the source α_n of the noise and the magnification factor F in observation. Then the noise (squared average) is

$$I_n^2 = 2F^2 \alpha_n \Delta f \qquad (29)$$

Given the transport mechanism (see Figs. 7 and 14 and related discussions), the dark current noise is generation-recombination (g–r) in nature. The noise current should be given by the standard g–r noise expression (Rose, 1963)

$$i_{n,\text{dark}}^2 = 4e g_{\text{noise}} I_{\text{dark}} \Delta f \qquad (30)$$

where g_{noise} is the noise gain, and I_{dark} is the device dark current. If we label the emission current (see the top part in Fig. 14) from one well as $i_{\text{em}}^{(1)}$, the dark current is

$$I_{\text{dark}} = i_{\text{em}}^{(1)}/p_c = i_{\text{em}}/(N p_c) \qquad (31)$$

where $i_{\text{em}} \equiv N \times i_{\text{em}}^{(1)}$ is the total emission current from all N wells. Equivalently, one can express the dark current in an alternative form $I_{\text{dark}} = i_{\text{trap}}^{(1)}/p_c$, where $i_{\text{trap}}^{(1)}$ is the trapping current per well and $i_{\text{trap}}^{(1)} = i_{\text{em}}^{(1)}$. The "g–r" (emission and trapping here) noise therefore consists of two contributions: fluctuations in i_{em} and i_{trap}. The magnification factor is $1/(Np_c)$ according to Eq. (31). Then we have

$$i_{n,\text{dark}}^2 = 2e\left(\frac{1}{Np_c}\right)^2 (i_{\text{em}} + i_{\text{trap}})\Delta f$$

$$= 4e\left(\frac{1}{Np_c}\right)^2 i_{\text{em}}\Delta f$$

$$= 4e\frac{1}{Np_c} I_{\text{dark}}\Delta f$$

$$\equiv 4eg_{\text{noise}} I_{\text{dark}}\Delta f \qquad (32)$$

where the noise gain is defined by $g_{\text{noise}} \equiv 1/(Np_c)$. In a conventional photoconductor, the noise gain equals the photoconductive gain $g_{\text{noise}} = g_{\text{photo}}$ (at least as a very good approximation for all practical purposes). Here we see that g_{noise} is different (Liu, 1992b) from the photoconductive [Eq. (23)] $g_{\text{photo}} = p_e/(Np_c)$. Experiments of Levine et al. (1992a), Levine et al. (1992b), and Xing et al. (1994) reported a gain derived from the ratio of the measured current responsivity and absorption (i.e., the photoconductive gain), and a gain derived from direct noise measurements (the noise gain). The ratio of the two measured gains gave the escape probability p_e, which approached unity as the bias voltage was increased. In the limit of $p_c \to 1$, Eq. (32) does not give the expected N full shot noise sources in series. Beck (1993) extended the model and derived a more general expression $g_{\text{noise}} = (1 - p_c/2)/(Np_c)$, which does give the expected result when $p_c \to 1$. Several other related discussions of gain in QWIPs are given by Shadrin et al. (1995a), Shadrin et al. (1995b), and Choi (1997).

An expression for photon noise current can be easily obtained by replacing I_{dark} in Eq. (32) with photocurrent. One of the most important sources of photon noise is caused by the background photons absorbed by a detector. The background photon noise usually determines a detector operating temperature.

b. *Detectivity and Blip Condition*

Analytical Expressions. We first list a few expressions and then provide the derivations. The goal here is to provide a simple analytical estimate that displays the key physical parameters. In doing so, we can easily identify the key parameters and point out possible directions for improvement.

The two most important IR photon detector characteristics are detectivity (D^*) and blip (background limited IR performance) temperature (T_{blip}). Here D^* is the signal- (per unit incident power) to-noise ratio appropriately normalized by the detector area and the measurement electrical bandwidth. The relevant noise contributions are from (1) the detector itself (i.e., dark current) and (2) the fluctuation of the photocurrent induced by background photons incident on the detector. The blip regime is defined as the regime where the dominant noise is caused by the background photons. It is always desirable to operate a detector under blip condition for maximal sensitivity.

The detector noise limited D^* as a function of detector parameters is given by

$$D^*_{\text{det}} = \frac{\lambda}{2hc} \frac{\eta}{\sqrt{N}} \sqrt{\frac{\tau_{\text{life}}}{N_{2D}}} \tag{33}$$

where λ is the wavelength, η is the absorption quantum efficiency, N is the number of quantum wells, and N_{2D} is given by

$$N_{2D} = \frac{m}{\pi\hbar^2} k_B T \exp(-hc/\lambda_c k_B T + E_f/k_B T) \tag{34}$$

where m is the effective mass in the well, T is the temperature, λ_c is the cutoff wavelength, and E_f is the Fermi energy. There is a simple relation between the Fermi energy and the well two-dimensional doping density $N_D = (m/\pi\hbar^2)E_f$ (assuming doping is in the well and the dopants are completely ionized). Some of the shortcomings of QWIP are seen from Eqs. (33) and (34). Since the lifetime is short for QWIPs, a lower detector dark current limited $D^*_{\text{det}} \propto \sqrt{\tau_{\text{life}}}$ is anticipated. In addition, an "extra" thermal excitation factor appears [$E_f/k_B T$ in the exponent of Eq. (34)]. This causes a larger dark current and hence results in a lower D^*_{det}. Note that this discussion is for a QWIP with its noise coming solely from its dark current (i.e., no background); and a similar discussion is presented for background photon noise limited detectivity.

From Eq. (33), the expected general behavior for a photoconductor is seen, such as (1) a higher η, longer τ_{life}, shorter λ_c or lower T lead to a higher D^*; and (2) λ_c and T are the most sensitive parameters, being in the exponent. Noting that η is proportional to the doping density and hence the Fermi energy ($\eta \propto E_f$), there is an optimum value for E_f. Since $D^* \propto E_f \exp(-E_f/2k_B T)$, the maximum occurs when $E_f = 2k_B T$ (Kane et al., 1992). This condition dictates an optimum value for N_D for maximizing

D^*. Figure 19 shows calculated D^* values versus wavelength and temperature. Typical values are used: $\eta = 25\%$, $N = 50$, $\tau_{\text{life}} = 5$ ps, and $m^* = 0.067$ for GaAs reduced effective mass ($m = m_e \times m^*$, where m_e is the free electron mass). The optimum doping or Fermi energy value $E_f = 2k_B T$ is assumed.

From Eq. (33), a point worth noting is that *provided* η can be made high, say close to 100%, one should use the *least* number of quantum wells in a QWIP. The limiting and the best case is a single well ($N = 1$) QWIP with $\eta = 100\%$. In this limit, a QWIP would in fact have a comparable performance to that of HgCdTe or InSb detectors. This limit is not completely impractical, for example, it may be achievable by a waveguide-grating coupler with a high quality factor. Further work in this direction would have a potentially high payoff.

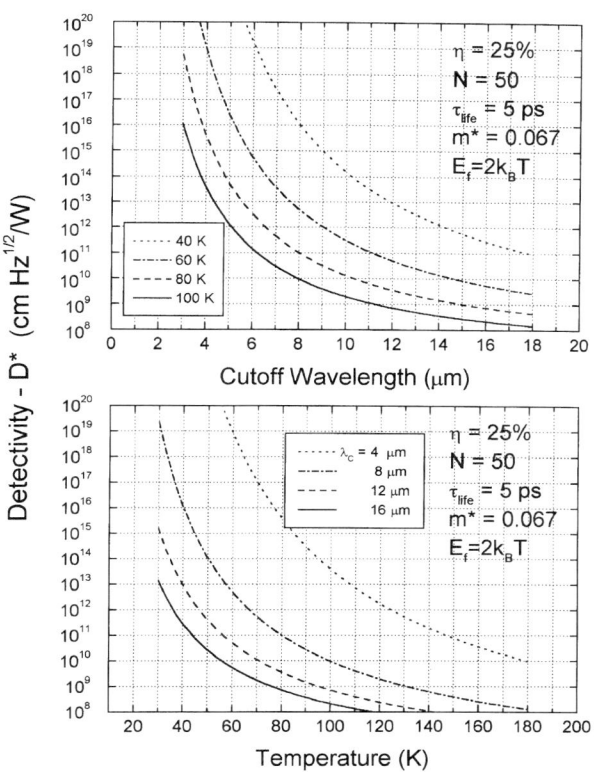

FIG. 19. Calculated detector dark current limited detectivity vs (upper) wavelength for different temperatures and vs (lower) temperature for different wavelengths.

For comparison, the background limited spectral peak D^* is given by

$$D^*_{\text{blip}} = \frac{\lambda_p}{2hc}\sqrt{\frac{\eta_p}{\phi_{B,\text{ph}}}} \qquad (35)$$

where λ_p is the peak detection wavelength, η_p is the peak absorption, and $\phi_{B,\text{ph}}$ is the integrated background photon number flux (per unit area) incident on the detector. One can see that for a given wavelength and if a detector is blip, D^* depends only on the absorption quantum efficiency and the background photon flux. The lifetime becomes irrelevant in this regime. Unlike a broadband photon detector with a cutoff, a QWIP spectral response is peaked and can be approximated by a Lorentzian line shape. In evaluating $\phi_{B,\text{ph}}$ therefore an integration is performed using the blackbody function and a Lorentzian with unity height peaked at λ_p. Another parameter, the full width of the response curve $\Delta\lambda$, must be specified for the integration. For a common QWIP the range is approximately $\Delta\lambda/\lambda_p = 10$–30% (Levine, 1993).

The blip temperature is determined by the following equation:

$$\eta^{(1)}\tau_{\text{life}}\phi_{B,\text{ph}} = \frac{m}{\pi\hbar^2} k_B T \exp(-hc/\lambda_c k_B T + E_f/k_B T) \qquad (36)$$

where $\eta^{(1)}$ is the peak absorption efficiency for one quantum well. Solving this transcendental equation for T gives T_{blip}. Figure 20 shows the calculated T_{blip} versus QWIP peak detection wavelength (λ_p) for different values of τ_{life}. Typical values have been used: response bandwidth $\Delta\lambda/\lambda_p = 20\%$, $\eta^{(1)} = 0.5\%$, 90° full cone field of view (FOV), and 300-K background temperature. Note that the cutoff wavelength is 10% larger than the peak in this calculated example (i.e., $\lambda_c = 1.10 \times \lambda_p$). Another shortcoming of QWIP is clearly displayed in Eq. (36). The short lifetime results in a lower T_{blip} and the effect of a finite E_f on the exponential also leads to a lower T_{blip}.

From Eq. (36) for T_{blip}, the most sensitive parameter is λ_c, being on the exponent. The short lifetime, although giving rise to a fast intrinsic response speed, is the cause for the low T_{blip}. It is interesting to note that T_{blip} depends on the *one* well absorption not the total absorption. It is also interesting to note that improving $\eta^{(1)}$ has the *same* effect as improving τ_{life}. Detectors made of HgCdTe or InSb have a performance similar to the solid line ($\tau_{\text{life}} = 1$ ns) or better in Fig. 20. The practical values for QWIPs τ_{life} fall in the range of 1–10 ps. It then follows that if $\eta^{(1)}$ can be enhanced by about 100 times, QWIP performance would be comparable to HgCdTe or InSb.

From Eq. (36), an optimum condition for the Fermi energy and hence the

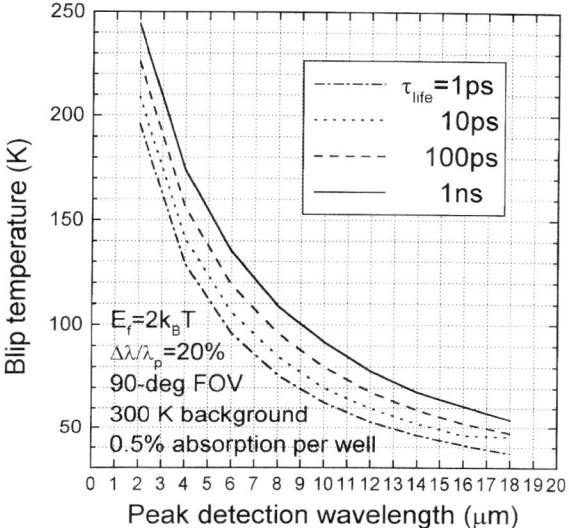

FIG. 20. Calculated background-limited infrared performance (blip) temperature vs peak detection wavelength for different carrier lifetimes (τ_{life}). For GaAs QWIPs, $\tau_{\text{life}} \approx 1$ ps.

doping density can be found. Given λ_c, τ_{life}, and $\phi_{B,\text{ph}}$, Eq. (36) can be rewritten as

$$E_f/k_B T \exp(-E_f/k_B T) = \text{Constant} \times \exp(-hc/\lambda_c k_B T) \qquad (37)$$

noting $\eta^{(1)} \propto N_D \propto E_f$. One can adjust E_f to maximize the left-hand side of the equation, which maximizes T_{blip}. The optimum condition is $E_f = k_B T$, which is different from the optimum condition for maximizing the detector limited detectivity by a factor of 2 (i.e., $E_f = 2k_B T$).

Derivations. Here we present the derivations of the analytical expressions Eqs. (33), (35), and (36).

As given before, the current noise power spectral density $S_i \equiv i_{n,\text{dark}}^2/\Delta f$ relates to dark current by

$$S_i = 4egI_{\text{dark}} \qquad (38)$$

where $g \approx 1/Np_c$ is the gain. We assume that we are in the regime where noise and photoconductive gains are the same. This is valid for structures where $p_c \ll 1$ and $p_e \approx 1$ (i.e., a nearly ideal photoconductor). The dark current can be estimated

$$I_{\text{dark}} = e \frac{N_{2D}}{L_p} v A \qquad (39)$$

where L_p is the period of the QWIP multiple quantum wells, v is the drift velocity, and A is the device area. The above-barrier two-dimensional electron density can be approximated by Eq. (34) in the low applied field regime where the thermionic emission is the dominant electron escape process. Note that the energy difference from the ground state to the top of the barrier corresponds approximately to the cutoff. Equation (39) is the same as Eq. (17).

As before [Eq. (26)], the current responsivity of a QWIP is given by

$$\mathcal{R}_i = e \frac{\lambda}{hc} \eta g \qquad (40)$$

The detector dark current limited D^* is given by

$$D^*_{\text{det}} = \frac{\mathcal{R}_i}{\sqrt{S_i/A}} \qquad (41)$$

Using relations $v = L_p/\tau_{\text{trans}}$ and $g = \tau_{\text{life}}/N\tau_{\text{trans}}$, where τ_{trans} is the transit time across one period, and substituting Eqs. (38), (39), and (40) into Eq. (41), we get Eq. (33).

The derivation of Eqs. (35) and (36) can be constructed similarly. Replacing I_{dark} in Eq. (38) with the photocurrent caused by the background photons one obtains the noise spectral density caused by the background. The detectivity follows the same definition of Eq. (41), leading to Eq. (35). The blip condition is defined when the photocurrent caused by the background equals the dark current. (This condition is equivalent to when background photon noise equals the dark current noise.) Using Eqs. (39) and (40), Eq. (36) is easily obtained.

For completeness, the ideal (blackbody) background photon flux is given by

$$\phi_{B,\text{ph}} = \int d\lambda \left(\pi \sin^2 \frac{\theta}{2} \right) \eta(\lambda) L_B(\lambda) \qquad (42)$$

where θ is the FOV full-cone angle, the photon irradiance is given by

$$L_B(\lambda) = \frac{2c}{\lambda^4} \frac{1}{e^{hc/\lambda k_B T_B} - 1} \qquad (43)$$

T_B is the background temperature, and the spectral lineshape of a QWIP is modeled by

$$\eta(\lambda) = \frac{1}{1 + (\Delta\lambda/2\lambda - \Delta\lambda/2\lambda_p)^2} \qquad (44)$$

4. Design of an Optimized Detector

In this subsection, we summarize the guidelines in designing an optimum photoconductive QWIP, which involves choosing the following parameters: well width L_w, barrier height, barrier width L_b, well doping density N_D, and number of wells. We use the simplest structure made of GaAs–AlGaAs square quantum wells. The well region is GaAs, and the barrier is $Al_xGa_{1-x}As$ so that its height is controlled by Al fraction x.

As discussed before, the optimum well shape is the one having the first excited state in resonance with the top of the barrier. Given this design rule, the well width and barrier height are fixed once an desired detection wavelength (peak wavelength λ_p) is chosen. Figure 21 shows these parameters for a range of λ_p for a GaAs–AlGaAs quantum well. The peak detection wavelength λ_p corresponds to the energy difference between the first excited

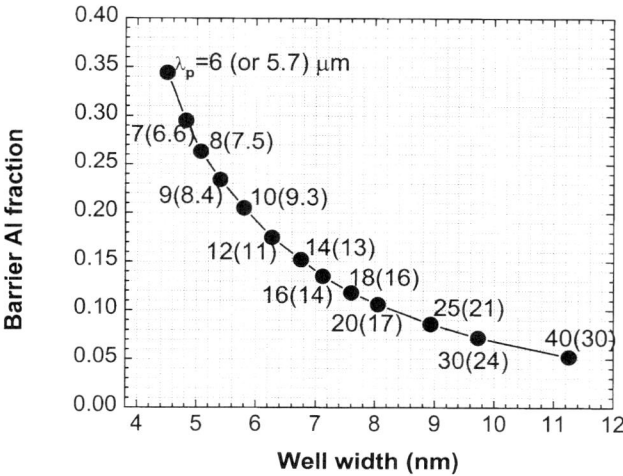

Fig. 21. Calculated parameters of barrier Al fraction and well width for a given peak detection wavelength (λ_p). Including many-body effects, the values in parentheses are for a quantum well with doping density of about 5×10^{11} cm^{-2}.

and the ground states. The calculation is a simple one-band effective mass model calculation. The difference in effective mass values between AlGaAs and GaAs is included. The barrier height V_b relates to Al fraction by $V_b = 0.87 \times x$ eV. All higher order effects have been neglected, such as, band nonparabolicity, Coulomb interaction between ionized donors and electrons (Hartree correction), exchange-correction effect, and depolarization-exciton effect. For a structure appropriate for $\lambda_p = 10$ μm these effects all lead to a modification of the transition energy in the few-percent range. A detailed discussion is given in the chapter by Helm in this volume.

As an example, the main effect of exchange-correlation and depolarization-exciton is to shift the transition energy to a higher value. The amount is about 10 meV for a 2D electron (or doping) density of $N_D = 5 \times 10^{11}$ cm^{-2}. Taking this into account, the renormalized peak wavelengths are shown in Fig. 21 in parentheses. The effect is not very important for short-wavelength structures, but is dramatic for long wavelengths. In general, longer than about 20 μm, Hartree, exchange-correlation, and depolarization-exciton effects must be included to obtain good fit to experiments. At wavelengths shorter than about 7 μm, the nonparabolicity effect becomes important.

The next parameter is the well doping. As discussed before, to maximize the detector limited detectivity, the doping density should be such that the Fermi energy is $E_f = 2k_BT$, where T is the desired operating temperature. On the other hand, to maximize the blip temperature, one should have $E_f = k_BT$. As before, the doping density relates to Fermi energy by $N_D = (m/\pi\hbar^2)E_f$. Figure 22 shows these two (trivial) relations for QWIPs with GaAs wells.

I have not found an intuitive way to choose the barrier width. The barrier width should be wide enough so that the interwell tunneling current is suppressed. There is a critical value beyond which any further increase in L_b does not lead to a lowering in dark current for the same applied field. This means that one should use thick barriers. On the other hand, the practical MBE growth is such that the thicker the total epilayer the lower the material quality. Furthermore, if the barrier width can be reduced, one can grow more repeats for higher absorption for a given total epilayer thickness. Empirically, a barrier width in the range of 30 to 50 nm seems to be sufficient. One can make some hand-waving arguments in estimating the critical barrier thickness. For example, if the operating field is below about 10 kV/cm, to ensure that the tunneling of electrons near the top of the barrier (e.g., within 20 meV) directly into the next well is suppressed, one needs a barrier width of more than about 20 nm. A systematic experimental and modeling study is desired.

Last, for the number of wells or repeats (N), since the absorption depends

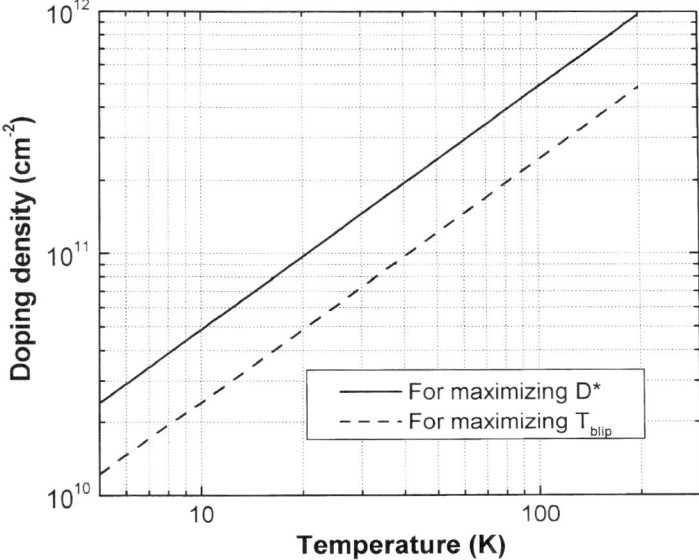

FIG. 22. Optimum doping density vs device operating temperature. The well material is GaAs.

strongly on the device geometry and optical coupling scheme, only a general guideline can be given: one should maximize the absorption with a minimum number of wells.

Note that these design guidelines are based on our current understanding and on our and other published experiments. There have not been systematic studies to reenforce–confirm these design "rules."

III. Novel Structures and Devices

1. MULTICOLOR AND MULTISPECTRAL DETECTORS

This part of the chapter focuses on novel structures and devices made possible by the quantum well approach for IR detection and imaging applications. Being based on thin multilayers grown by epitaxial techniques, the design of QWIPs is very flexible. This enables various implementations of multicolor and multispectral detectors. The approaches are divided into three basic categories: (1) multiple leads, (2) voltage switched, and (3) voltage tuned. The three cases are schematically shown in Fig. 23. The division here is somewhat arbitrary. In general, a multicolor or multispectral detector is a device having its spectral response varied with parameters like voltages or any other parameters such as pressure, magnetic field, filter position, and so on. The latter cases are not as desirable in practical applications and are not discussed here.

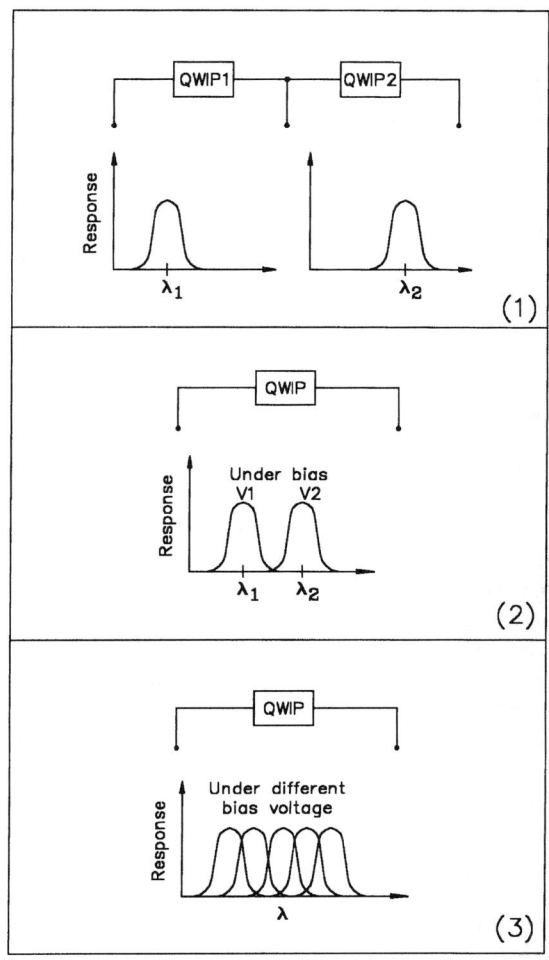

FIG. 23. Three different multicolor QWIP approaches. (1) Two QWIPS with different response wavelengths grown in a multistack. Intermediate leads are provided, so that each QWIP functions independently. (2) The QWIP has a response peaked at λ_1 at voltage V_1 and at λ_2 at voltage V_2. (3) The QWIP has a response that is continuously tuned by voltage.

Approach 1 is a direct one, which involves contacting each intermediate conducting layer separating one-color QWIPs grown in a multistack. This results in a separately readable and addressable multicolor QWIP with multiple electrical leads, and a two-color version has been demonstrated by Köck et al. (1992). The advantage of this approach is its simplicity in design and its negligible electrical crosstalk between colors. The drawback is the difficulty in fabricating a many color version because of the many separate leads required in contacting each intermediate layer. Bois et al. (1997) and Costard et al. (1998) have developed processing technologies for implementing two-color imaging arrays.

The second approach (middle part in Fig. 23) is to have a QWIP with a switchable response, for example, for an applied voltage V_1 the response is at λ_1 and for V_2 at λ_2. One such example is realized by stacking the usual one-color QWIPs separated by thin conducting layers (Liu et al., 1993d; Lenchyshyn et al., 1996). We rely on the highly nonlinear and exponential nature of the device dark current–voltage (I–V) characteristics. This implies that an applied voltage across the entire multistack would be distributed among the one-color QWIPs according to their values of DC resistances. Thus, when the applied voltage is increased from zero, most of the voltage will be dropped across the one-color QWIP with the highest resistance. As the voltage is further increased, an increasing fraction of the voltage will be dropped across the next highest resistance one-color QWIP, and so on. Since the detector responsivity of a one-color QWIP gradually turns on with applied voltage, we therefore can achieve a multicolor QWIP with spectral response peaks that turn on sequentially with applied voltage.

The band-edge profiles of a three-color version are schematically shown in Fig. 24. To quantitatively predict the magnitude of the photocurrent, we need to consider what is being measured. The use of a photoconductive detector usually involves applying a constant DC bias across detector in series with a load resistor R_s. The equivalent circuit of this three-color detector involves a network of photocurrent sources (i_{p1}, i_{p2}, and i_{p3}) and dynamic device resistances (r_1, r_2, and r_3), as shown in Fig. 25. Under the small-signal condition, the measured photoresponse current is

$$I_{\text{photo}} = \frac{i_{p1}r_1 + i_{p2}r_2 + i_{p3}r_3}{R_s + r_1 + r_2 + r_3} \tag{45}$$

The nonlinear nature of the dynamic resistances as a function of the voltage leads to a nonlinear weighting factor of the relative contributions among i_{p1}, i_{p2}, and i_{p3} to I_{photo}. This scheme of multicolor QWIP was first reported by Liu et al. (1993d) and subsequently studied in detail by Lenchyshyn et al.

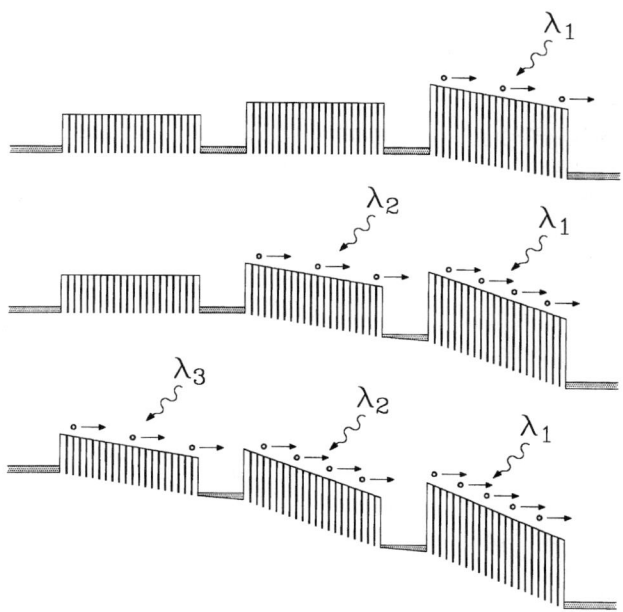

FIG. 24. Band-edge profile of a three-color detector at different bias voltages. The top part is for a small voltage where only the highest resistance one-color QWIP (at λ_1) is turned on, the middle part is appropriate for the situation where two of the three one-color QWIPs (at λ_1 and λ_2) contribute to the photocurrent, and in the lower part the applied voltage is high enough so that all three one-color QWIPs (at λ_1, λ_2, and λ_3) have nonnegligible electrical fields.

(1996). An example of a two-color QWIP is shown in Fig. 26. The advantage of this approach is that it is simple in fabrication (as it requires only two terminals) and suited for implementing a QWIP with many colors. The drawback is the difficulty to achieve a negligible electrical crosstalk between colors. A similar multicolor QWIP relying on high and low field domains was demonstrated by Gravé et al. (1992).

The last case (lower part in Fig. 23) is a QWIP with its response continuously tuned in a range of wavelengths. The demonstrated examples of approach 3 involve special shapes of quantum wells; for example, a stepped well by Martinet et al. (1992) so that the response spectrum shifts as a function of applied bias voltage. This provides a continuous tuning of the spectrum by moving the intersubband resonance position. A range from 8.5 to 13.5 μm has been achieved using stepped wells (Martinet et al., 1992). The large continuous tuning capability is the distinct feature of this

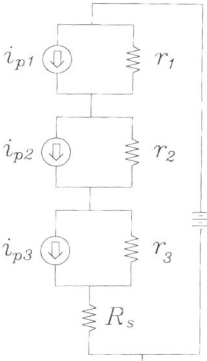

FIG. 25. Model equivalent circuit of a three-color detector biased through a series load resistor.

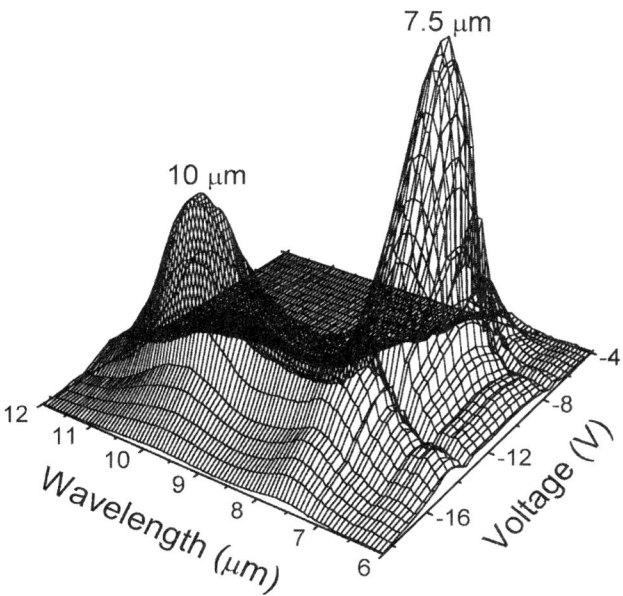

FIG. 26. A two-color QWIP response vs applied voltage and wavelength at 80 K. In the voltage range around -8 V, the dominant response is from the 7.5-μm QWIP; whereas at around -14 V, the 10-μm QWIP becomes strong. The full detail is found in Lenchyshyn et al. (1996).

approach. The difficulty is to ensure good QWIP performance for all voltages. To achieve good performance, the transition final state (usually the first excited state) must be close to the top of the barrier, as discussed before, providing a large intersubband transition strength and, at the same time, an easy escape for the excited carriers. These two conditions are difficult to fulfill for all voltages. Another factor, which may degrade the QWIP performance, is the use of relatively wide wells as in the case of a stepped well. This may lead to an enhanced trapping probability and hence a shorter carrier lifetime.

As an example, Fig. 27 depicts the stepped quantum well QWIP that we tested. The QWIP had 32 wells with the centers of the InGaAs layers doped with Si to $5 \times 10^{11} \, \text{cm}^{-2}$. The normalized spectra are given in Fig. 28. A large tuning in the range of 220 to 260 meV (i.e., 17%) was obtained. The "noise" in the -17-V curve was due to water absorption in the measurement optical path. Figure 29 shows a comparison between the experimental response peak positions (taken from Fig. 28) and calculated eigenenergy difference. A reasonable agreement was obtained. Also shown is the energy difference of the barrier height and the excited level $(V - E_1)$. It is clearly seen that the quantity $V - E_1$ varies substantially in the range between 0 and 50 meV. This may be a detrimental effect as discussed before.

A final comment on multicolor QWIPs is that it seems that for all the approaches, the device performance has not been fully optimized. Further work is therefore needed both in optimization and in new and better designs. Moreover, multicolor QWIPs require special gratings that have not

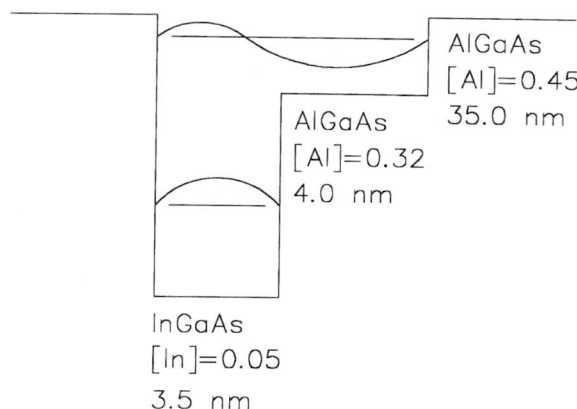

FIG. 27. Band-edge profile of stepped quantum well used for a voltage-tunable QWIP. The device had 32 repeats.

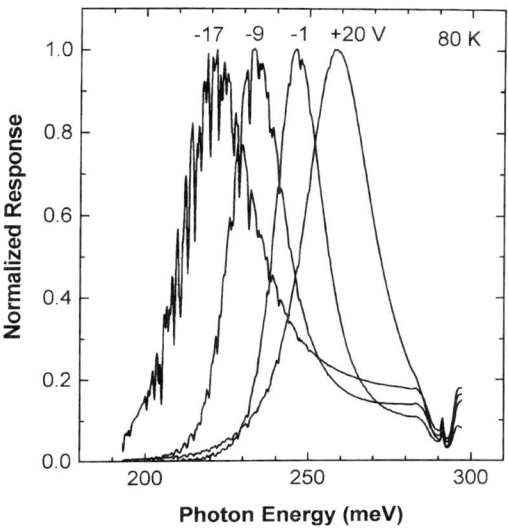

FIG. 28. Spectra at different applied voltages for sample in Fig. 27. (Data taken by E. Dupont.)

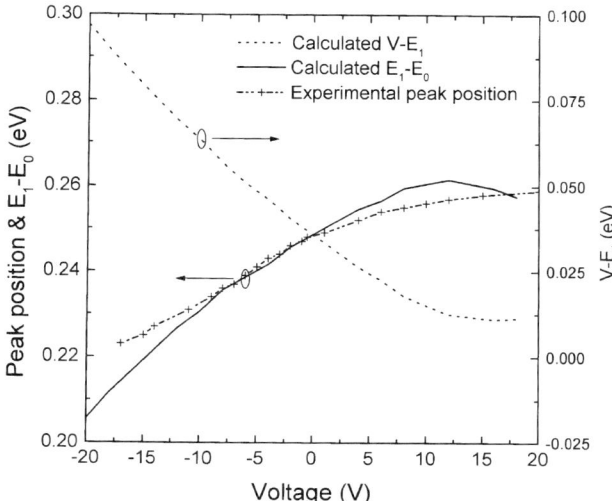

FIG. 29. Comparison with calculations. The experimental peak positions are taken from Fig. 27. The ground and first excited states are labeled by E_0 and E_1, respectively. Calculated energy difference between the top of the barrier (V) and E_1 is also shown. (Calculations made by G. C. Aers.)

been studied. One possibility is to use quasi-random gratings (Sarusi et al., 1994; Xing and Liu, 1996).

2. High-Frequency Detectors

One of the distinct advantages of QWIPs over standard detectors made of HgCdTe is their intrinsic high speed. High-frequency detectors may create new applications in, for example, environmental remote sensing of molecules and CO_2-laser-based (or other long-wavelength-laser-based) communication. The intrinsic high speed capability is related to the inherent short carrier lifetime $\tau_{life} \approx 5$ ps, as discussed before. In this section, we demonstrate this high-frequency capability and briefly review the three experimental techniques used to characterize the frequency capability: microwave rectification, optical heterodyne, and heterodyne and microwave mixing.

To characterize a QWIP frequency response, a standard microwave S parameter measurement can be used. Simpler than this, if one is only interested in a roll-off or maximum frequency, a microwave rectification measurement can be used. Complimentary to the optical heterodyne technique (discussed next), which involves generating a microwave signal within a QWIP at the difference frequency of two optical beams, we apply a microwave signal to the QWIP and measure the change in its DC biasing current. In the former case, the generated microwave propagates from QWIP to the measurement instrument, whereas for the latter, the applied microwave propagates from an external source to the QWIP. The rectification in a QWIP relies on its inherent nonlinear I–V characteristic. The small-signal rectified DC current is therefore given by

$$I_{rect} = \tfrac{1}{2} I'' V_\mu^2 \qquad (46)$$

where I'' is the second derivative of the I–V curve, and V_μ is the amplitude of the microwave voltage applied to the device. Both I'' and V_μ depend on the microwave frequency (ω). The dependence of I'' on ω reflects the frequency roll-off behavior of the intrinsic transport mechanism, and therefore is expected to behave as $1/[1 + (\omega\tau_{life})^2]$. Noting the fundamental relation between device dark current and photocurrent, the generation and recombination process of electrons in the quantum wells controls both currents. Based on this relation, we are inferring the photoconductive quantities, specifically the lifetime, by microwave rectification *without* applying any IR beams and with samples completely in dark.

Given the output power from a microwave source, V_μ varies as a function of frequency because of the device capacitance and differential resistance, and other parasitics. We then rewrite Eq. (46) separating out the frequency dependencies:

$$I_{\text{rect}} = \tfrac{1}{2} I_0'' V_{\mu 0}^2 \beta_1(\omega) \beta_2(\omega) \tag{47}$$

where

$$\beta_1(\omega) = \frac{1}{1 + (\omega \tau_{\text{life}})^2} \tag{48}$$

$\beta_2(\omega)$ is the circuit dependence, and I_0'' and $V_{\mu 0}$ are the low-frequency limiting values of I'' and V_μ, respectively.

The experiment is schematically shown in Fig. 30. Given the equivalent circuit of the device and its parasitic, the microwave voltage across the device is easily derived:

$$V_{\mu 0}^2 = 8 R_L P_{\text{out}} \tag{49}$$

and

$$\beta_2(\omega) = \frac{1}{(1 - \omega^2 LC)^2 + \omega^2 (R_L C + L/R)^2} \tag{50}$$

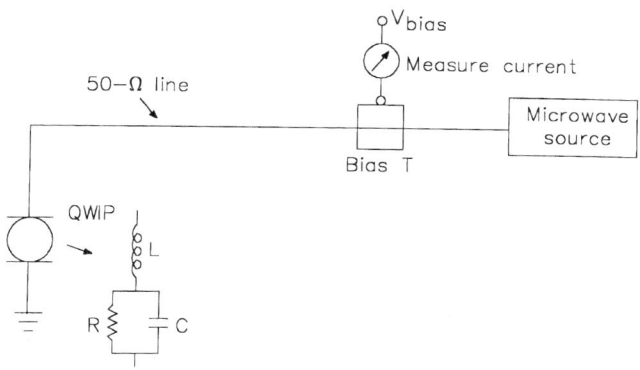

FIG. 30. Schematic of the microwave rectification experiment. The QWIP is mounted at the end of a 50-Ω transmission line by a short wirebond. The DC bias is applied through a bias T. The microwave is supplied by the microwave source with a frequency capability up to 40 GHz. The QWIP is modeled by a parallel resistance–capacitance equivalent circuit and the parasitic inductance is caused by the short wirebond.

where $R_L = 50\,\Omega$ is the transmission line impedance, P_{out} is the output power from the microwave source, C is the device capacitance, L is the parasitic inductance, and $R = 1/I'$ is device differential resistance. We have made the approximation $R + R_L \approx R$ because $R_L \ll R$ for a typical QWIP at 77 K. The $L \rightarrow 0$ limit of the above gives the usual RC roll-off: $1/[1 + (\omega R_L C)^2]$.

Liu et al. (1996) characterized a variety of devices with barrier widths from 234 to 466 Å and number of wells from 4 to 32. These experiments infer that the intrinsic photoconductive lifetime for these devices in the high biasing field regime is about 5 ps.

Figure 31 shows the measured (dots) and calculated (lines) rectified current versus frequency for three samples with different numbers of well. Taking appropriate values for device capacitance and parasitic inductance (mainly due to a short bond wire), the fitted $f_{max} = 1/(2\pi\tau_{life})$ is found to be 33 GHz, corresponding to $\tau_{life} = 4.8$ ps.

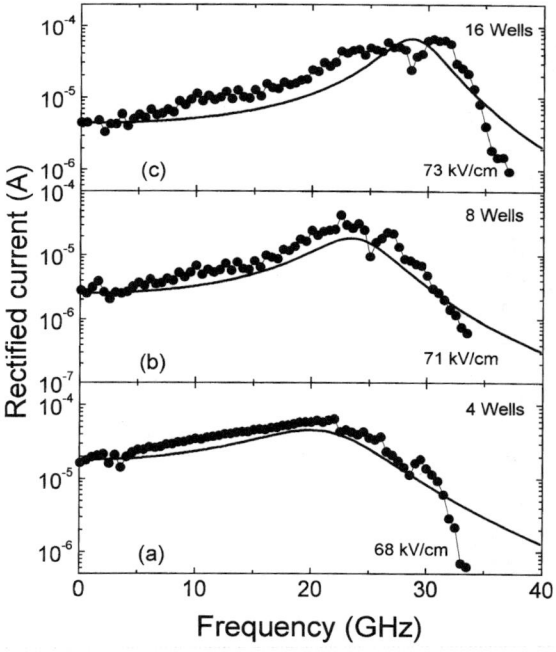

FIG. 31. Rectified current versus frequency for (a) the 4-well, (b) 8-well, and (c) 16-well samples (Fig. 11). The solid lines are calculated. The bias voltages were chosen to give approximately the same electric field of 70 kV/cm. All device areas were $10 \times 10\,\mu m^2$. The device temperature was 77 K. The microwave power was 0 dBm.

One possible use (Brown et al., 1993; Richards et al., 1998) of the high-frequency capability is optical heterodyne detection. Liu et al. (1995a) have carried out experiments with an intermediate frequency (IF) up to 26.5 GHz. A CO_2 laser and a lead-salt tunable diode laser (TDL) were used as the IR sources. The experiment is schematically shown in Fig. 32. The CO_2 laser frequency was fixed, while the TDL frequency was tuned by varying its operating temperature.

Much beyond 26.5 GHz, several factors combine to make a direct measurement of the IF signal very difficult. Two of the impediments to working with such high frequencies are (i) the lack of a sufficiently sensitive preamplifier and analyzer combination and (ii) the difficulty of making a low loss, broadband circuit that can be used to couple the weak IF signal from the QWIP. To extend the frequency range in optical heterodyne detection beyond that achievable in conventional CO_2 laser or TDL systems, we use microwave mixing techniques. The central idea here is to employ classical mixing in the QWIP to down-convert the IF signal generated by the optical heterodyne. This, as in the rectification experiment, relies on the nonlinear nature of the QWIP I–V characteristic. Specifically, as shown in Fig. 33, we apply to the QWIP not only the two IR beams, as in a conventional heterodyne experiment, but an additional microwave excitation. Let f_{IR1} and f_{IR2} denote the two IR frequencies, and let $f_{\mu wave}$ denote the microwave frequency. Then the IR heterodyne frequency is $f_{het} = |f_{IR1} - f_{IR2}|$, while the down-converted signal frequency is $|f_{het} - f_{\mu wave}|$. In this scheme, one can reach very high f_{het} frequencies using conventional electronics at the output of the QWIP. Heterodyne detection for a frequency separation of two IR beams up to 82.16 GHz has been achieved (Liu et al., 1995c).

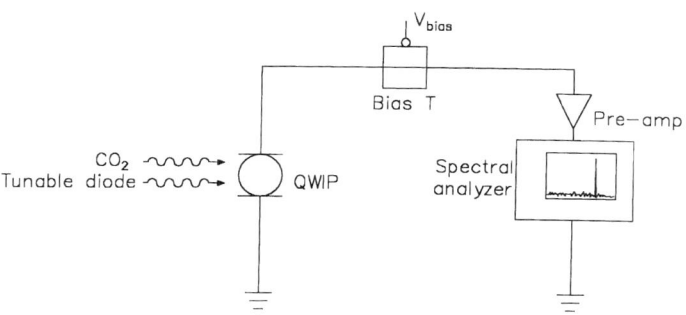

FIG. 32. Schematic of the optical heterodyne experiment. The difference frequency of the two infrared beams (CO_2 and tunable diode) is measured on the spectral analyzer.

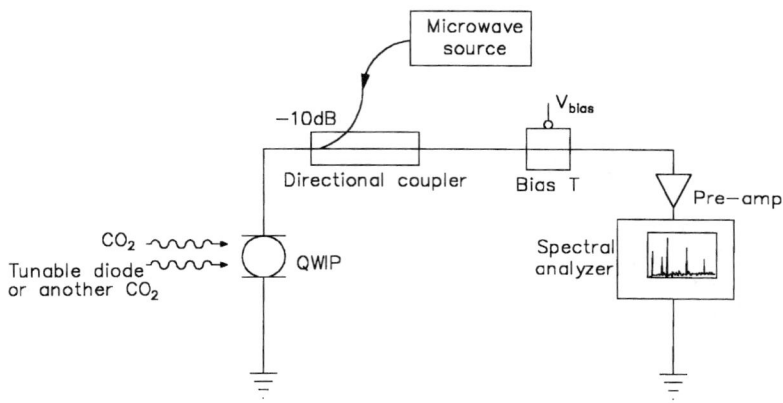

FIG. 33. Schematic of the infrared heterodyne and microwave mixing experiment. The measured signals, in general, consist of all possible mixing frequencies of the two infrared and the one microwave frequencies.

The device used in our experiments is a 100-well QWIP with a well width of 4.5 nm. The barriers are all 40-nm-thick $Al_{0.21}Ga_{0.79}As$. The center 2.5 nm of each well was doped with Si to 2.5×10^{18} cm^{-3}. The 75-μm-diameter mesa device was connected to the end of a 50-Ω microstrip transmission line by a short wirebond. The standard 45° facet geometry was used for optical coupling. The device was mounted on the cold finger of a liquid nitrogen dewar.

Shown in Fig. 34 is the measured optical heterodyne signal (squares), and up- and down-converted mixing signal (dots) vs heterodyne frequency (f_{het}) at a bias voltage of 2 V. The dashed and dotted lines are the expected roll-off behaviors due to the device RC time constant and the photocarrier lifetime. The optical heterodyne data were taken with a CO_2 laser and the TDL laser. We normalized the IF signal for a constant TDL power. When two CO_2 lasers and a microwave source were used for the mixing experiment, we normalized the signal for a constant incident power of about 0.2 mW from each of the lasers and 0.3 mW from the microwave source.

The five data points shown in Fig. 34 for the mixing experiment were taken using different sources and different microwave coupling schemes. For the first three points at $f_{het} = 1.83$, 5.37, and 15.5 GHz, we used one CO_2 laser, the TDL, and the microwave source at frequencies of $f_{\mu wave} = 9$, 15, and 10 GHz, respectively, and measured the sum (up-converted) signals of f_{het} and $f_{\mu wave}$. The microwave was coupled into the device through a directional coupler. The point at 41.42 GHz was obtained using two

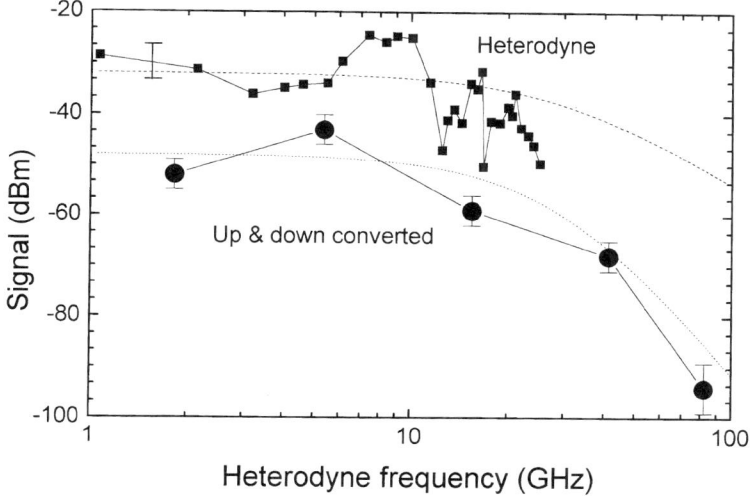

FIG. 34. Direct infrared heterodyne (squares) and mixed heterodyne frequency with microwave frequency (dots) signal vs heterodyne frequency for a bias voltage of 2 V. The dashed and dotted lines show the expected roll-off behaviors. The heterodyne frequency is defined as the difference between the two infrared frequencies. The incident powers from the two infrared sources and the microwave source are normalized to about 0.2 and 0.3 mW, respectively. The device temperature is 80 K.

CO_2 lasers separated by $f_{het} = 41.42$ GHz and the microwave source at $f_{\mu wave} = 20$ GHz, and measuring the difference (down-converted) signal at $41.42 - 20 = 21.42$ GHz. The point at 82.16 GHz was obtained using two CO_2 lasers separated by this frequency and a Gunn oscillator mechanically tuned in the range of 91–94 GHz. The difference signal in the range of 9–12 GHz was measured. For this data point, the millimeter-wave radiation was coupled into the device through free space and the power was estimated by using the QWIP as a rectifier (Liu et al., 1996). We estimate that only about 0.1% of the Gunn output power of 30 mW was coupled into the device. Note that the absolute value of the data point at 82.16 GHz in Fig. 34 has a relatively large uncertainty of about ±10 dB because of the difficulty in calibrating the actual microwave power on to the QWIP.

Neglecting the high-frequency roll-off, the amplitude of the heterodyne current is given by

$$i_{het} = 2\mathcal{R}_i \sqrt{P_{IR1} P_{IR2}} \tag{51}$$

where $\mathcal{R}_i$ is the detector current responsivity. Therefore, the power coupled to the 50-Ω-input spectrum analyzer is $P_{het} = i_{het}^2 R_L/2$, where $R_L = 50\,\Omega$. Taking $P_{IR1} = P_{IR2} = 0.2$ mW and $\mathscr{R}_i = 0.4$ A/W for polarized light for this device, the heterodyne power is expected to be $P_{het} = 0.64\,\mu$W $= -32$ dBm. The expected roll-off behavior is proportional to $1/[1 + (f/f_{RC})^2][1 + (f/f_{max})^2]$, where f is the IF frequency. The roll-off frequency limited by the RC effect is $f_{RC} = 28$ GHz, while the photocarrier lifetime-limited response frequency is $f_{max} = 1/\tau_{life} = 32$ GHz for $\tau_{life} = 5$ ps. In Fig. 34 we show the expected heterodyne power vs frequency of -32 dBm/$[(1 + (f/f_{RC})^2)(1 + (f/f_{max})^2)]$ with a dashed line.

There is no standard theory for the mixing results. We use the following simple analysis. The heterodyne current i_{het} is treated as a microwave or millimeter-wave signal which in turn mixes with the applied voltage $v_{\mu wave}$. We then have

$$i_{mix} = \tfrac{1}{2} I'' v_{\mu wave} v_{het} \tag{52}$$

where the heterodyne voltage v_{het} relates to i_{het} by $v_{het} = i_{het} R$ and $R = (I')^{-1}$. The mixing (up- and down-converted heterodyne) signal power in the absence of frequency roll-off is then given by

$$P_{mix} = \tfrac{1}{2} i_{mix}^2 R_L = 4 R_L^2 (I''/I')^2 \mathscr{R}_i^2 P_{IR1} P_{IR2} P_{\mu wave} \tag{53}$$

where we have used $v_{\mu wave}^2 = 8 R_L P_{\mu wave}$ and Eq. (51). Taking $P_{IR1} = P_{IR2} = 0.2$ mW, $P_{\mu wave} = 0.3$ mW, and $I''/I' = 0.9$ V^{-1} obtained by numerically differentiating the device I–V curve at the operating bias voltage, the power P_{mix} is calculated to be 1.6×10^{-8} W $= -48$ dBm. Following the same qualitative argument as in the optical heterodyne case, we expect the same roll-off behavior in P_{het} as before, and, since we are treating v_{het} as an input microwave signal which mixes with the applied microwave signal $v_{\mu wave}$, the QWIP as a mixing element should display a similar roll-off behavior. We then expect that P_{mix} should roll off faster than P_{het}. We plot in Fig. 34 the expected "doubled" roll-off behavior:

$$-48 \text{ dBm}/[(1 + (f/f_{RC})^2)(1 + (f/f_{max})^2)]^2$$

with a dotted line. The agreement between the expected and the measured frequency dependence is acceptable.

As a final note, the high-frequency capabilities have not been given much attention. Continued work both theoretically (Ershov, 1996; Ryzhii et al., 1997b; Ryzhii and Ryzhii, 1998) and experimentally (Ehret et al., 1997; Richards et al., 1998) may lead to some unique applications.

3. INTEGRATION WITH LEDs

To utilize the flexibility of quantum wells, we investigate the integration of QWIPs with light emitting diodes (LEDs). The devices described here are intended for thermal imaging applications requiring large-area devices in the mid- and far-infrared (M&FIR) region of wavelength longer than 2 μm. The conventional approach uses InSb or HgCdTe 2D detector arrays hybrid bonded to Si readout ICs (integrated circuits). The hybrid technology has also been applied to QWIP arrays with great success (see Chapter 4 by Gunapala and Bandara).

The basic idea (Liu *et al.*, 1995b) of an integrated QWIP–LED is to epitaxially grow a QWIP and a LED on top of each other. The QWIP can be either *n*- or *p*-type. Figure 35 shows a GaAs–AlGaAs *n*-QWIP with an InGaAs–GaAs quantum well LED. Only two contacts are made — to the heavily *p*-doped LED contact layer and the heavily *n*-doped QWIP emitter. A forward bias is applied to this serial QWIP–LED, which should be large enough to turn both the QWIP and the LED to their operating bias conditions. For concreteness, we assume that the QWIP detects M&FIR light of wavelength greater than 2 μm and the LED emits in the near-infrared (NIR) region of wavelength 800–1000 nm. The QWIP is a photoconductor so that under M&FIR light illumination its resistance decreases, which leads to an increase in the voltage drop across the LED and therefore an increase in the amount of NIR emission. This device is therefore an

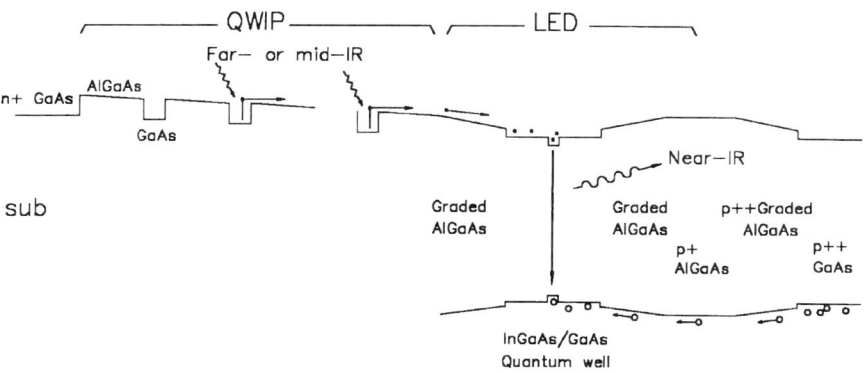

FIG. 35. Band-edge profile of the integrated QWIP with LED. For the QWIP part, only the conduction band-edge is shown. A forward bias is applied so that both the QWIP and the LED are in operating conditions. Photocurrent electrons from the QWIP recombine with injected holes in LED, giving rise to an increase in LED emission.

M&FIR to NIR converter. For 77-K operation, a well-designed QWIP can be very efficient (Andersson and Lundqvist, 1991) with easily higher than 10% absorption, and the LED technology is well developed. An optimized QWIP–LED therefore can be very efficient with little or no loss of performance compared with the QWIP alone used as an M&FIR detector. However, the advantage of this integrated QWIP–LED is of importance technologically. In this scheme, one can make large-format 2D imaging devices *without* the need of making any circuits on the device chip and *without* the need of hybrid bonding with another readout chip (normally a Si IC). The resulting NIR emission can be easily imaged using the well-developed Si CCD, which has a spectral response covering the NIR wavelength.

The first experimental demonstration of QWIP–LED was made by Liu *et al.* (1995b). The concept was independently proposed by Ryzhii *et al.* (1995). We briefly summarize the device details and the experimental results. A QWIP structure for a peak detection wavelength of 9 μm was grown and then an InGaAs–GaAs quantum well LED. Shown in Fig. 35, the QWIP was made of 50 repeat $Al_{0.25}Ga_{0.75}As$–GaAs quantum wells with 40-nm barriers and 5.9-nm wells. The well center regions were doped with Si to a density of 5×10^{11} cm^{-2}. Growth continued with the following LED layers: a 40-nm graded $Al_xGa_{1-x}As$ layer from $x = 0.25$ to 0.12, a 30-nm GaAs, a 7.0-nm $In_{0.2}Ga_{0.8}As$ well, a 30-nm GaAs, a 40-nm graded $Al_xGa_{1-x}As$ layer from $x = 0.12$ to 0.25, a 50-nm p^+–$Al_{0.25}Ga_{0.75}As$ with Be doping graded from 1 to 8×10^{18} cm^{-3}, a 40-nm graded p^{++}–$Al_xGa_{1-x}As$ layer from $x = 0.25$ to 0.12 with its doping increased from 8×10^{18} to 10^{19} cm^{-3}, and finally a 200-nm p^{++}–GaAs top contact layer with 10^{19} cm^{-3} doping. We used InGaAs as the well material so that the emitted light was not absorbed in any other layers. The design of the LED for this first test of the QWIP–LED concept borrowed that of a state-of-the-art quantum well laser.

The measurements were carried out on a mesa device with a top optical window for NIR emission and a 45° edge facet for IR coupling. The device was mounted in a liquid nitrogen optical cryostat. The LED emission was at a wavelength of 927 nm; while the QWIP response peaked at a wavelength of 9.2 μm. The LED emission power increased linearly with bias current. The measured external efficiency was about 1%, limited entirely by the device geometry. This implies that the LED internal efficiency is 100% (within the experimental uncertainty). This first demonstration (see Liu *et al.*, 1995b, for details) showed the potential for achieving a highly efficient up-conversion imaging device.

An intriguing possibility for QWIP–LED is to fabricate a continuously image conversion device — a pixelless large-area imaging device (Liu *et al.*,

1997a). The idea is that since the entire active layer of a QWIP–LED in the growth direction is very thin (normally less than 5 μm) the up-conversion process (i.e., photoexcitation in QWIP, carrier transport to LED, and radiative recombination in LED) should have little in-plane spreading. Theoretical analyses (Ryzhii et al., 1997a; Ershov et al., 1997a) show that the intrinsic spreading and crosstalk are negligible for a practical QWIP with a large number of wells ($N > 20$), supporting the pixelless concept. As an order-of-magnitude estimate, the in-plane spreading in QWIP is limited by the diffusion of excited carriers within the lifetime, and that in the LED is related to the bipolar diffusion and the radiative interband recombination lifetime. Taking some typical numbers (10 and 1 cm^2/s for the diffusion constants, and 5 ps and 1 ns for the lifetimes, in QWIP and LED, respectively) we get spreading lengths (square root of the product of diffusion constant and lifetime) of 0.1 and 0.3 μm, respectively. These values are negligible since they are much smaller than the wavelength of the IR to be imaged.

The first demonstration of the pixelless QWIP–LED used a p-type QWIP for the simplicity of avoiding the fabrication of a grating (Liu et al., 1997a; Allard et al., 1997). The proof-of-the-concept experiment was done with a 4 × 3 mm^2 device mounted in a liquid nitrogen dewar. For convenience, a "reflective" geometry was used (see Fig. 36); that is, both the incident IR light and the emission NIR light were through the front surface of device. A 1-mm-diameter aperture was illuminated from the back by a blackbody source with a temperature of 1000 K. A ZnSe lens mapped the aperture's image onto the QWIP–LED device. The bias across the device was kept at -3.0 V (-0.62 mA) during the experiment. The NIR photons generated by the device's LED portion were then directed toward a Si-CCD camera by a Ge beamsplitter coated to reflect the near IR. The beamsplitter coating also gave a high transmission in the long-wavelength infrared. Note that this is only an experimental setup. For practical applications, a "transmissive" geometry (discussed later) should be used.

Figure 37a maps out the NIR image. The background image (collected by blocking the aperture) was subtracted. A line profile of the center of the image is shown in Fig. 37b along with a line profile of the image of the aperture collected directly by the CCD (the smooth trace). The direct CCD image is identical to the thermal image of the aperture, confirmed by using a Mitsubishi thermal imaging camera. The widths at the base of the traces in Fig. 37b are essentially the same but differ by approximately 150 μm at half height. The discrepancy is due to an optical smearing mechanism in the geometry of this particular sample. Since the sample's thickness is 130–150 μm and the critical angle at a GaAs–vacuum interface is about 18°, any feature observed in LED emission is, to the lowest order, intrinsically

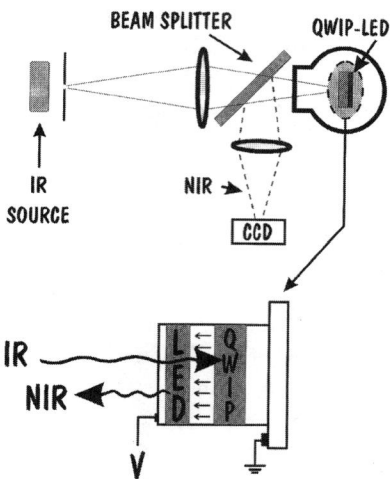

FIG. 36. Experimental setup employed to assess the QWIP–LED integrated device's functionality as an imaging detector. A 1-mm-diameter aperture is illuminated with a 1000-K blackbody source and imaged onto the device's face. The IR photons impinging on the QWIP section of the device generate a photocurrent (depicted by the left-pointing straight arrows found between the LED and the QWIP in the bottom drawing). The carriers move into the LED section of the device and subsequently give rise to near IR emission, which is then imaged onto a Si-CCD camera.

smeared over distances between 170 and 200 μm. Although this first experiment was only able to image a hot object with substantial smearing the basic concept is clearly demonstrated.

Apart from improving both the QWIP and the LED efficiencies, the main challenge for achieving a pixelless QWIP–LED imaging device with negligible distortion, smearing, and crosstalk seems to be related to extrinsic optical effects. One must ensure that both incident IR light and the emission NIR light have negligible crosstalk and smearing. There are many possible approaches such as removing the substrate and incorporating a distributed Bragg reflector.

Currently we are working toward an optimized large-area pixelless imaging device. The latest result is shown in Fig. 38 (Dupont et al., 1999). The device was made of a n-type QWIP on a heavily n-doped substrate. The QWIP was similar to the previous but with 40 quantum wells. The LED was also similar to the previous but with a 30 nm GaAs active region (instead of InGaAs). The heavily p-doped top contact layer used $Al_{0.15}Ga_{0.85}As$. The substrate absorption eliminates the crosstalk and smearing. Sharp, nearly ideal images were obtained (Dupont et al., 1999).

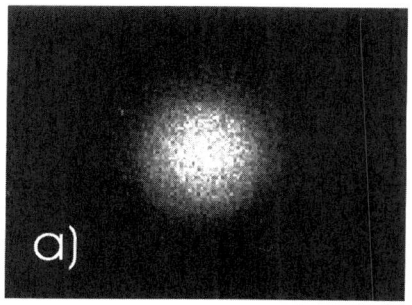

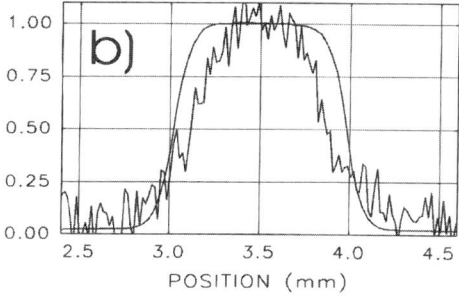

FIG. 37. (a) Up-converted image of a 1-mm-diameter aperture illuminated from the back by a 1000-K blackbody source. (b) Line profile of the center of the image in (a). The solid, noiseless curve is the line profile of the aperture's image collected directly by the CCD camera.

Note that this latest result still represents an intermediate step toward our final goal since the experimental geometry was reflective and the device performance was compromised due to the absorption in the substrate. Our current efforts concentrate on the implementation of a transmissive device. We are investigating the use of photon recycling effect (Schnitzer *et al.*, 1993) to improve the LED efficiency. There is a trade-off of LED efficiency and smearing if the photon recycling effect is used. Two of the envisioned final camera system layouts are sketched in Fig. 39. Another major issue relates to the materials quality. The pixelless device is more demanding on materials. Any leakage point in the QWIP would cause a hot spot in the LED emission, and any leakage point through the entire structure would render the whole device useless.

Besides the points already discussed, there are other potential advantages of this up-conversion approach. These include (a) the ease to implement multicolor imaging devices in a pixelless geometry, (b) the low cooling

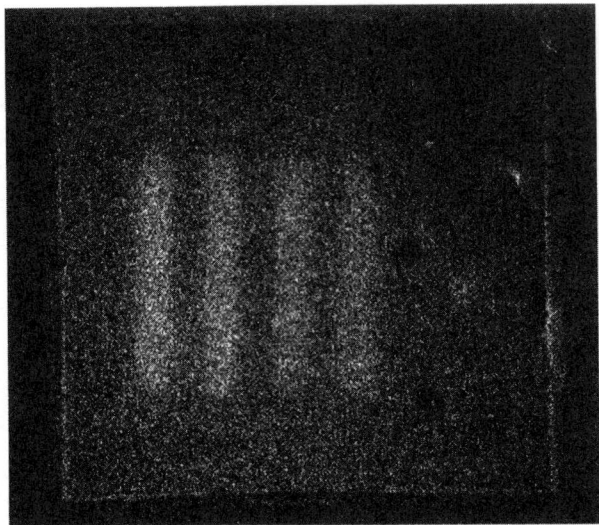

FIG. 38. A thermal image of a four-bar 1000-K infrared test target. The active device area is 1×1 cm^2. The image was taken at 8 frames per second. The device temperature was about 70 K.

power requirement since the readout chip (CCD) does not need to be cooled, and (c) the simplicity to fabricate ultra-large-area imaging devices. The concept of up-conversion may be expanded to cover the technological important region of 1.5 μm (Liu *et al.*, 1998d).

IV. Conclusions

1. CONCLUDING REMARKS

In this chapter, we briefly introduced the intersubband transition using the simplest ideal case of an infinitely high barrier quantum well. For a detailed and in-depth treatment of intersubband transition, the readers are referred to Chapter 1 of this volume by Helm. We then briefly discussed the wavelengths covered by the intersubband transition in GaAs-based quantum well structures. The main focus of the chapter was on the QWIP physics. Important areas such as dark current, photocurrent, and detector performance and design were discussed in great detail. The readers are also

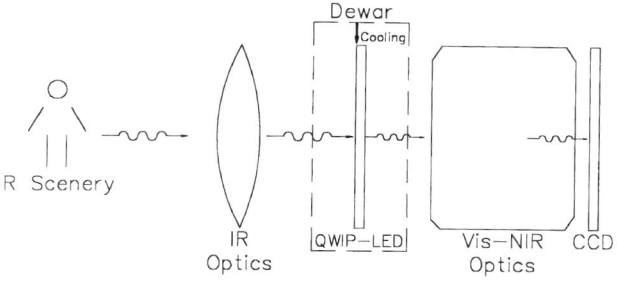

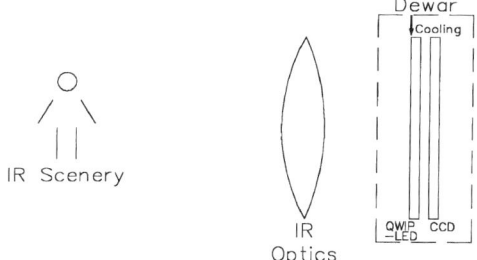

FIG. 39. Possible final imaging system configurations. Top: the IR scenery is mapped on the transmissive QWIP–LED device by the IR optics, the device is mounted in the dewar, the output emission image is mapped onto the CCD by visible or near-IR (Vis-NIR) optics. The Vis-NIR optics could be a lens or a fiber bundle. Bottom: the CCD is mounted in the dewar in proximity with QWIP–LED.

referred to Chapter 4 by Gunapala and Bandara for a review of various QWIP structures and of QWIP focal plane array technology and applications. Finally, we presented examples of novel structures and devices including multicolor and multispectral detectors, high-frequency QWIPs, and integrated QWIP–LEDs.

To conclude, there is no question that the new quantum well approach to IR detection has been one of the success stories in modern semiconductor research. Although there are many issues that need to be studied further, this new technology will find its way into the commercial market. In comparison with other technologies, the QWIP approach is favorable and attractive in several areas such as producibility, large arrays, multicolor and multiband, high speed, and integration.

2. LIST OF SYMBOLS

A	Normalization area
C	Device capacitance
c	Speed of light
D	Transmission coefficient
D^*	Detectivity
D^*_{det}	Detector-limited detectivity
D^*_{blip}	Background-limited detectivity
E	Total energy
E_{act}	Thermal activation energy
E_{ex}	Exchange energy
E_f	Fermi energy
E_g	Bandgap energy
E_n	Quantum well eigenenergy
E_0	Ground state eigenenergy
E_1	First excited state eigenenergy
ΔE	Broadening half width
ΔE_C	Conduction band offset
$\mathscr{E}$	Electric field
e	Elementary charge
F	Magnification factor in observation
f	Oscillator strength
f_{het}	Heterodyne frequency
f_{IR}	Frequency of infrared light
f_{max}	Lifetime limited maximum frequency
$f_{\mu\text{wave}}$	Microwave frequency
Δf	Measurement bandwidth
g	Gain
g_{photo}	Photoconductive gain
g_{noise}	Noise gain
HH1	Heavy hole ground state eigenenergy
h	Planck constant
$\hbar$	Reduced Planck constant
I'	First derivative of I–V
I''	Second derivative of I–V
I''_0	Second derivative of I–V in low-frequency limit
I_n	Noise current spectral density
I_{photo}	Photocurrent
I_{rect}	Rectified current
i_{em}	Emission current
$i^{(1)}_{\text{em}}$	Emission current from one well

i_{het}	Heterodyne current
i_{mix}	Mixing current
$i_{n,dark}$	Dark current noise spectral density
$i_{n,J}$	Johnson noise current spectral density
i_{p1}, i_{p2}, i_{p3}	Photocurrents from individual one-color QWIP in a multicolor stacked structure
$i_{photo}^{(1)}$	Direct photoemission current from one well
i_{trap}	Trapping current
$i_{trap}^{(1)}$	Trapping current into one well
J_{dark}	Dark current density
j_{em}	Emission current density from well
j_{trap}	Trapping current density into well
j_{3D}	Three-dimensional current density in barrier
k_B	Boltzmann constant
$\vec{k}_{xy}$	In-plane wavevector
L	Inductance
$\mathscr{L}$	Length
L_b	Barrier thickness
L_p	Length of one period
L_w	Well width
$L_B(\lambda)$	Background photon irradiance
m	Effective mass in well
m_b	Effective mass in barrier
m_e	Electron rest mass
m^*	Reduced effective mass in well
N	Number of wells
N_D	Two-dimensional doping density
N_{2D}	Total escaped electron density
n, n'	Wave function indices
n_r	Refractive index
n_{ex}	Number of photoexcited electrons in upper subband
n_{2D}	Two-dimensional electron density in well
n_{3D}	Three-dimensional electron density in barrier
P_{het}	Heterodyne signal power
P_{IR}	Infrared power
P_{mix}	Mixing signal power
P_{out}	Microwave output power
$P_{\mu wave}$	Microwave power
p_c	Trapping or capture probability
p_e	Escape probability
R	Device dynamic resistance
R_L	Transmission line impedance

R_s	Series load resistor
$\mathscr{R}_i$	Current responsivity
r_1, r_2, r_3	Dynamic resistances of individual one-color QWIP in a multicolor stacked structure
S_i	Noise current power spectral density
T	Temperature
T_{blip}	Temperature for background-limited infrared performance
V	Effective barrier height
V_b	Barrier height at zero bias
V_μ	Amplitude of microwave voltage
$V_{\mu 0}$	Amplitude of microwave voltage in low-frequency limit
V_1, V_2	Applied voltages
v	Drift velocity
v_{het}	Heterodyne signal voltage
v_{sat}	Saturated drift velocity
v_1	Velocity in first excited state
$v_{\mu wave}$	Microwave voltage
x	Aluminum mole fraction
z_c	Classical turning point
α	Absorption constant
α_n	Source of noise
$\beta_1(\omega), \beta_2(\omega)$	Intrinsic and circuit frequency roll-off factors
ε_0	Vacuum permittivity
η	Absorption quantum efficiency
$\eta^{(1)}$	Absorption quantum efficiency for one well
$\lambda, \lambda_1, \lambda_2$	Wavelengths
λ_c	Cutoff wavelength
λ_p	Peak detection wavelength
$\Delta\lambda$	Full width
μ	Mobility
μ_h	Hole mobility
ν	Photon frequency
ω	Photon angular frequency
Φ	Photon number per unit time incident on detector
$\phi_{B,ph}$	Integrated background photon number flux density incident on detector
ψ_n	Envelope wave function
θ	Internal angle of incidence
τ_{esc}	Escape time
τ_{life}	Lifetime from continuum to quantum well
τ_{relax}	Intersubband relaxation time
τ_{scatt}	Scattering time from subband to continuum

τ_{trans}	Transit time across one well
$\tau_{trans,tot}$	Total transit time
τ_{tunnel}	Tunneling time

ACKNOWLEDGMENTS

I have benefited greatly from many discussions over the years with fellow researchers in the field, especially G. C. Aers of NRC, J. Y. Andersson of Industrial Microelectronics Center (Sweden), E. R. Brown and K. L. Wang of UCLA, G. J. Brown and F. Szmulowicz of the U.S. Air Force Research Laboratory, B. F. Levine of Bell Laboratories, K. K. Choi and R. Q. Yang of the U.S. Army Research Laboratory, S. D. Gunapala and S. V. Bandara of NASA JPL, D. Jiang and H. Zheng of Chinese Academy of Sciences, S. S. Li of University of Florida, H. Ohno of the Tohoku University (Japan), A. G. U. Perera and M. Ershov of Georgia State University, E. Rosencher and B. Vinter of Thomson-CSF (France), V. Ryzhii of University of Aizu (Japan), and H. Schneider of Fraunhofer Institute (Germany). Special thanks are due to M. Buchanan for fabricating all the devices, to Z. R. Wasilewski for growing the highest quality wafers by MBE, to all the research (former and present) associates, L. B. Allard, E. Dupont, L. C. Lenchyshyn, J. Li, L. Li, J. R. Thompson, A. Shen, C.-Y. Song, and A. G. Steele, to all my former students M. G. Boudreau, M. Lamm, M. Paton, P. H. Wilson, and B. Xing, and to E. V. Kornelsen for supporting this work in its earliest stages. I am also extremely grateful to the funding provided by DND-DREV and to the technical monitors, P. Chevrette, D. Faubert, R. Corriveau, who recognized the potential value of QWIPs.

REFERENCES

Allard, L. B., Liu, H. C., Buchanan, M., and Wasilewski, Z. R. (1997). *Appl. Phys. Lett.* **70**, 2784.
Andersson, J. Y. (1995). *J. Appl. Phys.* **78**, 6298.
Andersson, J. Y., and Lundqvist, L. (1991). *Appl. Phys. Lett.* **59**, 857.
Ando, T., Fowler, A. B., and Stern, F. (1982). *Rev. Mod. Phys.* **54**, 437.
Andrews, S. R., and Miller, B. A. (1991). *J. Appl. Phys.* **70**, 993.
Baier, J., Bayanov, I. M., Plödereder, U., and Seilmeier, A. (1996). *Superlattices Microstructures* **19**, 9.
Bandara, K. M. S. V., Coon, D. D., Byungsung, O., Lin, Y. F., and Francombe, M. H. (1988). *Appl. Phys. Phys.* **53**, 1931; erratum **55**, 206 (1989).
Beck, W. (1993). *Appl. Phys. Lett.* **63**, 3589.
Blom, P. W. M., Smit, C., Haverkort, J. E. M., and Wolter, J. H. (1993a). *Phys. Rev. B* **47**, 2072.
Blom, P. W. M., Haverkort, J. E. M., van Hall, P. J., and Wolter, J. H. (1993b). *Appl. Phys. Lett.* **62**, 1490.

Bois, Ph., Costard, E., Duboz, J. Y., and Nagle, J. (1997). *Proc. SPIE* **3061**, 764.
Brown, E. R., McIntosh, K. A., Smith, F. W., and Manfra, M. J. (1993). *Appl. Phys. Lett.* **62**, 1513.
Brown, G. J., and Szmulowicz, F. (1996). In *Long Wavelength Infrared Detectors*, ed. M. Razeghi (Gordon and Breach, Amsterdam), pp. 271–333.
Chang, L. L., Esaki, L., and Sai-Halaz, G. A. (1977). *IBM Tech. Discl. Bull.* **20**, 2019.
Choi, K. K. (1997). *The Physics of Quantum Well Infrared Photodetectors.* World Scientific, Singapore.
Choi, K. K., Levine, B. F., Malik, R. J., Walker, J., and Bethea, C. G. (1987). *Phys. Rev. B***35**, 4172.
Chu, C. H., Hung, C. I., Wang, Y. H., and Houng, M. P. (1987). *IEEE Photon. Technol. Lett.* **9**, 1262.
Chui, L. C., Smith, J. S., Margalit, S., Yariv, A., and Cho, A. Y. (1983). *Infrared Phys.* **23**, 93.
Coon, D. D., and Karunasiri, R. P. G. (1984). *Appl. Phys. Lett.* **45**, 649.
Coon, D. D., Karunasiri, R. P. G., and Liu, L. Z. (1985). *Appl. Phys. Lett.* **47**, 289.
Coon, D. D., Karunasiri, R. P. G., and Liu, H. C. (1986). *Appl. Phys. Lett.* **60**, 2636.
Costard, E., Bois, P., Audier, F., and Herniou, E. (1998). *Proc. SPIE* **3436**, 228.
Dereniak, E. L., and Crowe, D. G. (1984). *Optical Radiation Detectors.* John Wiley & Sons, New York.
Dupont, E., Liu, H. C., Buchanan, M., Wasilewski, Z. R., St-Germain, D., and Chevrette, P. (1999). *Appl. Phys. Lett.* **75**, 563.
Ehret, S., Schneider, H., Fleissner, J., Koidl, P., and Böhm, G. (1997). *Appl. Phys. Lett.* **71**, 641.
Ershov, M. (1996). *Appl. Phys. Lett.* **69**, 3480.
Ershov, M. (1998). *Appl. Phys. Lett.* **73**, 3432.
Ershov, M., Ryzhii, V., and Hamaguchi, C. (1995). *Appl. Phys. Lett.* **67**, 3147.
Ershov, M., Hamaguchi, C., and Ryzhii, V. (1996). *Jpn. J. Appl. Phys.* **35**, 1395.
Ershov, M., Liu, H. C., and Schmitt, L. M. (1997a). *J. Appl. Phys.* **82**, 1446.
Ershov, M., Liu, H. C., Li, L., Buchanan, M., Wasilewski, Z. R., and Ryzhii, V. (1997b). *Appl. Phys. Lett.* **70**, 1828.
Ershov, M., Liu, H. C., Buchanan, M., Wasilewski, Z. R., and Ryzhii, V. (1997c). *Appl. Phys. Lett.* **70**, 414.
Esaki, L., and Sakaki, H. (1977). *IBM Tech. Discl. Bull.* **20**, 2456.
Faist, J., Capasso, F., Sirtori, C., Sivco, D. L., Hutchinson, A. L., Chu, S. N. G., and Cho, A. Y. (1993). *Appl. Phys. Lett.* **63**, 1354.
Goodnick, S. M., and Lugli, P. (1988). *Phys. Rev. B***37**, 2578.
Goossen, K. W., and Lyon, S. A. (1985). *Appl. Phys. Lett.* **47**, 1257.
Goossen, K. W., and Lyon, S. A. (1988). *J. Appl. Phys.* **63**, 5149.
Gravé, I., Shakouri, A., Kruze, N., and Yariv, A. (1992). *Appl. Phys. Lett.* **60**, 2362.
Gunapala, S. D., Levine, B. F., Pfeiffer, L., and West, W. (1991). *J. Appl. Phys.* **69**, 6517.
Harwit, A., and Harris, J. S. (1987). *Appl. Phys. Lett.* **50**, 685.
Hasnain, G., Levine, B. F., Gunapala, S., and Chand, N. (1990). *Appl. Phys. Lett.* **57**, 608.
Kane, M. J., Millidge, S., Emeny, M. T., Lee, D., Guy, D. R. P., and Whitehouse, C. R. (1992). In *Intersubband Transitions in Quantum Wells*, eds. E. Rosencher, B. Vinter, and B. Levine (Plenum, New York), pp. 31–42.
Kinch, M. A., and Yariv, A. (1989). *Appl. Phys. Lett.* **55**, 2093; see also a comment by B. F. Levine and the reply in **56**, 2354–2356 (1990).
Kingston, R. H. (1978). *Detection of Optical and Infrared Radiation.* Springer-Verlag, New York.
Köck, A., Gornik, E., Abstreiter, G., Böhm, G., Walther, M., and Weimann, G. (1992). *Appl. Phys. Lett.* **60**, 2011.
Lenchyshyn, L. C., Liu, H. C., Buchanan, M., and Wasilewski, Z. R. (1996). *J. Appl. Phys.* **79**, 8091.

Levine, B. F. (1993). *J. Appl. Phys.* **74**, R1.
Levine, B. F., Choi, K. K., Bethea, C. G., Walker, J., and Malik, R. J. (1987). *Appl. Phys. Lett.* **50**, 1092.
Levine, B. F., Bethea, C. G., Hasnain, G., Shen, V. O., Pelve, E., Abbott, R. R., and Hsieh, S. J. (1990). *Appl. Phys. Lett.* **56**, 851.
Levine, B. F., Zussman, A., Kuo, J. M., and de Jong, J. (1992a). *J. Appl. Phys.* **71**, 5130.
Levine, B. F., Zussman, A., Gunapala, S. D., Asom, M. T., Kuo, J. M., and Hobson, W. S. (1992b). *J. Appl. Phys.* **72**, 4429.
Li, S. S., and Su, Y.-K. (editors). (1998). *Intersubband Transitions in Quantum Wells: Physics and Devices*. Kluwer, Boston.
Liu, H. C. (1992a). *Appl. Phys. Lett.* **60**, 1507.
Liu, H. C. (1992b). *Appl. Phys. Lett.* **61**, 2703.
Liu, H. C. (1993). *J. Appl. Phys.* **73**, 3062.
Liu, H. C. (1996). In *Long Wavelength Infrared Detectors*, ed. M. Razeghi (Gordon and Breach, Amsterdam), pp. 1–59.
Liu, H. C., Aers, G. C., Buchanan, M., Wasilewski, Z. R., and Landheer, D. (1991a). *J. Appl. Phys.* **70**, 935.
Liu, H. C., Buchanan, M., Wasilewski, Z. R., and Chu, H. (1991b). *Appl. Phys. Lett.* **58**, 1059.
Liu, H. C., Li, J., Buchanan, M., Wasilewski, Z. R., and Simmons, J. G. (1993a). *Phys. Rev. B* **48**, 1951.
Liu, H. C., Wasilewski, Z. R., Buchanan, M., and Chu, H. (1993b). *Appl. Phys. Lett.* **63**, 761.
Liu, H. C., Steele, A. G., Buchanan, M., and Wasilewski, Z. R. (1993c). *J. Appl. Phys.* **73**, 2029.
Liu, H. C., Li, J., Thompson, J. R., Wasilewski, Z. R., Buchanan, M., and Simmons, J. G. (1993d). *IEEE Electron. Dev. Lett.* **14**, 566.
Liu, H. C., Levine, B. F., and Andersson, J. Y. (editors). (1994). *Quantum Well Intersubband Transition Physics and Devices*. Kluwer, Dordrecht, Netherlands.
Liu, H. C., Jenkins, G. E., Brown, E. R., McIntosh, K. A., Nichols, K. B., and Manfra, M. J. (1995a). *IEEE Electron. Dev. Lett.* **16**, 253.
Liu, H. C., Li, J., Wasilewski, Z. R., and Buchanan, M. (1995b). *Electron. Lett.* **31**, 832.
Liu, H. C., Li, J., Brown, E. R., McIntosh, K. A., Nichols, K. B., and Manfra, M. J. (1995c). *Appl. Phys. Lett.* **67**, 1594.
Liu, H. C., Li, J., Buchanan, M., and Wasilewski, Z. R. (1996). *IEEE J. Quantum Electron.* **32**, 1024.
Liu, H. C., Allard, L. B., Buchanan, M., and Wasilewski, Z. R. (1997a). *Electron. Lett.* **33**, 379.
Liu, H. C., Li, L., Buchanan, M., and Wasilewski, Z. R. (1997b). *J. Appl. Phys.* **82**, 889.
Liu, H. C., Li, L., Buchanan, M., Wasilewski, Z. R., Brown, G. J., Szmulowicz, F., and Hegde, S. M. (1998a). *J. Appl. Phys.* **83**, 585.
Liu, H. C., Buchanan, M., and Wasilewski, Z. R. (1998b). *J. Appl. Phys.* **83**, 6178.
Liu, H. C., Buchanan, M., and Wasilewski, Z. R. (1998c). *Appl. Phys. Lett.* **72**, 1682.
Liu, H. C., Gao, M., Buchanan, M., Wasilewski, Z. R., and Poole, P. (1998d). *Proc. SPIE* **3491**, 214.
Long, D. (1980). In *Optical and Infrared Detectors*, ed. R. J. Keyes (Springer-Verlag, Berlin), Ch. 4.
Man, P., and Pan, D. S. (1995). *Appl. Phys. Lett.* **66**, 192.
Martinet, E., Rosencher, E., Luc, F., Bois, Ph., Costard, E., and Delaitre, S. (1992). *Appl. Phys. Lett.* **61**, 246.
Meshkov, S. V. (1986). *Sov. Phys. JETP* **64**, 1337.
Muraki, K., Fujiwara, A., Fukatsu, S., Shiraki, Y., and Takahashi, Y. (1996). *Phys. Rev. B* **53**, 15477.

Perera, A. G. U., Shen, W. Z., Matsik, S. G., Liu, H. C., Buchanan, M., and Schaff, W. J. (1998). *Appl. Phys. Lett.* **72**, 1596.
Petrov, A. G., and Shik, A. (1991). *Semicond. Sci. Technol.* **6**, 1163.
Petrov, A. G., and Shik, A. (1998). *J. Appl. Phys.* **83**, 3203.
Richards, R. K., Hutchinson, D. P., Bennett, C. A., Simpson, M. L., Liu, H. C., and Buchanan, M. (1998). *ECS Proc.* **98-21**, 95.
Rose, A. (1963). *Concepts in Photoconductivity and Allied Problems*. Interscience Publishers, John Wiley & Sons, New York.
Rosencher, E., Vinter, B., and Levine, B. (editors). (1992a). *Intersubband Transitions in Quantum Wells*. Plenum, New York.
Rosencher, E., Luc, F., Bois, Ph., and Delaitre, S. (1992b). *Appl. Phys. Lett.* **61**, 468.
Ryzhii, V. (1997). *J. Appl. Phys.* **81**, 6442.
Ryzhii, M., and Ryzhii, V. (1998). *Appl. Phys. Lett.* **72**, 842.
Ryzhii, V., Ershov, M., Ryzhii, M., and Khmyrova, I. (1995). *Jpn. J. Appl. Phys.* **34**, L38.
Ryzhii, V., Liu, H. C., Khmyrova, I., and Ryzhii, M. (1997a). *IEEE J. Quantum Electron.* **33**, 1527.
Ryzhii, V., Khmyrova, I., and Ryzhii, M. (1997b). *Jpn. J. Appl. Phys.* **36**, 2596.
Ryzhii, M., Ryzhii, V., and Willander, M. (1998). *J. Appl. Phys.* **84**, 3403.
Sarusi, G., Levine, B. F., Pearton, S. J., Bandara, K. M. S., and Leibenguth, R. E. (1994). *Appl. Phys. Lett.* **64**, 960.
Schnitzer, I., Yablonovitch, E., Caneau, C., and Gmitter, T. J. (1993). *Appl. Phys. Lett.* **62**, 131.
Schönbein, C., Schneider, H., Rehm, R., and Walther, M. (1998). *Appl. Phys. Lett.* **73**, 1251, see a comment by B. F. Levine in **74**, 892 (1999).
Sclar, N. (1984). *Prog. Quant. Elect.* **9**, 149.
Serzhenko, F. L., and Shadrin, V. D. (1991). *Sov. Phys. Semicond.* **25**, 953.
Shadrin, V. D., Mitin, V. V., Kochelap, V. A., and Choi, K. K. (1995a). *J. Appl. Phys.* **77**, 1771.
Shadrin, V. D., Mitin, V. V., Choi, K. K., and Kochelap, V. A. (1995b). *J. Appl. Phys.* **78**, 5765.
Smith, J. S., Chui, L. C., Margalit, S., Yariv, A., and Cho, A. Y. (1983). *J. Vac. Sci. Technol. B* **1**, 376.
Sotirelis, P., von Allmen, P., and Hess, K. (1993). *Phys. Rev. B* **47**, 12744.
Steele, A. G., Liu, H. C., Buchanan, M., and Wasilewski, Z. R. (1991). *Appl. Phys. Lett.* **59**, 3625.
Steele, A. G., Liu, H. C., Buchanan, M., and Wasilewski, Z. R. (1992). *J. Appl. Phys.* **73**, 1062.
Szmulowicz, F., and Brown, G. J. (1995). *Phys. Rev. B* **51**, 13203.
Tatham, M. C., Ryan, J. F., and Foxon, C. T. (1989). *Phys. Rev. Lett.* **63**, 1637.
Thibaudeau, L., Bois, P., and Duboz, J. Y. (1996). *J. Appl. Phys.* **79**, 446.
Wasilewski, Z. R., Liu, H. C., and Buchanan, M. (1994). *J. Vac. Sci. Technol. B* **12**, 1273.
West, L. C., and Eglash, S. J. (1985). *Appl. Phys. Lett.* **46**, 1156.
Whitney, R. L., Cuff, K. F., and Adams, F. W. (1993). In *Semiconductor Quantum Wells and Superlattices for Long-Wavelength Infrared Detectors*, ed. M. O. Manasreh (Artech House, Boston), pp. 55–108.
Xing, B., and Liu, H. C. (1996). *J. Appl. Phys.* **80**, 1214.
Xing, B., Liu, H. C., Wilson, P. H., Buchanan, M., Wasilewski, Z. R., and Simmons, J. G. (1994). *J. Appl. Phys.* **76**, 1889.
Xu, Y., Shakouri, A., Yariv, A., Krabach, T., and Dejewski, S. (1995). *Electron. Lett.* **31**, 320.

CHAPTER 4

Quantum Well Infrared Photodetector (QWIP) Focal Plane Arrays

S. D. Gunapala and S. V. Bandara

JET PROPULSION LABORATORY
CALIFORNIA INSTITUTE OF TECHNOLOGY
PASADENA, California

I. INTRODUCTION	198
II. COMPARISON OF VARIOUS TYPES OF QWIPs	201
1. n-Doped Bound-to-Bound QWIPs	201
2. n-Doped Bound-to-Continuum QWIPs	202
3. n-Doped Bound-to-Quasibound QWIPs	203
4. n-Doped Broadband QWIPs	205
5. n-Doped Bound-to-Bound Miniband QWIPs	206
6. n-Doped Bound-to-Continuum Miniband QWIPs	207
7. n-Doped Bound-to-Miniband QWIPs	208
8. n-Doped Asymmetrical $GaAs-Al_xGa_{1-x}As$ QWIPs	209
9. p-Doped QWIPs	210
10. Single-Quantum-Well Infrared Photodetectors	211
11. Indirect Bandgap QWIPs	212
12. n-Doped $In_{0.53}Ga_{0.47}As-In_{0.52}Al_{0.48}As$ QWIPs	213
13. n-Doped $In_{0.53}Ga_{0.47}As-InP$ QWIPs	214
14. InGaAsP Quaternary QWIPs	215
15. n-Doped $GaAs-Ga_{0.5}In_{0.5}P$ QWIPs	216
16. n-Doped $GaAs-Al_{0.5}In_{0.5}P$ QWIPs	217
17. n-Doped $In_{0.15}Ga_{0.85}As-GaAs$ QWIPs	217
18. p-Doped $In_{0.53}Ga_{0.47}As-InP$ QWIPs	218
III. FIGURES OF MERIT	220
1. Absorption Spectra	221
2. Dark Current	222
3. Responsivity	225
4. Dark Current Noise	228
5. Noise Gain and Photoconductive Gain	229
6. Quantum Efficiency	232
7. Detectivity	234
IV. LIGHT COUPLING	237
1. Random Reflectors	238
2. Two-Dimensional Periodic Gratings	239

3. Corrugated Structure 240
4. Microlenses 242
V. IMAGING FOCAL PLANE ARRAYS 246
   1. Effect of Nonuniformity 246
   2. 128 × 128 VLWIR Imaging Camera 249
   3. 256 × 256 LWIR Imaging Camera 251
   4. 640 × 486 LWIR Imaging Camera 254
   5. Dualband (MWIR and LWIR) Detectors 260
   6. Dualband (LWIR and VLWIR) Detectors 263
   7. High-Performance QWIPs for Low-Background Applications 264
   8. Broadband QWIPs for Thermal Infrared Imaging Spectrometers 267
VI. APPLICATIONS 270
   1. Fire Fighting 270
   2. Volcanology 271
   3. Medicine 271
   4. Defense 273
   5. Astronomy 274
VII. SUMMARY 275
   REFERENCES 278

I. Introduction

Intrinsic infrared detectors in the long-wavelength range (6–20 μm) are based on interband transition, which promotes an electron across the bandgap (E_g) from the valence band to the conduction band, as shown in Fig. 1. These photoexcited electrons can be collected efficiently, thereby producing a photocurrent in the external circuit. Since the incoming photon must excite an electron from the valence band to the conduction band, the energy of the photon (hv) must be higher than the E_g of the photosensitive material. Therefore, the spectral response of the detectors can be tuned by controlling the E_g of the photosensitive material. Examples for such materials are $Hg_{1-x}Cd_xTe$ and $Pb_{1-x}Sn_xTe$ in which the energy gap can be controlled by varying x. This means detection of very long wavelength infrared (VLWIR; >12 μm) radiation up to 20 μm requires small bandgaps down to 62 meV. It is well known that these low-bandgap materials are more difficult to grow and process than large-bandgap semiconductors such as GaAs. These difficulties motivate the exploration of utilizing the intersubband transitions in multi-quantum-well (MQW) structures made of large-bandgap semiconductors (Fig. 2).

The idea of using MQW structures to detect infrared radiation can be explained by using the basic principles of quantum mechanics. The quantum well is equivalent to the well-known particle in a box problem in quantum mechanics, which can be solved by the time-independent Schrödinger

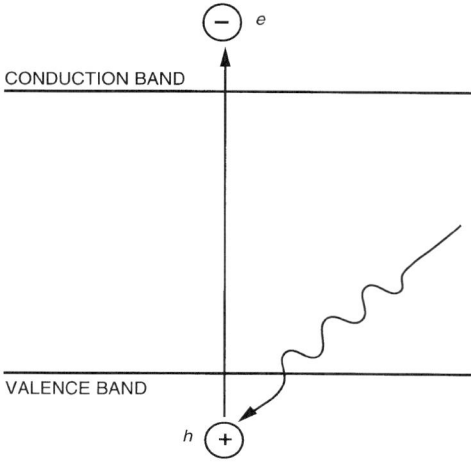

FIG. 1. Band diagram of conventional intrinsic infrared photodetector (from Gunapala and Bandara, 1995).

equation. The solutions to this problem are the eigenvalues that describe energy levels inside the quantum well in which the particle is allowed to exist. The positions of the energy levels are primarily determined by the quantum well dimensions (height and width). For infinitely high barriers and parabolic bands, the energy levels in the quantum well are given by

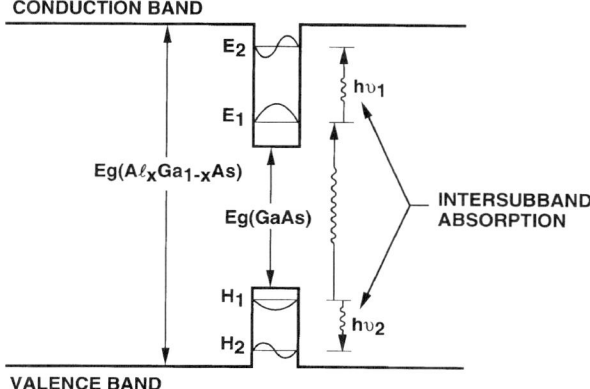

FIG. 2. Schematic band diagram of a quantum well. Intersubband absorption can take place between the energy levels of a quantum well associated with the conduction band (n-doped) or the valence band (p-doped) (from Levine, 1993).

(Weisbuch, 1987)

$$E_j = \left(\frac{\hbar^2\pi^2}{2m^*L_w^2}\right)j^2 \qquad (1)$$

where L_w is the width of the quantum well, m^* is the effective mass of the carrier in the quantum well, and j is an integer. Thus the intersubband energy between the ground and the first excited state is

$$(E_2 - E_1) = (3\hbar^2\pi^2/2m^*L_w^2) \qquad (2)$$

The quantum well infrared photodetectors (QWIPs) discussed in this chapter utilize the photoexcitation of an electron (hole) between the ground state and the first excited state in the conduction (valence) band quantum well (see Fig. 2). The quantum well structure is designed so that these photoexcited carriers can escape from the quantum well and be collected as photocurrent. In addition to larger intersubband oscillator strength, these detectors afford greater flexibility than extrinsically doped semiconductor infrared detectors because the wavelength of the peak response and cutoff can be continuously tailored by varying layer thickness (quantum well width) and barrier composition (barrier height).

The lattice matched GaAs–Al_xGa_{1-x}As material system is a very good candidate to create such a quantum well structure, because the bandgap of Al_xGa_{1-x}As can be changed continuously by varying x (and hence the height of the quantum well). Thus, by changing the quantum well width L_w and the barrier height (Al molar ratio of Al_xGa_{1-x}As alloy), this intersubband transition energy can be varied over a wide range, from short-wavelength infrared (SWIR; 1–3 μm) and midwavelength infrared (MWIR; 3–5 μm), through long-wavelength infrared (LWIR; 8–12 μm) and into the VLWIR (>12 μm). It is important to note that unlike intrinsic detectors, which utilize interband transition, the quantum wells of these detectors must be doped since the photon energy is not sufficient to create photocarriers ($hv < E_g$).

The possibility of using GaAs–Al_xGa_{1-x}As MQW structures to detect infrared radiation was first suggested by Esaki and Sakaki (1977), experimentally investigated by Smith et al. (1983), and theoretically analyzed by Coon and Karunasiri (1984). The first experimental observation of the strong intersubband absorption was performed by West and Eglash (1985), and the first QWIP was demonstrated by Levine et al. (1987b) at Bell Laboratories. Levine et al. (1988c) also introduced QWIPs involving bound-to-continuum intersubband transitions with wider Al_xGa_{1-x}As barriers and demonstrated dramatically improved detectivity. Recent developments in these detectors have already led to the demonstration of large (up to 640 × 486) high-sensitivity staring arrays by several groups (Bethea *et al.*, 1991, 1993; Bethea and Levine, 1992; Kozlowski *et al.*, 1991; Beck *et al.*, 1994; Gunapala *et al.*, 1997a, 1997b, 1998a; Andersson *et al.*, 1997; Choi *et al.*, 1998b; Breiter *et al.*, 1998).

II. Comparison of Various Types of QWIPs

1. *n*-DOPED BOUND-TO-BOUND QWIPs

As mentioned previously, the first *bound-to-bound* state QWIP was demonstrated by Levine *et al.* (1987b), which consisted of 50 periods of $L_w = 65$ Å GaAs and $L_b = 95$ Å $Al_{0.25}Ga_{0.75}As$ barriers sandwiched between top (0.5-μm-thick) and bottom (1-μm-thick) GaAs contact layers. The center 50 Å of the GaAs wells were doped to $N_D = 1.4 \times 10^{18}$ cm^{-3} and the contact layers were doped to $N_D = 4 \times 10^{18}$ cm^{-3}. This structure was grown by molecular beam epitaxy (MBE). These thicknesses and compositions were chosen to produce only two states in the quantum well with energy spacing giving rise to a peak wavelength of 10 μm. The measured (Levine *et al.*, 1987b) absorption spectra peaked at $\lambda_p = 10.9$ μm, with a full width at half maximum of $\Delta v = 97$ cm^{-1}. The peak absorbance $a = -\log(\text{transmission}) = 2.2 \times 10^{-2}$ corresponds to a net absorption of 5% (i.e., $a = 600$ cm^{-1}).

After the absorption of infrared photons, the photoexcited carriers can be transported either along the plane of quantum wells (with an electric field along the quantum wells) or perpendicular to the wells (with an electric field perpendicular to the epitaxial layers). As far as the infrared detection is concerned, perpendicular transport is superior to parallel transport (Wheeler and Goldberg, 1975) since the difference between the excited-state and ground-state mobilities is much larger in the latter case, and consequently, transport perpendicular to the quantum wells (i.e., growth direction) gives a substantially high photocurrent. In addition, the heterobarriers block the transport of ground-state carriers in the quantum wells, and thus lower dark current. For these reasons, QWIPs are based on escape and perpendicular transport of photoexcited carriers, as shown in Fig. 3.

In the latter versions of the bound-to-bound state QWIPs, Choi *et al.* (1987c) used slightly thicker and higher barriers to reduce tunneling-induced dark current. When they increased the barrier thickness from $L_b = 95$ to 140 Å and $Al_xGa_{1-x}As$ barrier height from $x = 0.25$ to 0.36, the dark

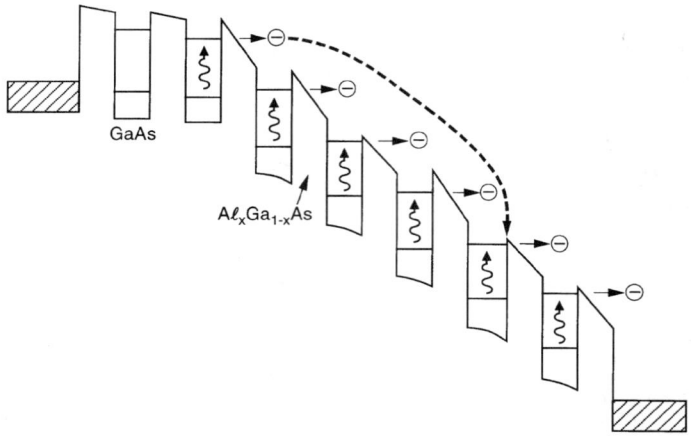

FIG. 3. Conduction-band diagram for a bound-to-bound QWIP, showing the photoexcitation (intersubband transition) and tunneling out of well (from Levine, 1993).

current (also the photocurrent) was significantly reduced. The nonlinear behavior of the responsivity and the dark current versus bias voltage observed in the bound-to-bound QWIPs is due to the complex tunneling process associated with the high-field domain formation (Choi *et al.*, 1987c).

2. *n*-DOPED BOUND-TO-CONTINUUM QWIPs

In the previous section, we mentioned QWIP containing two bound states. By reducing the quantum well width, it is possible to push the strong bound-to-bound intersubband absorption into the continuum, resulting in a strong bound-to-continuum intersubband absorption. The major advantage of the bound-to-continuum QWIP is that the photoexcited electron can escape from the quantum well to the continuum transport states without tunneling through the barrier, as shown in Fig. 4. As a result, the bias required to efficiently collect the photoelectrons can be reduced dramatically, and hence lower the dark current. Due to the fact that the photoelectrons do not have to tunnel through the barriers, the $Al_xGa_{1-x}As$ barrier thickness of bound-to-continuum QWIP can be increased without reducing the photoelectron collection efficiency. Increasing the barrier width from a few hundred angstrom to 500 Å can reduce the ground state sequential tunneling by an order of magnitude. By making use of these improvements, Levine *et al.* (1990a) has successfully demonstrated the first

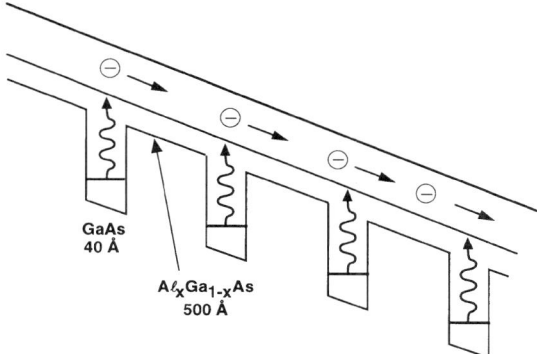

FIG. 4. Conduction-band diagram for a bound-to-continuum QWIP, showing the photo-excitation and hot-electron transport process (from Levine, 1993).

bound-to-continuum QWIP with a dramatic improvement in the performance (i.e., detectivity 3×10^{10} cm $\sqrt{\text{Hz}}/\text{W}$ at 68 K for a QWIP, which had cutoff wavelength at 10 μm).

3. n-DOPED BOUND-TO-QUASIBOUND QWIPs

Improving QWIP performance depends largely on minimizing the parasitic current (i.e., dark current) that plagues all light detectors. The dark current is the current that flows through a biased detector in the dark (i.e., with no photons impinging on it). As Gunapala and Bandara (1995) have discussed elsewhere, at temperatures above 45 K (typical for $\lambda < 14\,\mu\text{m}$), the dark current of the QWIP is entirely dominated by classical thermionic emission of ground-state electrons directly out of the well into the energy continuum. Minimizing the dark current is critical to the commercial success of the QWIP as it enables the highly desirable high-temperature detector operation.

Therefore, Gunapala and Bandara (1995) have designed the *bound-to-quasibound* quantum well by placing the first excited state exactly at the well top, as shown in Fig. 5. The best previous QWIPs pioneered by Levine *et al.* (1988c) at AT&T Bell Laboratories were of the bound-to-continuum variety, so called because the first excited state was a continuum energy band above the quantum well top (typically 10 meV). Dropping the first excited state to the quantum well top causes the barrier to thermionic emission (roughly the energy height from the ground state to the well top)

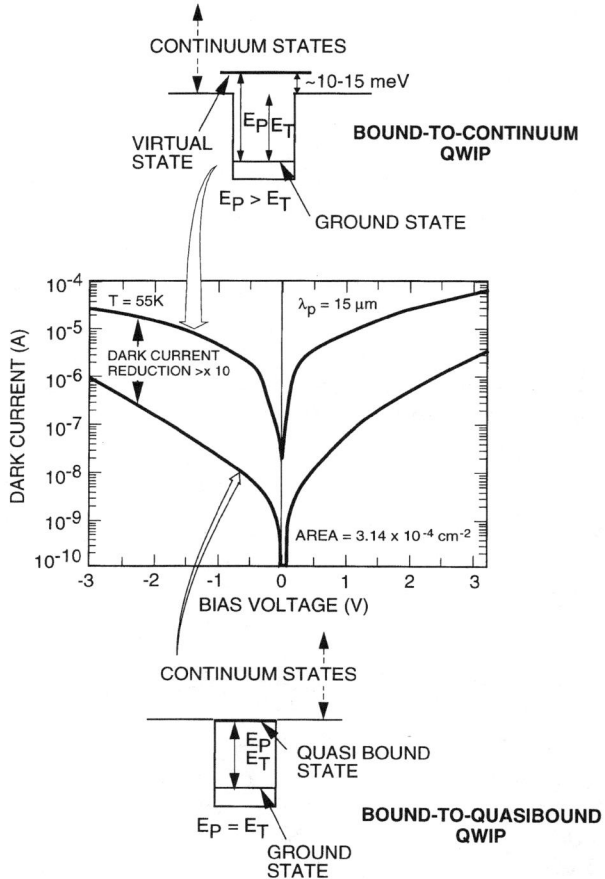

FIG. 5. Comparison of dark currents of bound-to-continuum and bound-to-quasibound VLWIR QWIPs as a function of bias voltage at temperature $T = 55$ K (from Gunapala et al., 1997a).

to be ~ 10 meV more in bound-to-quasibound QWIP than in the bound-to-continuum one, causing the dark current to drop significantly at elevated operating temperatures. The advantage of the bound-to-quasibound QWIP over the bound-to-continuum QWIP is that in the case of bound-to-quasibound QWIP the energy barrier for the thermionic emission is the same as it is for the photoionization, as shown in Fig. 5 (Gunapala and Bandara, 1995). In the case of a bound-to-continuum QWIP the energy barrier for the thermionic emission is 10–15 meV less than the photoionization energy. Thus, the dark current of bound-to-quasibound QWIPs is reduced by an order of magnitude (i.e., $I_d \propto e^{-\Delta E/kT} \approx e^{-2}$ for $T = 55$ K), as shown in Fig. 5.

4. n-Doped Broadband QWIPs

A broadband MQW structure can be designed by repeating a unit of several quantum wells with slightly different parameters such as quantum well width and barrier height. The first device structure (shown in Fig. 6), demonstrated by Bandara et al. (1998a), has 33 repeated layers of GaAs three-quantum-well units separated by $L_B \sim 575$ Å thick $Al_xGa_{1-x}As$ barriers (Bandara et al., 1998b). The well thickness of the quantum wells of three-quantum-well units are designed to respond at peak wavelengths around 13, 14, and 15 μm, respectively. These quantum wells are separated by 75-Å-thick $Al_xGa_{1-x}As$ barriers. The Al mole fraction (x) of barriers throughout the structure was chosen such that the $\lambda_p = 13$ μm quantum well operates under bound-to-quasibound conditions. The excited state energy level broadening was further enhanced due to overlap of the wave functions associated with excited states of quantum wells separated by thin barriers. Energy band calculations based on a two-band model shows excited state energy levels spreading about 28 meV. An experimentally measured responsivity curve at $V_B = -3$ V bias voltage has shown broadening of the spectral response

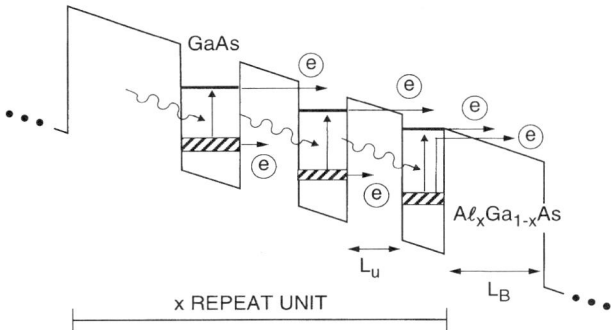

FIG. 6. Schematic diagram of the conduction band in broadband QWIP in an externally applied electric field. The device structure consists of 33 repeated layers of three-quantum-well units separated by thick $Al_xGa_{1-x}As$ barriers. Also shown are the possible paths of dark current electrons and photocurrent electrons of the device under a bias (from Bandara et al., 1998b).

up to $\Delta\lambda \sim 5.5\,\mu m$; that is, the full width at half maximum from 10.5 to 16 μm. This broadening $\Delta\lambda/\lambda_p \sim 42\%$ is about a 400% increase compared to a typical bound-to-quasibound QWIP.

5. n-DOPED BOUND-TO-BOUND MINIBAND QWIPs

The superlattice miniband detector uses the concept of infrared photoexcitation between minibands (ground state and first excited state) and transport of these photoexcited electrons along the excited-state miniband. When the carrier de Broglie wavelength becomes comparable to the barrier thickness of the superlattice, the wave functions of the individual wells tend to overlap due to tunneling, and energy minibands are formed. The miniband occurs when the bias voltage across one period of the superlattice becomes smaller than the miniband width (Capasso et al., 1986).

Experimental work on infrared detectors involving the miniband concept was initially carried out by Kastalsky et al. (1988). The spectral response of this GaAs–AlGaAs detector was in the range 3.6–6.3 μm and indicated that low-noise infrared detection was feasible without the use of external bias. O et al. (1990) reported experimental observations and related theoretical analysis for this type of detectors with absorption peak in the LWIR spectral range (8–12 μm). Both these detectors consist of a bound-to-bound miniband transition (i.e., two minibands below the top of the barrier) and a graded barrier between the superlattice and the collector layer as a blocking barrier for ground-state miniband tunneling dark current.

To further reduce ground-state miniband tunneling dark current, Bandara et al. (1992) used a square step barrier at the end of the superlattice. This structure, illustrated in Fig. 7a, was grown by MBE and consists of 50 periods of 90-Å GaAs quantum wells and 45-Å $Al_{0.21}Ga_{0.79}As$ barriers. A 600-Å $Al_{0.15}Ga_{0.85}As$ blocking layer was designed so that it has two minibands below the top of the barrier with the top of the step blocking barrier being lower than the bottom of the first excited-state miniband, but higher than the top of the ground-state miniband. The spectral photoresponse was measured at 20 K with a 240-mV bias voltage across the detector and at 60 K with a 200-mV bias voltage. The experimental response band of this detector was in the VLWIR range with peak response at 14.5 μm. The rapid fall-off in the photocurrent at higher bias voltage values was observed and attributed to the progressive decoupling of the miniband as well as the rapid decrease in the impedance of the detector.

The peak responsivity of this detector at 20 and 60 K are 97 and 86 mA/W for unpolarized light. Based on these values and noise measurements, the estimated detectivity at 20 and 60 K are 1.5×10^9 and 9×10^8 cm $\sqrt{Hz}/W$,

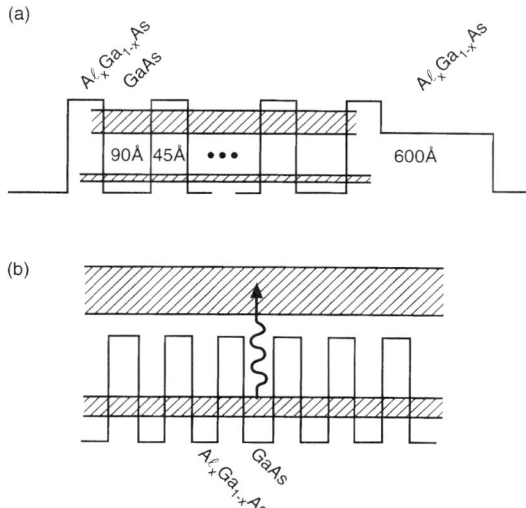

FIG. 7. (a) Parameters and band diagram for LWIR GaAs–Al$_x$Ga$_{1-x}$As superlattice miniband detector with $\lambda_c \sim 15\,\mu$m. (Bandara et al., 1988); (b) device structure of bound-to-continuum miniband (from Gunapala et al., 1991b).

respectively. Although this detector operates under a modest bias and power conditions, the demonstrated detectivity is relatively lower than the responsivity of usual QWIPs. This is mainly due to lower collection efficiency. Although there is enough absorption between minibands, only photoexcited electrons in few quantum wells near the collector contact contribute to the photocurrent.

6. n-DOPED BOUND-TO-CONTINUUM MINIBAND QWIPs

It is anticipated that placing the excited state miniband in the continuum levels would improve the transportation of the photoexcited electrons (i.e., the responsivity of the detector). This is the same as in the case of the wide-barrier bound-to-continuum detectors discussed previously. A detector based on photoexcitation from a single miniband below the top of the barriers to one above the top of the barriers is expected to show a higher performance. Gunapala et al. (1991b) proposed and demonstrated this type of bound-to-continuum miniband photoconductor based on GaAs–Al$_x$Ga$_{1-x}$As superlattice operating in the 5- to 9-μm spectral range. Their structure shows more than an order of magnitude improvement in electron transport and detector performance, compared with previous bound-to-bound state miniband detectors.

Device structures (as shown in Fig. 7b) studied by Gunapala et al. (1991b) consisted of 100 periods of GaAs quantum wells of either $L_b = 30$ Å or $L_b = 45$ Å barriers of $Al_{0.28}Ga_{0.72}As$, and $L_w = 40$ Å GaAs wells (doped $N_D = 1 \times 10^{18}$ cm^{-3}) sandwiched between doped GaAs contact layers. The absolute values for the peak absorption coefficients are $\alpha = 3100$ cm^{-1} and $\alpha = 1800$ cm^{-1} for the $L_b = 30$ and 45 Å structures, respectively. The structure with narrower barrier ($L_b = 30$ Å) has a higher peak absorption coefficient as well as a broader spectrum, resulting in significantly larger integrated absorption strength. Also, the dark current of this structure ($L_b = 30$ Å) is much larger than that of the other ($L_b = 45$ Å) structure.

Detectivities at peak wavelength for the preceding miniband detectors were calculated using measured responsivities and dark currents. For the $L_b = 30$ Å structure the result was $D^* = 2.5 \times 10^9$ and 5.4×10^{11} cm$\sqrt{Hz}$/W for $T = 77$ and 4 K at -80 mV bias. For $L_b = 45$ Å structure, they obtained $D^* = 2.0 \times 10^9$ and 2.0×10^{10} cm$\sqrt{Hz}$/W for $T = 77$ and 4 K at -300 mV bias. These values are significantly larger than the previous bound-to-bound miniband results. Although the responsivity is improved by placing the excited state in the continuum, it also increases the thermionic dark current because of the lower barrier height. This fact is more critical for LWIR detectors because the photoexcitation energy becomes even smaller; that is, the detector operating temperature will be lowered.

7. n-Doped Bound-to-Miniband QWIPs

Yu and Li (1991) and Yu et al. (1991) proposed and demonstrated a miniband transport QWIP that contained two bound states with higher energy level being resonance with the ground-state miniband in the superlattice barrier (see Fig. 8a). In this approach, infrared radiation is absorbed in the doped quantum wells, exciting an electron into the miniband and transporting it in the miniband until it is collected or recaptured into another quantum well. Thus, the operation of this miniband QWIP is analogous to that of a weakly coupled MQW bound-to-continuum QWIP. In this device structure, the continuum states above the barriers are replaced by the miniband of the superlattice barriers. These miniband QWIPs show lower photoconductive gain than bound-to-continuum QWIPs because the photoexcited electron transport occurs in the miniband where electrons must transport through many thin heterobarriers, resulting in lower mobility. The bandwidth of the absorption spectrum is controlled by the position

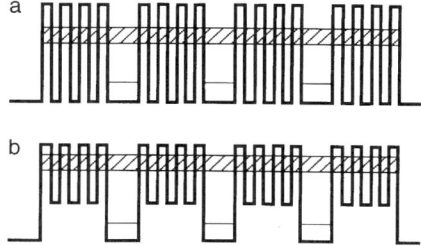

FIG. 8. Band diagrams for (a) bound-to-miniband and (b) step bound-to-miniband QWIP structures (from Levine, 1993).

of the miniband, relative to the barrier threshold, as well as the width of the miniband, which is exponentially dependent on the thickness of the superlattice barriers. Faska *et al.* (1992) adopted this bound-to-miniband approach and demonstrated excellent LWIR images from a 256×256 focal plane array (FPA) camera. These bound-to-miniband QWIPs have been demonstrated using a GaAs–Al_xGa_{1-x}As material system. To further improve performance (by decreasing dark current) of these miniband QWIPs, Yu *et al.* (1992) proposed a step bound-to-miniband QWIP, which is shown in Fig. 8b. This structure consists of GaAs–Al_xGa_{1-x}As superlattice barriers and $In_{0.07}Ga_{0.93}$As strained quantum wells, which are deeper than superlattice barrier wells, as shown in Fig. 8b (Li *et al.*, 1993).

8. *n*-Doped Asymmetrical GaAs–Al_xGa_{1-x}As QWIPs

For typical QWIP structures with symmetrical rectangular wells, the electric field shift of intersubband absorption is relatively small, since the linear shift term is forbidden by symmetry. This was demonstrated by Harwit and Harris (1987) who obtained a shift of less than 2 meV for a field of 36 kV/cm across a symmetrical QWIP structure. Large stark shifts can be obtained by designing QWIP structures with asymmetrical quantum wells such as stepped well. Martinet *et al.* (1992) have demonstrated a linear stark shift of ~ 10 meV for an electric field shift of ~ 15 kV/cm, while Mii *et al.* (1989) measure a shift of 8 meV at 18 kV/cm. These shifts are an order of magnitude larger than that for the symmetrical rectangular well in the same electrical field. In addition to absorption spectral changes due to asymmetrical wells or barriers in the structure, there will be much stronger bias-dependent behavior in escape probability, photoconductive gain, and the responsivity spectrum.

By introducing a QWIP structure with asymmetrical barriers, Levine et al. (1990c) demonstrated an electrically tunable photodetector without sacrificing the responsivity. Lacoe et al. (1992) demonstrated a large difference in spectral width and cutoff wavelength by using graded barrier QWIP with much stronger (81 meV) grading in the barrier. They also find that the responsivity spectral line width is much narrower ($\Delta\lambda < 1\,\mu$m) and the cutoff wavelength much shorter ($\lambda_c = 7.6\,\mu$m) in the negative bias direction where the QWIP operates on a bound-to-bound transition, compared to the broad ($\Delta\lambda = 6\,\mu$m) longer-wavelength cutoff ($\lambda_c = 11.4\,\mu$m) positive bias bound-to-continuum transition. Levine et al. (1991c) have also discussed LWIR photoinduced charge polarization and storage in graded asymmetrical quantum wells. They have studied the long-wavelength photo-induced charge polarization and electron storage produced by infrared-induced intersubband absorption in highly asymmetrical graded quantum wells. At zero bias a large photovoltage is observed, while at high bias the charge transfer probability approached unity.

9. p-DOPED QWIPs

Levine et al. (1991b) experimentally demonstrated the first QWIP that used hole intersubband absorption in the GaAs valence band. The strong mixing (Chang and James, 1989; Chiu et al., 1983; Karunasiri et al., 1990; Pinczuk et al., 1986; Wieck et al., 1984) between the light and heavy holes (at $k \neq 0$) allows the desired normal incidence illumination geometry to be used. The samples were grown on a (100) semi-insulating GaAs substrate, using gas source MBE, and consisted of 50 periods of $L_w = 30\,\text{Å}$ (or $L_w = 40\,\text{Å}$) quantum wells (doped $N_D = 4 \times 10^{18}\,\text{cm}^{-3}$ with Be) separated by $L_b = 300\,\text{Å}$ barriers of $Al_{0.3}Ga_{0.7}As$, and capped by $N_D = 4 \times 10^{18}\,\text{cm}^{-3}$ contact layers.

The unpolarized responsivity spectra of the $L_w = 40\,\text{Å}$ QWIP are compared for the two geometries (i.e., 45° and normal incidence, as discussed in Section IV). The two spectra are essentially identical (peak wavelength $\lambda_p = 7.2\,\mu$m and long wavelength cutoff $\lambda_c = 7.9\,\mu$m) and the normal incidence responsivity is larger than that of 45° illumination, which is consistent with both polarizations contributing to the photoresponse. The peak unpolarized responsivities (for the $L_w = 40\,\text{Å}$ sample at $\lambda_p = 7.2\,\mu$m and $V_b = 4\,\text{V}$) are $R_p = 39$ and $35\,\text{mA/W}$ for normal and 45° incidence, respectively, which are approximately an order of magnitude smaller than the responsivities for n-QWIPs. As a further test of polarization behavior, Levine et al. (1991b) found that by using the 45° geometry, s-polarized light had twice the photoresponse of p polarization. This is again in strong contrast with n-QWIPs for which the s-polarized photoresponse is forbidden by symmetry. A comparison between the $L_w = 30$ and $40\,\text{Å}$ detectors shows that, as expected, the narrower well QWIP has a broader spectral response ($\lambda_c = 8.6\,\mu\text{m}$) due to the excited state being pushed farther up into the continuum and thereby broadening the absorption (Levin et al., 1989). The 30-Å well detector also has a slightly longer peak wavelength ($\lambda_p = 7.4\,\mu\text{m}$), consistent with the ground state being pushed up even farther than the excited state. Similar lineshape effects (Levine et al., 1989) have been seen in n-QWIPs. Peak detectivity $D_\lambda^* = 1.7 \times 10^{10}\,\text{cm}\sqrt{\text{Hz}}/\text{W}$ and $D_\lambda^* = 3.5 \times 10^9\,\text{cm}\sqrt{\text{Hz}}/\text{W}$ were measured at $T = 77\,\text{K}$ for $L_w = 30$ and $40\,\text{Å}$ detectors, respectively.

The experimentally measured photoconductive gains are $g = 3.4 \times 10^{-2}$ and $g = 2.4 \times 10^{-2}$ for $L_w = 30$ and $40\,\text{Å}$, respectively. This yields hot hole mean free path $L = 510$ and $360\,\text{Å}$ for the two samples (i.e., photoexcited carrier travels only one or two periods before being recaptured). It should be noted that both g and L are over an order of magnitude smaller than the corresponding values for n-QWIPs (where photoconductive gains of $g \approx 1$ have been obtained; Levine et al., 1990a). Since $L = v_s \tau_r$ (where τ_r is the well recapture time; Levine et al., 1992b), we can see that the lower g is due to a lower velocity associated with the higher hole effective mass, and a shorter lifetime due to increased scattering between the light and heavy hole bands. Having now obtained g, we can relate it to the peak responsivity, and thus using the measured values for R_p and g directly determine the low-temperature quantum efficiency. This yields double-pass values of $\eta = 17\%$ and 28% for the $L_w = 30$ and $40\,\text{Å}$ QWIPs, respectively. Thus, in spite of the larger effective mass of the holes compared to that of electrons, the different symmetry (normal incidence) of the intersubband absorption, and the much smaller photoconductive gain and mean free path, the quantum efficiency (and hence escape probability) for bound-to-continuum n- and p-QWIPs are similar.

10. SINGLE-QUANTUM-WELL INFRARED PHOTODETECTORS

Although there has been extensive research on MQW infrared photodetectors that typically contain many (~ 50) periods, there has been relatively limited experimental work (Liu et al., 1991a, 1991b; Rosencher et al., 1992) done on QWIPs containing only a single quantum well. Bandara et al. (1993a, 1993b, 1993c) have performed a complete series of experiments on single-quantum-well structures with an n-type doped well, and undoped well, and a p-type doped well. These doped single-well detectors are, in fact, particularly interesting since they have exceptionally high photoconductive gain compared to MQW detectors. In addition, their simple band structures enable accurate calculations of the bias voltage dependence of the potential profiles for each of the two barriers, band-bending effects in the contacts, as well as charge accumulation (or depletion) in the quantum well.

11. Indirect Bandgap QWIPs

Infrared detectors operating in the MWIR spectral range ($\lambda \sim 3$–$5\,\mu$m) are also of interest due to the atmospheric window in this wavelength range. However, the short wavelength limit in the GaAs–Al_xGa_{1-x}As materials system imposed by keeping the Al_xGa_{1-x}As barriers direct is $\lambda = 5.6\,\mu$m. That is, if the Al concentration x is increased beyond $x = 0.4$ the indirect X valley becomes the lowest bandgap. This has been thought to be highly undesirable since Γ–X scattering together with GaAs X-barrier trapping can result in inefficient carrier collection and, thus, poor responsivity. In view of technology advantages in the more mature GaAs–Al_xGa_{1-x}As material system, however, it would be highly desirable to design detectors using indirect Al_xGa_{1-x}As. Levine et al. (1991a) successfully demonstrated thye first bound-to-continuum GaAs–Al_xGa_{1-x}As indirect bandgap QWIP operating at $\lambda_p = 4.2\,\mu$m.

A single period of the structure (Levine et al., 1991a) consists of $L_w = 30$ Å of GaAs (doped $n = 1 \times 10^{18}$ cm^{-3}) and 500 Å of undoped $Al_{0.55}Ga_{0.45}$As. This is repeated 50 times and is sandwiched between a 0.5-μm top and 1-μm bottom contact layer (also doped $n = 1 \times 10^{18}$ cm^{-3}). The conduction band diagram under bias for the structure is shown in Fig. 9. The solid line in Fig. 9 is the Γ-valley band edge and the dotted line is that of the X valley. The Γ-barrier height of $E_b^\Gamma = 452$ meV results in a calculated single bound Γ state at an energy $E_0^\Gamma = 165$ meV, and an optical absorption peak (from E_0^Γ to the excited continuum state E_1^Γ) at an energy of $\Delta E_{01}^\Gamma = 289$ meV (i.e., 2 meV above E_b^Γ) corresponding to a calculated absorption peak at $\lambda_p = 4.3\,\mu$m.

The bias dependence of R at $\lambda_p = 4.2\,\mu$m has also been measured (Levine et al., 1991a) and found to have significant zero bias photovoltaic effect with $R_p = 0.05$ A/W at $V_b = 0$. This yields the photoconductive gain of $g = 3.0$ and 1.0 for $V_b = 4$ and -4 V, respectively, and (using the superlattice length $l = 2.65\,\mu$m) the corresponding values of the hot electron mean free path $L = 8.0$ and $2.65\,\mu$m. These large values for g and L are comparable to (or even larger than) the best results in the usual Γ direct gap devices (Hasnain et al., 1990b; Levine et al., 1990a) and thus confirm excellent transport and efficient carrier collection expected in this indirect barrier QWIP. That is, this large mean free path is due to the difficulty of an X electron being

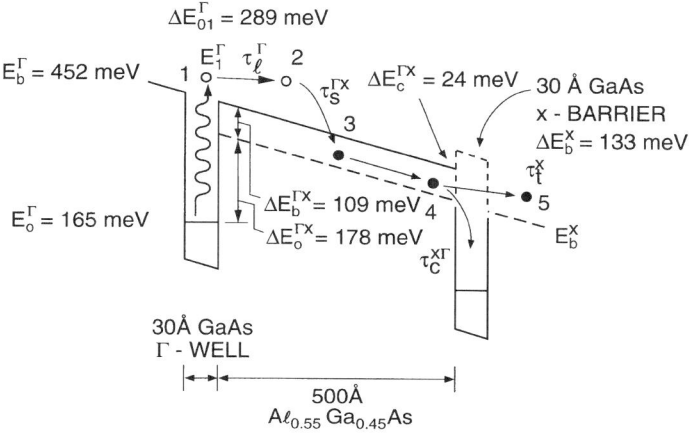

FIG. 9. Schematic conduction band diagram of indirect $Al_{0.55}Ga_{0.45}As$ barrier QWIP. The solid lines are the direct Γ-valley band edge, while the dashed lines are the indirect X-band edge (from Levine et al., 1991a).

captured by a Γ well. The peak detectivity D_λ^* at $V_b = -1$ V and $T = 77$ K is $D_\lambda^* = 4 \times 10^{10}$ cm $\sqrt{Hz}/W$. However, an even larger value $D_\lambda^* = 1.1 \times 10^{12}$ cm $\sqrt{Hz}/W$ is obtained at zero bias [determined by the Johnson noise generated by QWIP differential resistance (Levine et al., 1991a)]. This large D^* is comparable to that obtained using the direct-gap material $In_{0.53}Ga_{0.47}As-In_{0.52}Al_{0.48}As$, which is discussed in the next subsection.

12. n-Doped $In_{0.53}Ga_{0.47}As-In_{0.52}Al_{0.48}As$ QWIPs

To shift the intersubband absorption resonance into the higher energy spectral region ($\lambda = 3-5$ μm), Levine et al. (1988a) have investigated lattice matched quantum well superlattices of $In_{0.53}Ga_{0.47}As-In_{0.52}Al_{0.48}As$ grown using MBE on an InP substrate and reported intersubband absorption in this heterosystem. This direct gap heterostructure has conduction band discontinuity of 500 meV, which is significantly higher than that of the direct gap $GaAs-Al_xGa_{1-x}As$ system, therefore allowing for shorter wavelength operation.

A 50-period MQW superlattice consisting of $In_{0.53}Ga_{0.47}As$ wells (doped $N_D = 1 \times 10^{18}$ cm^{-3}) having a width $L_w = 50$ Å, and 150-Å barriers of $In_{0.52}Al_{0.48}As$, was grown on an InP substrate. The experimental absorption peak is at $\lambda = 4.4$ μm, in good agreement with the theoretical estimation of the energy separation of bound states (see Levine et al., 1988a, for details). To achieve higher performances in the MWIR range, Hasnain et al. (1990a) designed a MQW structure of the same materials system involving bound-to-continuum intersubband absorption. This structure consisting of 50 periods of 30-Å $In_{0.53}Ga_{0.47}As$ wells (doped $N_D = 2 \times 10^{18}\,cm^{-3}$) and 300-Å $In_{0.52}Al_{0.48}As$ barriers, was grown by MBE on an InP substrate. The absorption spectrum is peaked at 279 meV ($\lambda = 4.4\,\mu m$) with a full width at half maximum of 93 meV. Although the peak absorption of this bound-to-continuum detector is 4.2 times lower than that of a bound-to-bound detector (Levine et al., 1988a) the line width is five times greater. Thus, it has comparable (20% higher) absorption strength covering the full 3- to 5-μm MWIR band. The noise measured in these MQW detectors at 500 Hz corresponds to the shot noise of the dark current resulting peak detectivity at 77 K of $D^* = 1.5 \times 10^{12}\,cm\,\sqrt{Hz}/W$ with a background limited (for a 180° field of view) $D^*_{BL} = 2.3 \times 10^{10}\,cm\,\sqrt{Hz}/W$ at 120 K and lower temperatures. These values are comparable to those demonstrated with the Pt–Si devices (Sheperd 1988) presently used in MWIR band.

13. n-Doped $In_{0.53}Ga_{0.47}As$–InP QWIPs

An InGaAs–InP materials system has been used extensively for optical communication devices and therefore has a highly developed growth and processing technology. Since the quality of barriers is extremely important for optimum QWIP performance, and InP is binary, whereas $Al_xGa_{1-x}As$ is a ternary alloy, Gunapala et al. (1991a) investigated the hot electron transport and performance of detectors fabricated from these two materials. Two structures were grown by metal–organic molecular beam epitaxy (MOMBE) with arsine and phosphine as group V sources, trimethylindium and trimethylgallium as group III sources, and elemental Sn as n-type dopant sources (Gunapala et al., 1991a; Ritter et al., 1991). The first structure consisted of 20 periods of $L_w = 60$ Å $In_{0.53}Ga_{0.47}As$ quantum wells lattice matched to 500-Å InP barriers. A second sample contained 50 periods of $L_w = 50$ Å $In_{0.53}Ga_{0.47}As$ wells separated by 500-Å InP barriers. These MQWs were doped $N_D = 5 \times 10^{17}\,cm^{-3}$, and had top and bottom 0.4-μm contact layers of $N_D = 1 \times 10^{18}\,cm^{-3}$ doped $In_{0.53}Ga_{0.47}As$. The intersubband absorption was measured on a 45° multipass waveguide. The peak ($\lambda_p = 8.1\,\mu m$) room temperature absorption coefficient $\alpha = 950\,cm^{-1}$ is expected to increase by a factor of 1.3 at $T = 77\,K$, resulting in a low-temperature quantum efficiency of $\eta = 12\%$ (Gunapala et al., 1991a). For the 50-Å well QWIP, the corresponding value is $\eta = 11\%$. These values are quite comparable to those of GaAs sample, when the lower doping level of $N_D = 5 \times 10^{17}$ cm^{-3} in the wells is taken into account.

The bias dependence of the responsivities (which was essentially independent of temperature $T = 10$–80 K) was measured. Extremely large values of the responsivity, reaching $R = 6.5$ A/W at $V_b = 3.5$ V and $R = 3.5$ A/W at $V_b = -3.5$ V has been observed. This responsivity is five times larger than that of similar GaAs–Al$_x$Ga$_{1-x}$As QWIPs which demonstrates the excellent transport in this materials system. These large responsivity values yield very large values of photoconductive gain of $g = 9.0$ for $V_b = 3.5$ V and $g = 4.8$ for $V_b = -3.5$ V (Gunapala et al., 1991a). The corresponding value of the hot electron mean free path is $L = 10$ μm at $V_b = 3.5$ V, which is five times larger than that for similar GaAs–Al$_x$Ga$_{1-x}$As QWIPs. This excellent transport may be associated with the high-quality binary InP barriers, and higher mobility of InP compared with Al$_x$Ga$_{1-x}$As. The calculated peak detectivity $D_\lambda^* = 9 \times 10^{10}$ cm $\sqrt{\text{Hz}}$/W based on noise current and measured responsivity at $V_b = 1.2$ V and $T = 77$ K of 60-Å QWIP compares favorably with GaAs–Al$_x$Ga$_{1-x}$As QWIPs operating at this wavelength. Jelen et al. (1998a, 1998b) also investigated lattice matched InGaAs–InP QWIPs and observed large photoconductive gain indicating improved transport properties in the binary InP barrier (Jelen et al., 1998a, 1998b).

14. INGAASP QUATERNARY QWIPS

In the GaAs–Al$_x$Ga$_{1-x}$As system, transport properties of Al$_x$Ga$_{1-x}$As can be affected by oxygen related defects, and sometimes preferred elevated growth temperatures could result in undesirable dopant diffusion. Also, aluminum oxidation restricts fabrication methods such as epitaxial regrowth. Gunapala et al. (1991c, 1992a) investigated QWIP structures based on quaternary, Ga$_{1-x}$In$_x$As$_y$P$_{1-y}$ materials grown in a VG-V80H system modified for metal organic molecular beam epitaxy on semi-insulating (100) InP wafers. These structures consisted of InGaAsP quantum wells, lattice matched to 500-Å InP barriers. The quantum wells were n-type doped with $N_D = 5 \times 10^{17}$ cm^{-3} and sandwiched between 0.4-μm In$_{0.53}$Ga$_{0.47}$As contact layers of $N_D = 1 \times 10^{18}$ cm^{-3}. The responsivity spectra peaked at wavelengths of $\lambda_p = 8.6$, $\lambda_p = 11.0$ μm ($L_w = 54$ Å), and $\lambda_p = 12.1$ μm ($L_w = 63$ Å) with $\lambda_c = 9.4$ μm, $\lambda_c = 12.5$ and 13.2 μm, respectively, have been observed. The responsivities were measured and results are approximately twice as large as the responsivity of similar GaAs–Al$_x$Ga$_{1-x}$As QWIPs. These results clearly demonstrate the excellent hot electron transport in this materials system. Note also that this large maximum responsivity is still two times smaller than that of InGaAs–InP QWIPs previously discussed. This reduction of responsivity can be attributed to an increase in scattering and hence reduction in photoconductive gain of the quaternary GaInAsP material. The peak detectivities are comparable to GaAs–Al$_x$Ga$_{1-x}$As detectors operating at similar wavelengths. Hoff et al. (1995a, 1995b) also, investigated p-doped QWIP based on quaternary, Ga$_{1-x}$In$_x$As$_y$P$_{1-y}$ materials. Extensive theoretical modeling and experimental measurements have been carried out on QWIPs with different combinations of these materials system. Both p-doped binary–quaternary QWIP (GaAs–Ga$_{0.71}$In$_{0.29}$As$_{0.39}$P$_{0.61}$) and p-doped quaternary–ternary QWIP (Ga$_{0.87}$In$_{0.13}$As$_{0.74}$P$_{0.26}$/Ga$_{0.51}$In$_{0.49}$P) shows spectral coverage from 4 to 6 μm and background limited performances at $T = 100$ K (Hoff et al., 1995a, 1995b).

15. n-Doped GaAs–Ga$_{0.5}$In$_{0.5}$P QWIPs

In a GaAs–Ga$_{0.5}$In$_{0.5}$P materials system, Ga$_{0.5}$In$_{0.5}$P acts as the barrier material for the transport of electrons with an effective mass similar to that of Al$_{0.3}$Ga$_{0.7}$As. Therefore, Gunapala et al. (1990a) investigated this lattice-matched GaAs–Ga$_{0.5}$In$_{0.5}$P MQW structure grown on a GaAs substrate as an alternative system to GaAs–Al$_x$Ga$_{1-x}$As for LWIR detection. In addition, these infrared photoconductive measurements have enabled us to accurately determine the GaAs–Ga$_{0.5}$In$_{0.5}$P band offset.

The QWIP structure consisting of 10 periods of 40-Å GaAs quantum wells (doped $N_D = 2 \times 10^{18}$ cm^{-3}) and 300 Å of undoped Ga$_{0.5}$In$_{0.5}$P barriers grown by atmospheric pressure metal–organic vapor phase epitaxy (MOVPE) at a substrate temperature of 675°C on a semi-insulating undoped GaAs substrate. The absorption spectrum is peaked at 8 μm (155 meV) with a full width at half maximum of 82 meV (i.e., $\Delta\lambda/\lambda = 53\%$). The measured responsivity spectrum is also peaked at 8 μm, where the room temperature absorption peaked, but the width decreases from $\Delta\lambda/\lambda = 53$ to 21%. The measured peak responsivity is 0.34 A/W corresponding to a photoconductive gain of $g = 0.86$.

To determine the lattice-matched GaAs–Ga$_{0.5}$In$_{0.5}$P conduction band offset ΔE_c, we calculated the position and bandwidth of the low temperature absorption spectrum and adjusted the quantum well depth (i.e., ΔE_c) to fit the experiment. The theory includes both nonparabolicity as well as the exchange interaction. The exchange effect is significant and lowers the bound-state energy level (Bandara et al., 1988; Choe et al., 1990) by 19 meV for a doping density of $N_D = 2 \times 10^{18}$ cm^{-3}. From these results the conduction band offset ΔE_c was determined to be 221 ± 15 meV. From the known bandgaps of GaAs and Ga$_{0.5}$In$_{0.5}$P, the bandgap difference $\Delta E_g = 483$ meV was obtained, and thus from $\Delta E_c + \Delta E_v = \Delta E_g = 483$ meV, we determine that $\Delta E_v = 262 \pm 15$ meV. Also, Jelen et al. (1997) have demonstrated a

15-μm cutoff wavelength (13-μm detection peak) QWIP fabricated using n-type GaAs–Ga$_{0.5}$In$_{0.5}$P materials and fabrication of a preliminary FPA camera has been carried out.

16. n-Doped GaAs–Al$_{0.5}$In$_{0.5}$P QWIPs

The GaAs–(Al$_x$Ga$_{1-x}$)$_{0.5}$In$_{0.5}$P quaternary has a direct Γ-valley bandgap from $x = 0$ to $x = 0.7$, which then becomes an indirect X-valley conduction band from $x = 0.7$ to $x = 1$ (Watanabe and Ohba, 1987). The limiting composition GaAs–Al$_{0.5}$In$_{0.5}$P heterobarrier (although having an indirect gap) has a very large Γ-valley conduction band discontinuity of $\Delta E_c \sim 0.5$ eV (Watanabe and Ohba, 1987). This makes it an interesting system for short-wavelength QWIPs. A MQW structure was grown on a GaAs substrate via gas-source MBE. It consisted of 20 periods of 30-Å doped ($N_D = 1 \times 10^{18}$ cm^{-3}) quantum wells of GaAs and 500-Å barriers of undoped Al$_{0.5}$In$_{0.5}$P, surrounded by GaAs contact layers (doped $N_D = 2 \times 10^{18}$ cm^{-3}). The responsivity measured on a 45° polished QWIP had a narrow spectral shape ($\Delta\lambda/\lambda = 12\%$ indicating a nearly resonant bound-to-bound transition) and peaked at $\lambda_p = 3.25$ μm. The fact that this $\lambda \approx 3$ μm QWIP is grown lattice matched to a GaAs substrate means that it can be integrated with a long-wavelength GaAs–Al$_x$Ga$_{1-x}$As QWIP ($\lambda = 6$–20 μm), grown on the same wafer, allowing for the fabrication of monolithic multicolor infrared detectors.

17. n-Doped In$_{0.15}$Ga$_{0.85}$As–GaAs QWIPs

For all of the GaAs-based QWIPs that have been demonstrated thus far, GaAs is the low-bandgap well material and the barriers are lattice-matched Al$_x$Ga$_{1-x}$As, Ga$_{0.5}$In$_{0.5}$P, or Al$_{0.5}$In$_{0.5}$P. It is interesting, however, to consider GaAs as the barrier material since the transport in binary GaAs is expected to be superior to that of a ternary alloy, as was previously found to be the case in the In$_{0.53}$Ga$_{0.47}$As–InP binary barrier structures (Gunapala et al., 1991a, 1992a). To achieve this, Gunapala et al. (1994a) have used the lower bandgap non-lattice-matched alloy In$_x$Ga$_{1-x}$As as well material, together with GaAs barriers. Band-edge discontinuities and critical thicknesses of quantum well structures of this materials system have been studied earlier (Andersson et al., 1988; Yao et al., 1991). It has been demonstrated (Elman et al., 1989; Zhou et al., 1989) that strain layer heterostructures can be grown for lower In concentrations ($x < 0.2$), which results in lower barrier heights. Therefore, this heterobarrier system is very suitable for very long wavelength ($\lambda > 14$ μm) QWIPs.

To further investigate this materials system, Gunapala et al. (1994b) have used three samples that have been designed to give a very wide variation in QWIP absorption and transport properties. Detector A was designed to have an intersubband infrared absorption transition occurring between a single localized bound state in the quantum well and a delocalized state in the continuum (Levine et al., 1990a; Steele et al., 1991). Detector C was designed with a wider well width $L_w = 70$ Å, yielding two bound states in the well. Therefore, the intersubband transition is from the bound ground state to the bound excited state and requires electric field assisted tunneling for the photoexcited carrier to escape into the continuum (Choi et al., 1987b; Levine et al., 1987a). Due to the low effective mass of the electrons in the GaAs barriers, the electric field required for the field assisted tunneling is expected to be smaller in these structures in comparison with the usual GaAs–Al$_x$Ga$_{1-x}$As QWIP structures. Detector B was designed such that the second bound level is resonant with the conduction band of the GaAs barrier. Thus, the intersubband transition is from the bound ground state to the quasibound excited state which is intermediate between a strongly bound excited state and a weakly bound continuum state.

The responsivities of samples A, B, and C peak at 12.3, 16.0, and 16.7 μm, respectively. The absolute peaked responsivities (R_p) of the detectors were A, B, and C are 293, 510, and 790 mA/W, respectively, at bias $V_b = 300$ mV. As expected, the responsivity spectra of the bound-to-continuum QWIP (detector A) is much broader than the bound-to-bound (detector C) or bound-to-quasibound (detector B) QWIPs. Correspondingly, the magnitude of the peak absolute responsivity (R_p) is significantly lower than that of the bound-to-bound or bound-to-quasibound QWIPs, due to reduction of the absorption coefficient α. This reduction in the absorption coefficient is a result of the conservation of oscillator strength. These peak wavelengths and spectral widths are in good agreement with theoretical estimates of bound-to-continuum and bound-to-bound intersubband transition based on the 55% conduction band offset ($\Delta E_c/E_g$) of the GaAs–In$_{0.2}$Ga$_{0.8}$As materials system (i.e., $\Delta E_v/E_g = 45\%$). The large responsivity and detectivity D_λ^* values are comparable to those achieved with the usual lattice-matched GaAs–Al$_x$Ga$_{1-x}$As materials system (Levine et al., 1992a). The high photoconductive gains and the small carrier capture probabilities demonstrate the excellent carrier transport of the GaAs barriers and the potential of this heterobarrier system for VLWIR ($\lambda > 14 \mu$m) QWIPs.

18. p-Doped In$_{0.53}$Ga$_{0.47}$As–InP QWIPs

In Subsection 9 of this section we discussed the GaAs–Al$_x$Ga$_{1-x}$As QWIPs that are based on hole intersubband absorption in p-doped GaAs quantum wells. Due to the complex GaAs valence band interactions at the nonzero wave vector, the infrared absorption at normal incidence is allowed. Also, in Subsections 13 and 14 we have indicated that due to the high quality InP barriers, n-doped lattice-matched $In_{0.53}Ga_{0.47}As$–InP and InGaAsP–InP QWIPs have even larger responsivities than $GaAs$–$Al_xGa_{1-x}As$ n-QWIPs. Therefore, Gunapala et al. (1992b) have investigated the p-QWIPs in a lattice-matched $In_{0.53}Ga_{0.47}As$–InP materials system.

Since most of the $GaAs$–$Al_xGa_{1-x}As$ bandgap discontinuity is in the conduction band (i.e., $\Delta E_c/\Delta E_g = 65\%$), whereas in $In_{0.53}Ga_{0.47}As$–InP most of the bandgap difference is in the valence band (i.e., $\Delta E_v/\Delta E_g = 60\%$), the $In_{0.53}Ga_{0.47}As$–InP p-QWIP intersubband absorption occurs at a much shorter wavelength. The GaAs p-QWIPs discussed previously (Subsection 9), operated with a peak wavelength of $\lambda_p = 7.2$–$7.4\,\mu m$ and a cutoff wavelength of $\lambda_c = 7.9$–$8.6\,\mu m$. In strong contrast, the $In_{0.53}Ga_{0.47}As$–InP p-QWIPs discussed here have a responsivity peaked at $\lambda_p = 2.7\,\mu m$. This is, in fact, the shortest wavelength QWIP ever reported.

The devices discussed in the following were grown using MOMBE and consisted of 20 periods of 25-Å quantum wells of $In_{0.53}Ga_{0.47}As$ (doped $N_D = 2 \times 10^{18}\,cm^{-3}$ with Be) and 500-Å slightly p-doped barriers of InP. This slight p-doping is necessary to compensate for the n-background ($N_D \approx 4 \times 10^{16}$) of the InP barriers. Otherwise, n-InP and p-InGaAs result in a series of p-n junctions, which would significantly reduce the available carriers in the quantum well, and thus lower the infrared absorption. In addition, the p-n junctions would increase the series resistance and thus impede the transport. These MQWs were sandwiched between $In_{0.53}Ga_{0.47}As$ contact layers ($0.4\,\mu m$ top and bottom) having the same doping as the wells.

The responsivity spectrum demonstrates the shortest wavelength QWIP ever reported with a responsivity peak at $\lambda_p = 2.7\,\mu m$ and a cutoff wavelength of $\lambda_c = 3.0\,\mu m$ (the long-wavelength side where the responsivity drops to half of its peak value). The peak unpolarized responsivity, at normal incidence, is $R_p = 29\,mA/W$ at $V_b = 3\,V$ (corresponding to a double optical pass through the QWIP). This value is essentially unchanged from $T = 20$ to 80 K. The unpolarized responsivity measured using the 45° polished substrate control detector was nearly identical, with $R_p = 31\,mA/W$, for the same conditions. This is in good agreement with the $GaAs$–$Al_xGa_{1-x}As$ p-QWIPs discussed in Subsection 9, where both the magnitude and spectral shape of the normal incidence and 45° responses were also very similar. It is worth noting that there is a zero bias responsivity of $R_p = 6\,mA/W$, which also peaked at $\lambda_p = 2.7\,\mu m$. The peak detectivity for this unoptimized device was $D_\lambda^* = 3 \times 10^{10}\,cm\,\sqrt{Hz}/W$ at $T = 77\,K$. Later, Sengupta et al. (1996) have shown similar device performance of QWIPs based on the same material system.

III. Figures of Merit

Now we discuss and compare the optical and transport properties of bound-to-continuum QWIPs, bound-to-bound QWIPs, and bound-to-quasibound QWIPs. The structures of the six samples to be discussed are listed in Table I. These n-doped QWIPs were grown using MBE and the wells and contact layers were doped with Si. The quantum well widths L_w range from 40 to 70 Å, while the barrier widths are approximately constant at $L_b = 500$ Å. The Al molar fraction, in the $Al_xGa_{1-x}As$ barriers, varies from $x = 0.10$ to 0.31 (corresponding to cutoff wavelengths of $\lambda_c = 7.9$–19 μm). The photosensitive doped MQW region containing 25 to 50 periods is sandwiched between similarly doped top (0.5-μm) and bottom (1-μm) ohmic contact layers. These structural parameters have been chosen to give a very wide variation in QWIP absorption and transport properties (Levine et al., 1992b). In particular, samples A through D are n-doped with intersubband infrared transition occurring between a single localized bound state in the well and a delocalized state in the continuum (denoted B–C in Table I) (Levine et al., 1987d, 1990a, 1991e; Andersson and Lundqvist, 1991; Andrews and Miller, 1991; Janousek et al., 1990; Kane et al., 1989; Steele et al., 1991; Wu et al., 1992). Sample E has a high Al concentration $x = 0.26$ coupled with a wide well $L_w = 50$ A, yielding two bound states in the well. Thus, the intersubband transition from the bound ground state to the bound first excited state (denoted B–B in Table I), and therefore requires electric field assisted tunneling for the photoexcited carrier to escape into the continuum, as discussed in the previous section (Levine et al., 1987b, 1989;

TABLE I
STRUCTURE PARAMETERS FOR SAMPLES A–F

Sample	L_w(Å)	L_b(Å)	x	N_D(Å) (10^{18} cm^{-3})	Doping type	Periods	Intersubband transition
A	40	500	0.26	1.0	n	50	B–C
B	40	500	0.25	1.6	n	50	B–C
C	60	500	0.15	0.5	n	50	B–C
D	70	500	0.10	0.3	n	50	B–C
E	50	500	0.26	1.4	n	25	B–B
F	45	500	0.30	0.5	n	50	B–QB

Parameters include quantum well width L_w, barrier width L_b, $Al_xGa_{1-x}As$ composition x, doping density N_D, doping type, number of MQW periods, and type of intersubband transition bound-to-continuum (B–C), bound-to-bound (B–B), and bound-to-quasibound (B–QB) (Levine et al., 1992a).

Choi et al., 1987b). Sample F was designed to have a quasibound excited state (denoted B–QB in Table I; Kiledjian et al., 1991; Gunapala and Bandara, 1995), which is intermediate between a strongly bound excited state and weakly bound continuum state. It consists of a $L_w = 45$ Å doped quantum well and 500 Å of an $Al_xGa_{1-x}As$ barrier with $x = 0.3$. These quantum well parameters result in a first excited state in resonance with the barriers, and is thus expected to have an intermediate behavior.

1. ABSORPTION SPECTRA

The infrared absorption spectra for samples A–F were measured at room temperature, using a 45° multipass waveguide geometry (except for sample D, which was at such a long wavelength that the substrate multiphonon absorption obscured the intersubband transition). As can be readily seen in Fig. 10, the spectra of the bound-to-continuum QWIPs (samples A, B, and C) are much broader than the bound-to-bound or bound-to-quasibound QWIPs (samples E and F or the QWIPs discussed in the previous section). Correspondingly, the magnitude of the absorption coefficient α for the continuum QWIPs (left-hand scale) is significantly lower than that of the bound-to-bound QWIPs (right-hand scale), due to the conservation of oscillator strength. That is, $\alpha_p(\Delta\lambda/\lambda)/N_D$ is a constant, as was previously found (Gunapala et al., 1990b). The values of the peak room temperature absorption α_p, peak wavelength λ_p, cutoff wavelength λ_c (long wavelength λ

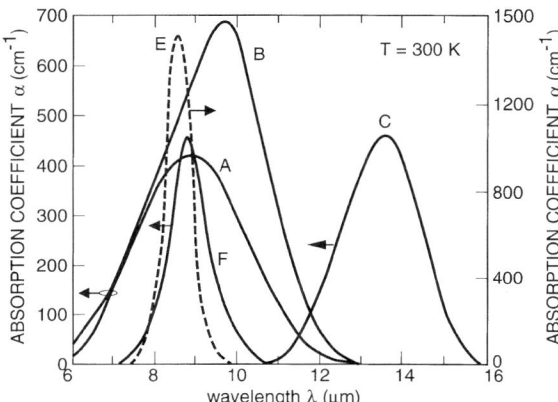

FIG. 10. Absorption coefficient spectra vs wavelength measured at $T = 300$ K for samples A, B, C, E, and F (from Levine et al., 1992a).

TABLE II

Optical Absorption Parameters for Samples A, B, C, E, and F

Sample	$\lambda_p(\mu m)$	$\lambda_c(\mu m)$	$\Delta\lambda(\mu m)$	$\Delta\lambda/\lambda$ (%)	α_p(300 K) (cm^{-1})	η_a(300 K) (%)	η_a(77 K) (%)	η_{max} (%)
A	9.0	10.3	3.0	33	410	10	13	16
B	9.7	10.9	2.9	30	670	15	19	25
C	13.5	14.5	2.1	16	450	11	14	18
E	8.6	9.0	0.75	9	1490	17	20	23
F	8.9	9.4	1.0	11	451	11	14	20

Parameters include peak absorption wavelength λ_p, long wavelength cutoff λ_c, spectral width $\Delta\lambda$, fractional spectral width $\Delta\lambda/\lambda$, peak room temperature absorption coefficient α_p(300 K), peak room temperature absorption quantum efficiency η_a(300 K), $T = 77$ K absorption quantum efficiency η_a(77 K), and maximum high bias net quantum efficiency η_{max} (Levine et al., 1992a).

for which α drops to half α_p) and spectral width $\Delta\lambda$ (full width at half α_p) are given in Table II. The room temperature absorption quantum efficiency η_a(300 K) evaluated from α_p(300 K) using

$$\eta_a = \tfrac{1}{2}(1 - e^{-2\alpha_p \ell}) \tag{3}$$

where η_a is the unpolarized double-pass absorption quantum efficiency, ℓ is the length of the photosensitive region; and the factor of 2 in the denominator is a result of the quantum mechanical selection rules, which allows only the absorption of radiation polarized in the growth direction. The low-temperature quantum efficiency η_a(77) was obtained by using α_p(77 K) $\approx$ 1.3α_p(300 K), as previously discussed. The last column containing η_{max} is discussed later.

To clearly compare the line shapes of the bound, quasibound, and continuum QWIPs, the absorption coefficients for samples A, E, and F were normalized to unity and plotted as $\tilde{\alpha}$ in Fig. 11, and the wavelength scale has been normalized by plotting the spectra against $\Delta\lambda \equiv (\lambda - \lambda_p)$, where λ_p is the wavelength at the absorption peak. The very large difference in spectral width is apparent with the bound and continuum excited state transitions ($\Delta\lambda/\lambda = 9$–11%) being 3–4 times narrower than for the continuum excited state QWIPs ($\Delta\lambda/\lambda = 33$%).

2. Dark Current

To measure the dark current–voltage curves, 200-μm-diameter mesas were fabricated, as described elsewhere (Gunapala and Bandara, 1995), and the results are shown in Fig. 12 for $T = 77$ K. Note that the asymmetry in

4 QUANTUM WELL INFRARED PHOTODETECTOR FOCAL PLANE ARRAYS 223

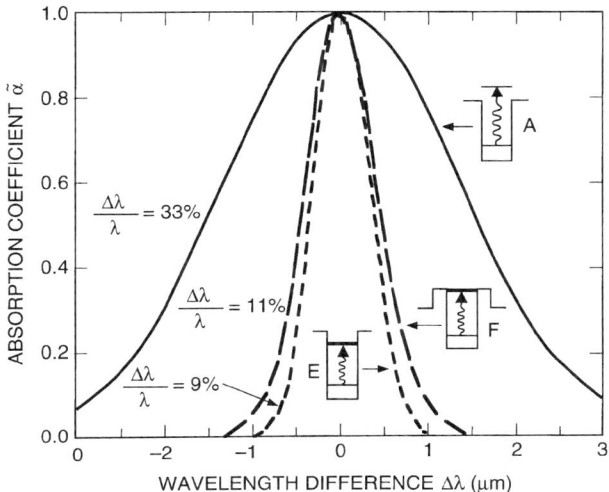

FIG. 11. Normalized absorption spectra vs wavelength difference $\Delta\lambda = (\lambda - \lambda_p)$. The spectral widths $\Delta\lambda/\lambda$ are also given. The inserts show the schematic conduction band diagram for sample A (bound-to-continuum), sample E (bound-to-bound), and sample F (bound-to-quasibound) (from Levine et al., 1992a).

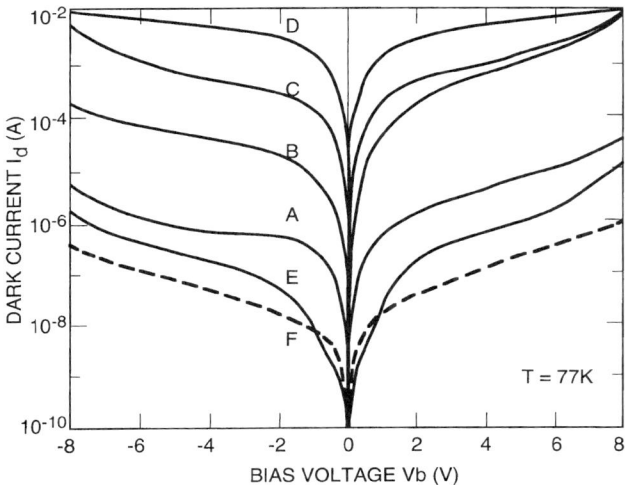

FIG. 12. Dark current I_d as a function of bias voltage V_b at $T = 77$ K for samples A–F (from Levine et al., 1992a).

the dark current (Zussman et al., 1991) with I_d being larger for positive bias (i.e., mesa top positive) than for negative bias. This can be attributed to the dopant migration in the growth direction (Liu et al., 1993), which lowers the barrier height of the quantum wells in the growth direction compared to the quantum well barriers in the other direction (which are unaffected). Note that, as expected, the dark current I_d increases as the cutoff wavelength λ_c increases. At bias $V_b = -1$ and 1 V, the curves for samples E and F cross. This is due to the fact that even though sample E has a shorter cutoff wavelength than sample F; it is easy for the excited electrons to tunnel out at sufficiently high bias. In contrast, sample F has a quasibound excited state, which is in resonance with the $L_b = 500$ Å thick barrier top.

Levine et al. (1990a) analyzed the origin of the dark current in detail and showed that thermionic-assisted tunneling is a major source of dark current (Gunapala et al., 1990b; Kinch and Yariv, 1989; Pelve et al., 1989; Zussman et al., 1991; Andrews and Miller, 1991). In that analysis, they first determine the effective number of electrons $n(V)$, which are thermally excited out of the well into the continuum transport states, as a function of bias voltage V:

$$n(V) = \left(\frac{m^*}{\pi\hbar^2 L_p}\right) \int_{E_0}^{\infty} f(E) T(E, V) \, dE \qquad (4)$$

where the first factor containing the effective mass m^* is obtained by dividing the two-dimensional density of states by the superlattice period L_p (to convert it into an average three-dimensional density), and where $f(E)$ is the Fermi factor $f(E) = [1 + \exp(E - E_0 - E_F)/kT]^{-1}$, E_0 is the ground state energy, E_F is the two-dimensional Fermi level, and $T(E, V)$ is the bias-dependent tunneling current transmission factor for a single barrier, which can be calculated using Wentzel–Kramers–Brillouin (WKB) approximation to a biased quantum well. Equation (4) accounts for both thermionic emission above the energy barrier E_b (for $E > E_b$) and thermionically assisted tunneling (for $E < E_b$). Then they calculated the bias dependent dark current $I_d(V)$ using $I_d(V) = n(V)ev(V)A$, where e is the electronic charge, A is the area of the detector, and v is the average transport velocity given by

$$v(V) = \mu F[1 + (\mu F/v_s)^2]^{1/2} \qquad (5)$$

where μ is the mobility, F is the average electric field, and v_s is the saturated drift velocity. Good agreement is achieved as a function of both bias voltage and temperature over a range of eight orders of magnitude in dark current (Levine et al., 1990a).

3. RESPONSIVITY

The responsivity spectra $R(\lambda)$ were measured (Zussman et al., 1991) on 200-μm-diameter mesa detectors using a polished 45° incident facet on the detector, together with a globar source and a monochromator. A dual lock-in ratio system with a spectrally flat pyroelectric detector was used to normalize the system spectral response due to wavelength dependence of the blackbody, spectrometer, filters, etc. The absolute magnitude of the responsivity was accurately determined by measuring the photocurrent I_p with a calibrated blackbody source. This photocurrent is given by

$$I_p \int_{\lambda_1}^{\lambda_2} R(\lambda)P(\lambda)\, d\lambda \tag{6}$$

where λ_1 and λ_2 are the integration limits that extend over the responsivity spectrum, and $P(\lambda)$ is the blackbody power per unit wavelength incident on the detector, which is given by

$$P(\lambda) = W(\lambda) \sin^2(\theta/2) AF \cos\phi \tag{7}$$

where A is the detector are, ϕ is the angle of incidence, θ is the optical field of view angle — that is, $\sin^2(\theta/2) = (4f^2 + 1)^{-1}$, where f is the f number of the optical system; in this case θ is defined by the radius ρ of the blackbody opening at a distance D from the detector, so that $\tan(\theta/2) = (\rho/D)$, F represents all coupling factors and $F = T_f(1 - r)C$, where T_f is the transmission of filters and windows, $r = 28\%$ is the reflectivity of the GaAs detector surface, C is the optical beam chopper factor ($C = 0.5$ in an ideal optical beam chopper), and $W(\lambda)$ is the blackbody spectral density given by the following equation (i.e., the power radiated per unit wavelength interval at wavelength λ by a unit area of a blackbody at temperature T_B).

$$W(\lambda) = (2\pi c^2 h/\lambda^5)(e^{hc/\lambda k T_B} - 1)^{-1} \tag{8}$$

By combining Eqs. (6) and (7) and using $R(\lambda) = R_p \tilde{R}(\lambda)$, where R_p is the peak responsivity and $\tilde{R}(\lambda)$ is normalized (at peak wavelength λ_p) experimental spectral responsivity, we can rewrite the photocurrent I_p as

$$I_p = R_p G \int_{\lambda_1}^{\lambda_2} \tilde{R}(\lambda) W(\lambda)\, d\lambda \tag{9}$$

where G represents all the coupling factors and is given by $G = \sin^2(\theta/2) AF \cos\phi$. Thus, by measuring the $T_B = 1000$ K blackbody photocurrent, R_p can be accurately determined.

The normalized responsivity spectra $\tilde{R}(\lambda)$ are given in Fig. 13 for samples A–F, where we again see that the bound and quasibound excited state QWIPs (samples E and F) are much narrower $\Delta\lambda/\lambda = 10$–$12\%$ than the continuum QWIPs $\Delta\lambda/\lambda = 19$–$28\%$ (samples A–D). Table III gives the responsivity peak λ_p and cutoff wavelengths λ_c as well as the responsivity spectral width $\Delta\lambda$. These responsivity spectral parameters are given in Table III and are similar to the corresponding absorption values listed in Table II.

The absolute peak responsivity R_p can be written in terms of quantum efficiency η and photoconductive gain g as

$$R_p = (e/h\nu)\eta g \tag{10}$$

Responsivity versus bias voltage curves for the bound, quasibound, and continuum, samples are shown in Fig. 14. Note that at low bias, the responsivity is nearly linearly dependent on bias and it saturates at high bias. This saturation occurs due to the saturation of drift velocity. For the longest wavelength sample D, where $\lambda_c = 19\ \mu m$, the dark current becomes too large at high bias to observe the saturation in R_p. The quasibound QWIP (sample F) behaves quite similarly to the bound QWIPs. The fully

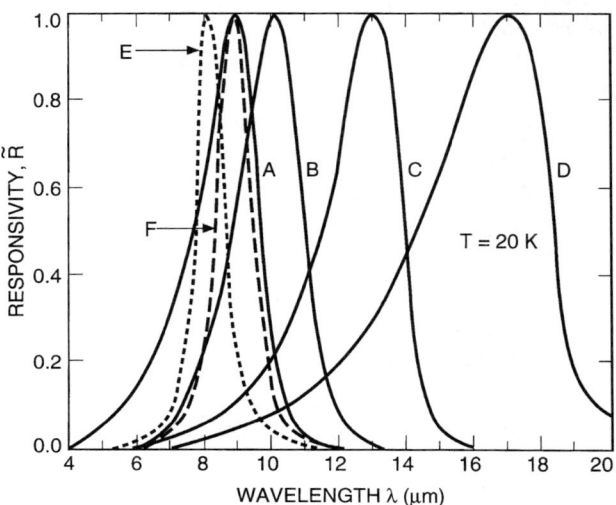

FIG. 13. Normalized responsivity spectra vs wavelength measured at $T = 20$ K for samples A–F (from Levine et al., 1992a).

TABLE III
RESPONSIVITY SPECTRAL PARAMETER FOR SAMPLES A–F

Sample	$\lambda_p^0(\mu m)$	$\lambda_c(\mu m)$	$\Delta\lambda(\mu m)$	$\Delta\lambda/\lambda(\%)$
A	8.95	9.8	2.25	25
B	9.8	10.7	2.0	20
C	13.2	14.0	2.5	19
D	16.6	19.0	4.6	28
E	8.1	8.5	0.8	10
F	8.4	8.8	1.0	12

Parameters include peak responsivity wavelength λ_p, long wavelength cutoff λ_c, spectral width $\Delta\lambda$, and fractional spectral width $\Delta\lambda/\lambda$ (Levine et al., 1992a).

bound sample E has a significantly lower responsivity. The responsivity does not start out linearly with bias but is in fact zero for finite bias. That is, there is a zero bias offset, due to the necessity of field assisted tunneling for the photoexcited carrier to escape from the well (Levine et al., 1987b, 1988b; Choi et al., 1987b; Vodjdani et al., 1991).

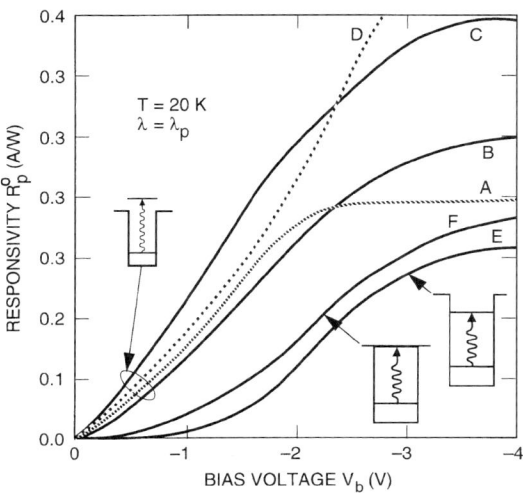

FIG. 14. Bias-dependent peak ($\lambda = \lambda_p$) responsivity R_p^0 measured at $T = 20$ K for samples A–F. The inserts show the conduction band diagrams (from Levine et al., 1992a).

4. Dark Current Noise

The dark current noise i_n was measured on a spectrum analyzer for all of the samples at $T = 77$ K as a function of bias voltage (Zussman et al., 1991). The result for sample B is shown in Fig. 15. The solid circles were measured for negative bias (mesa top negative), while the open circles are for positive bias. The smooth curves are drawn through the experimental data. Note that the current shot noise for positive bias is much larger than that for negative bias (e.g., at $V_b = 3.5$ V, it is four times larger). Also that near $V_b = 4$ V there is a sudden increase in the noise due to a different mechanism (possibly due to the avalanche gain (Levine et al., 1987c) process). This asymmetry in the dark current noise is due to the previously mentioned asymmetry in I_d. The photoconductive gain g can now be obtained using the current shot noise expression (Levine et al., 1990a; Liu, 1992a, 1992b; Beck, 1993; Hasnain et al., 1990b)

$$i_n = \sqrt{4eI_d g_n \Delta f} \tag{11}$$

where Δf is the band width (taken as $\Delta f = 1$ Hz).

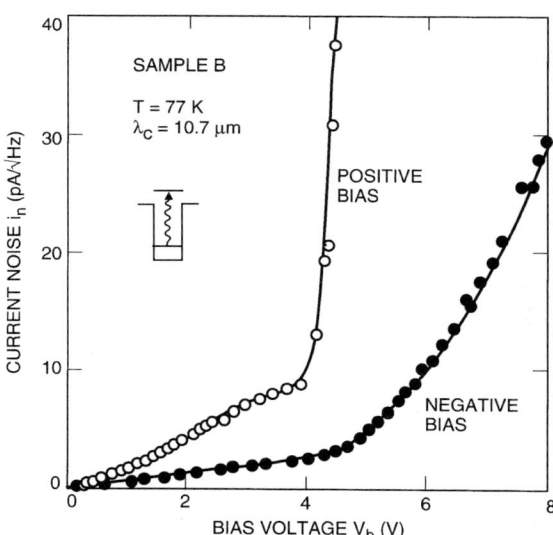

FIG. 15. Dark current noise i_n (at $T = 77$ K) vs bias voltage V_b for sample B. Both positive (open circles) and negative (solid circles) bias are shown. The smooth curves are drawn through the measured data. The insert shows the conduction band diagram (from Levine et al., 1992a).

5. Noise Gain and Photoconductive Gain

For a typical QWIP, where the dark current is dominated by thermionic emission, noise gain g_n and photoconductive gain g can be written in terms of well capture probability p_c (capture probability of electrons by the next period of the MQW) and the number of quantum wells N in the MQW region (Levine et al., 1992a; Liu, 1992a, 1992b; Beck, 1993; Choi, 1996),

$$g_n = \frac{1 - p_c/2}{N p_c} \qquad (12)$$

$$g = \frac{1 - p_c}{N p_c} \qquad (13)$$

where p_c is the capture probability of electrons by the next period of the MQW. Combining i_n from Fig. 15 and I_d from Fig. 12 enables the experimental determination of g as shown in Fig. 16. The solid circles are

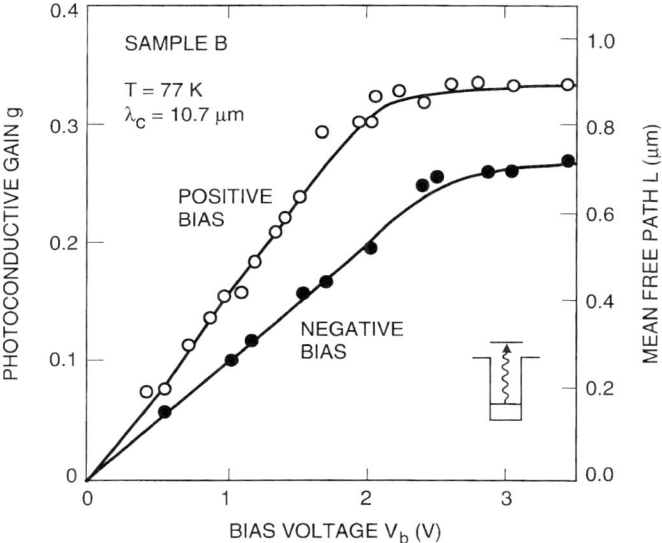

FIG. 16. Photoconductive gain g (left-hand scale) and hot electron mean free path L (right-hand scale) vs bias voltage V_b for sample B at $T = 77$ K. Both positive (open circles) and negative (solid circles) bias are shown. The smooth curves are drawn through the measured data. The insert shows the conduction band diagram (from Levine et al., 1992a).

for negative bias, while the open circles are for positive bias, and the smooth curves are drawn through the experimental points. As shown in Fig. 16, the photoconductive gain increases approximately linearly with the bias at low voltage and saturates near $V_b = 2$ V (due to velocity saturation) at $g \sim 0.3$. It is worth noting that, in spite of the large difference between the noise current i_n for positive and negative bias (as shown in Fig. 15), the photoconductive gains are quite similar. This demonstrates that, the asymmetry in i_n is due quantitatively to the asymmetry in I_d. It further shows that, although the number of carriers that escape from the well and enter the continuum (transmission factor γ; Choi, 1996) is strongly dependent on bias direction (due to the asymmetrical growth interfaces), the continuum transport (i.e., photoconductive gain), is less sensitive to the direction of carrier motion. The reason for this difference is that the transmission factor γ (and hence I_d) depends exponentially on the bias, whereas photoconductive gain is only linearly dependent on bias V_b. The photoconductive gain of QWIPs can be written as (Kastalsky et al., 1988; Hasnain et al., 1990b)

$$g = L/\ell \tag{14}$$

where L is the hot carrier mean free path and ℓ is the superlattice length ($\ell = 2.7\,\mu$m for sample B). Therefore, we can evaluate L as shown on the right-hand scale of Fig. 16. Thus, for this device L saturates at $\sim 1\,\mu$m. As mentioned earlier, the dramatic increase in i_n near $V_b = 4$ V in Fig. 15 is due to additional noise mechanisms and, therefore, should not be attributed to a striking increase in g.

From the strong saturation of R_p for sample A (in Fig. 14) we would also expect the photoconductive gain to be completely saturated for $|V_b| > 2$ V, which is how we have drawn the smooth curve in Fig. 17. However, if one obtains g by simply substituting the measured i_n into Eq. (11), the result is the dotted line in Fig. 17. Interpreting the large excess noise i_n above 4 V in Fig. 15 as I_d shot noise would lead to an incorrectly large gain. Likewise, interpreting the low i_n above $V_b = -2$ V in Fig. 17 as due to a low gain would also be incorrect. Levine et al. (1992b) have explained this by attributing the excess noise at high bias to the ground-state sequential tunneling, which is increasing the dark current I_d above that due to thermionic emission and thermionically assisted tunneling through the tip of the barriers. That is,

$$I_{d,m} = I_{d,\text{th}} + I_{d,\text{tu}} \tag{15}$$

where $I_{d,m}$ is the total measured current, $I_{d,\text{th}}$ is the usual thermionic contribution, and $I_{d,\text{tu}}$ is the ground state tunneling current. Note (Levine et

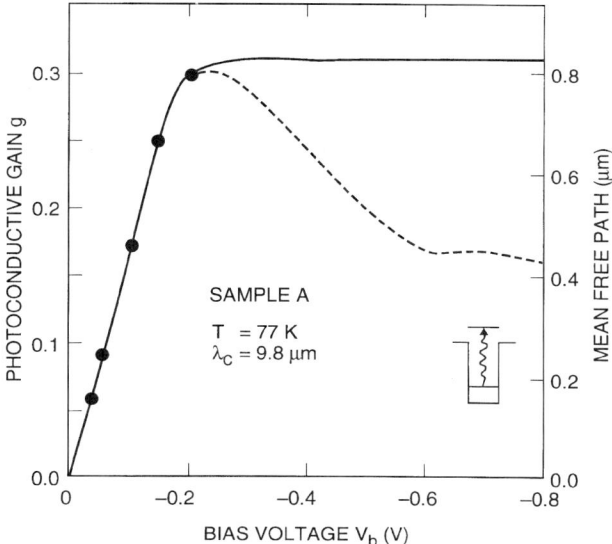

FIG. 17. Photoconductive gain g (left-hand scale) and hot electron mean free path L (right-hand scale) vs bias voltage V_b for sample A at $T = 77$ K. The solid curve drawn through the points is the correct interpretation; the dashed line is not. The insert shows the conduction band diagram (from Levine et al., 1992a).

al., 1990a) that electrons near the top of the well, which contribute to the dark current $I_{d,\text{tu}}$, are the same as those that can transport in the continuum and thus contribute to the photocurrent I_p. In contrast, the electrons that contribute to the ground state tunneling current $I_{d,\text{tu}}$ do not enter the continuum, but sequentially tunnel from one well to the next (Choi et al., 1987a). The gain associated with this process is $g_{\text{tu}} = L_p/\ell$ (where L_p is the superlattice period) and is very small compared with the usual continuum transport gain $g = L/\ell$. These two current processes lead to two contributions to the shot noise.

$$i_{d,m}^2 = i_{d,\text{th}}^2 + i_{d,\text{tu}}^2 \qquad (16)$$

By combining Eqs. (11), (15), and (16) and using the fact that $g_{\text{tu}} \ll g$ we can write

$$g_m = (1 - f)g \qquad (17)$$

where g_m is the measured gain and $f \equiv I_{d,\text{tu}}/I_{d,m}$. Therefore, at high bias when the current contribution from sequential tunneling increases, the measured gain decreases, exactly as found in Fig. 17. A more general formula for noise gain derived by Choi (1996) and Beck (1993) incorporated tunneling and thermionic contributions of the dark current. Bandara et al. (1998b) have observed two different gain mechanisms associated with photocurrent electrons and dark current electrons, which transported in two difference paths in a QWIP structure.

6. Quantum Efficiency

By using Eq. (10), the bias dependent photoconductive gain (Figs. 16 and 17), and the responsivity (Fig. 14), we can now determine the total measured quantum efficiency η (Levine et al., 1992a). The results of the continuum samples A–D are shown in Fig. 18. It is important to note that the total quantum efficiency does not vanish at zero bias but has a substantial value ranging from $\eta_0 = 3.2$–13%, corresponding to a finite probability of escaping from the quantum well. As the bias is increased, quantum efficiency increases approximately linearly and then saturates at high bias reaching maximum values of $\eta_{\text{max}} = 8$–25%. The saturation values of the total

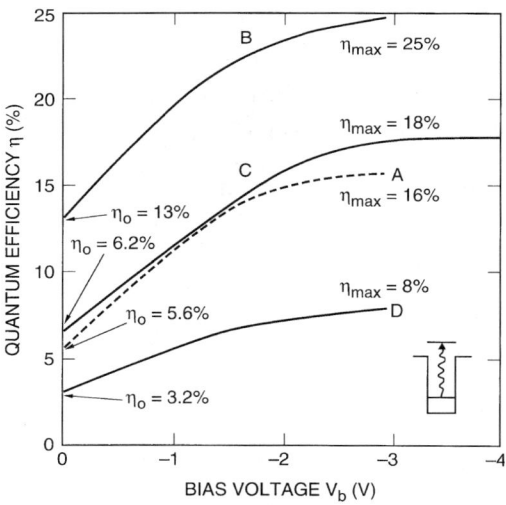

Fig. 18. Quantum efficiency η vs bias voltage V_b (negative) for samples A–D. The zero bias quantum efficiencies η_0 and the maximum quantum efficiencies η_{max} are shown. The insert shows the conduction band diagram (from Levine et al., 1992a).

quantum efficiencies are listed in Table II, where they appear to be comparable with the values obtained from the zero bias absorption measurements η_a (77 K). The difference in these values [i.e., between η_{max} and η_a (77 K)] can be attributed to lower measured gain as described in Subsection 5 of this section.

We now consider the bound and quasibound QWIPs (samples E and F). The photoconductive gains are plotted in Fig. 19, where they appear to be quite similar to the continuum QWIPs shown in Figs. 17, and the quantum efficiencies are shown in Fig. 20. For these QWIPs (sample E and F), η is quite similar to that of continuum QWIPs (Fig. 18), having a saturation values at high bias voltages. However, the required bias voltage to reach the saturation value is higher for bound and quasibound QWIPs than that for the continuum QWIPs. Also, zero bias quantum efficiency of these QWIPs (sample E and F) have much lower values. These differences are due to the necessity of field-assisted tunneling for the bound state excited photoelectrons to escape from the quantum well. This transmission factor (γ) can be included in the effective quantum efficiency (η) such that (Levine et al., 1992b; Choi, 1996)

$$\eta = \gamma \eta_a \qquad (18)$$

where η_a is the absorption quantum efficiency of the MQW structure. In this case, responsivity (R) and photoconductive gain (g) are related by $R = e\eta g/h\nu$, where $h\nu$ is the photoexcitation energy. Also, this transmission factor

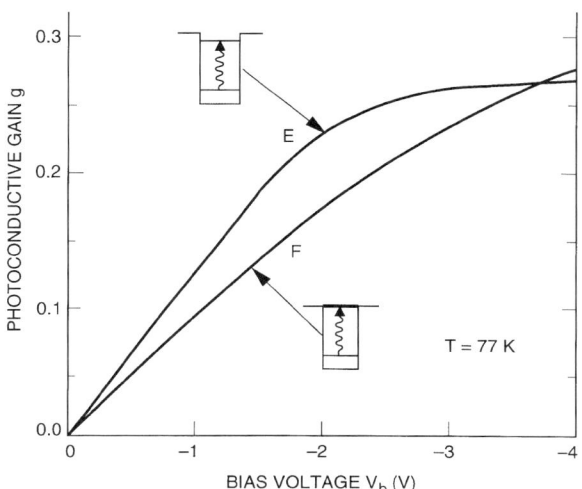

FIG. 19. Photoconductive gain g vs bias voltage V_b (negative) for samples E and F at $T = 77$ K. The inserts show the conduction band diagram (from Levine et al., 1992a).

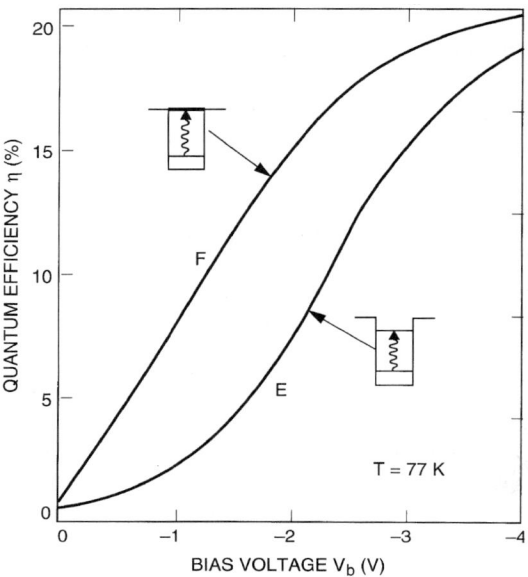

FIG. 20. Quantum efficiency η vs bias voltage V_b (negative) for samples E and F. The zero bias quantum efficiencies η_0 and the maximum quantum efficiencies η_{max} are shown. The inserts show the conduction band diagram (from Levine et al., 1992a).

can be included in the optical gain (g_o) which is defined (Choi, 1996) as $g_o = \gamma g$, and as a result, $R_p = e\eta_a g_o/h\nu$.

Typically, the γ of bound-to-continuum detectors is larger than the γ of bound-to-bound detectors and has a weaker bias voltage dependence. This is to be expected since the photoexcited carriers in bound-to-continuum QWIPs are above the top of the barriers and, thus, readily escape before being recaptured. However, the bound-to-bound QWIP is quite different due to the necessity of field-assisted tunneling for excited photoelectrons to escape from the quantum well (Levine et al., 1992b; Choi, 1996).

7. DETECTIVITY

We can now determine the peak detectivity D_λ^* defined as (Levine et al., 1990a; Zussman et al., 1991)

$$D_\lambda^* = R_p^0 \frac{\sqrt{A\Delta f}}{i_n} \tag{19}$$

where A is the detector area and $\Delta f = 1$ Hz. This is done as a function of bias for a continuum (A), a bound (E), and a quasibound (F) QWIP in Fig. 21. (The dashed lines near the origin are extrapolations.) For all three samples, D^* has a maximum value at a bias between $V_b = -2$ and -3 V. Since these QWIPs all have different cutoff wavelengths, these maximum D^* values cannot be simply compared. To facilitate this comparison, we note that the dark current has been demonstrated to follow an exponential law (Levine et al., 1990a, 1992a; Zussman et al., 1991) $I_d \propto e^{-(E_c - E_f)/kT}$ (where E_c is the cutoff energy $E_c = hc/\lambda_c$) over a wide range of both temperature and cutoff wavelength. Thus using $D^* \propto (R_p/i_n)$, we have

$$D^* = D_0^\lambda e^{E_c/2kT} \tag{20}$$

To compare the performance of these different QWIPs Levine et al. (1992b) plotted D^* against E_c on a log scale, as shown in Fig. 22 (Levine et al., 1990a, 1991b; Gunapala et al., 1991a; Zussman et al., 1991). The straight line fit the data very well, which is satisfying considering the samples have different doping densities N_D, different methods of crystal growth, different spectral widths $\Delta\lambda$, different excited states (bound, quasibound, and con-

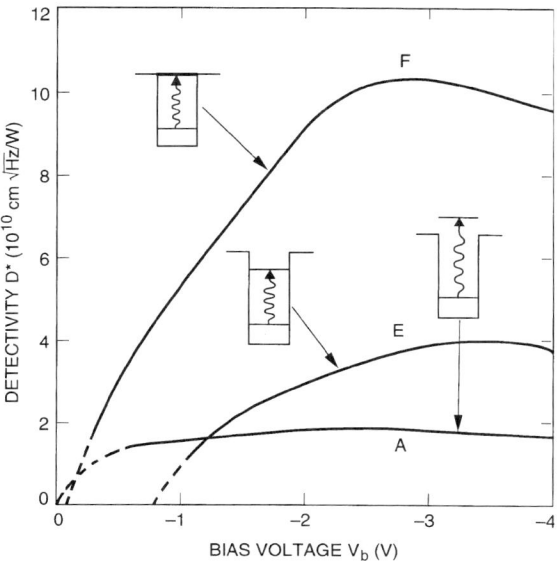

FIG. 21. Detectivity D^* (at $T = 77$ K) vs bias voltage V_b for samples A, E, and F. The inserts show the conduction band diagram (from Levine et al., 1992a).

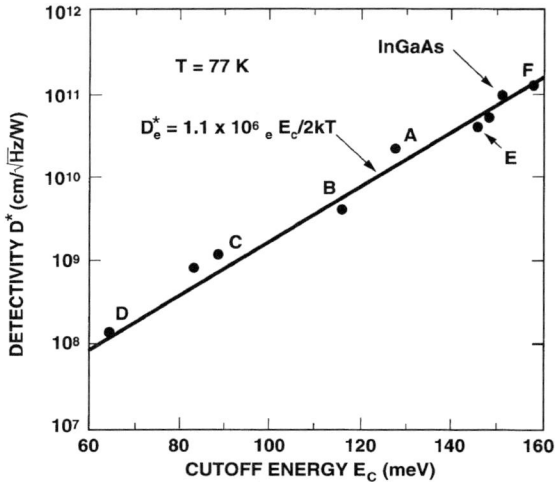

FIG. 22. Detectivity D^* (at $T = 77$ K) vs cutoff energy E_c for n-doped QWIPs. The straight line is the best fit to the measured data (from Levine et al., 1992a).

tinuum) and even, in one case, a different materials system (InGaAs) (Gunapala et al., 1991a). The best fit for $T = 77$ K detectivities of n-doped QWIPs is

$$D_e^* = 1.1 \times 10^6 e^{E_c/2kT} \text{ cm } \sqrt{\text{Hz}}/\text{W} \qquad (21)$$

Another useful figure of merit is blackbody responsivity and detectivity R_B and D_B^* and can be written as

$$D_B^* = R_B \frac{\sqrt{A\Delta f}}{i_n} \qquad (22)$$

with

$$R_B = \frac{\int_{\lambda_1}^{\lambda_2} R(\lambda) W(\lambda) \, d\lambda}{\int_{\lambda_1}^{\lambda_2} W(\lambda) \, d\lambda} \qquad (23)$$

It is worth noting that, for most applications, the blackbody responsivity R_B is reduced only a relatively small amount from the peak value R_p. Also note that, since QWIP dark current is mostly due to thermionic emission and thermionically assisted tunneling, unlike other detectors, QWIP detectivity increases nearly exponentially with the decreasing temperature as shown in Fig. 23.

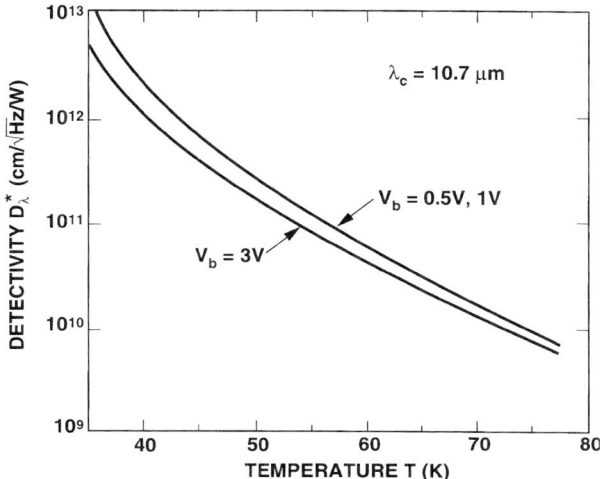

FIG. 23. Peak detectivity D_λ^* for a QWIP having cutoff wavelength of $\lambda_c = 10.7\,\mu$m as a function of temperature T for several bias voltages V_b (from Gunapala et al., 1990b).

IV. Light Coupling

QWIPs do not absorb radiation incident normal to the surface since the light polarization must have an electric field component normal to the superlattice (i.e., growth direction) to be absorbed by the confined carriers (Levine, 1993; Gunapala and Bandara, 1995). When the incoming light contains no polarization component along the growth direction, the matrix element of the interaction vanishes [i.e., $\vec{\varepsilon} \cdot \vec{p}_z = 0$ where $(\vec{\varepsilon})$ is the polarization and $(\vec{p}_z)$ is the momentum along growth direction (z)]. As a consequence, these detectors have to be illuminated through a 45° polished facet (Levine, 1993; Gunapala and Bandara, 1995). Clearly, this illumination scheme limits the configuration of detectors to linear arrays and single elements. For imaging, it is necessary to be able to couple light uniformly to two-dimensional arrays of these detectors.

Several different monolithic grating structures, such as linear gratings (Goosen and Lyon, 1985; Hasnain et al., 1989) two-dimensional (2-D) periodic gratings (Andersson and Lundqvist, 1991; Andersson et al., 1991a, 1991b; Sarusi et al., 1994b; Bandara et al., 1997), and random-reflectors (Sarusi et al., 1994a; Xing and Liu, 1996), have demonstrated efficient light coupling to QWIPs, and have made two-dimensional QWIP imaging arrays feasible (see Fig. 24). These gratings deflect the incoming light away from

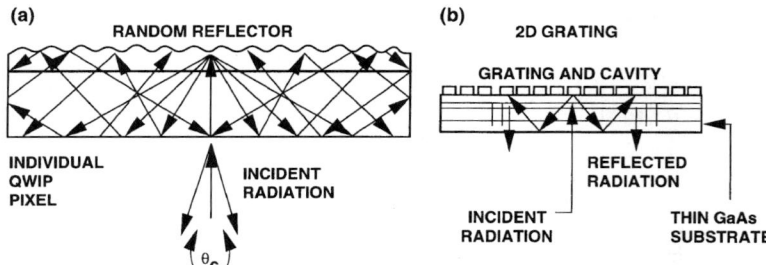

FIG. 24. (a) Schematic side view of a thin QWIP pixel with a random grating reflector. Ideally all the radiation is trapped except for a small fraction, which escapes through the escape cone. (b) Schematic diagram of 2-D periodic grating specifications. The grating features are spaced periodically along the x and y directions.

the direction normal to the surface, enabling intersubband absorption. These gratings were made of metal on top of each detector or crystallographically etched through a cap layer on top of the MQW structure. Normal incident light-coupling efficiency comparable to the light coupling efficiency of a 45° polished facet illumination was demonstrated using linear gratings (Goossen and Lyon, 1985; Hasnian et al., 1989).

1. RANDOM REFLECTORS

Random reflectors have demonstrated excellent optical coupling for individual QWIPs as well as for large area FPAs (Sarusi et al., 1994a; Xing and Liu, 1996; Gunapala et al., 1997a, 1997b). It has been shown that many more passes of infrared light (Fig. 24a), and significantly higher absorption, can be achieved with a randomly roughened reflecting surface. By careful design of surface texture randomization (with a three-level random reflector), an enhancement factor of 8 in responsivity compared to 45° illumination was demonstrated experimentally (Sausi et al., 1994a). The random structure on top of the detector prevents the light from being diffracted normally backward after the second bounce, as happens in the case of 2-D periodic grating (see Fig. 24b). Naturally, thinning down the substrate enables more bounces of light and, therefore, higher responsivity (Sarusi et al., 1994a; Gunapala et al., 1997a).

All these gratings were fabricated on the detectors by using standard photolithography and selective dry etching. The advantage of the photolithograpic process is its ability to accurately control the feature size and preserve the pixel-to-pixel uniformity, which is a prerequisite for high-sensitivity imaging FPAs. However, the resolution of photolithography and accuracy of etching processes become key issues in producing smaller grating feature sizes. These feature sizes are proportionally scaled with the peak response wavelength of the QWIP. The minimum feature size of random reflectors implemented in 15 and 9 μm cutoff FPAs were 1.25 and 0.6 μm, respectively (Gunapala et al., 1997a, 1997b). Thus, random reflectors of the 9-μm cutoff FPA were less sharp and had fewer scattering centers compared to random reflectors of the 15-μm cutoff FPA. These less sharp features in random gratings lowered the light coupling efficiency than expected. Therefore, it could be advantageous to utilize a 2-D periodic grating for light coupling in shorter wavelength QWIPs. However, one can avoid this problem by designing a random reflector which has features similar to 2-D periodic gratings (Xing and Liu, 1996).

2. TWO-DIMENSIONAL PERIODIC GRATINGS

Detailed theoretical analysis has been carried out on both linear and 2-D periodic gratings on QWIPs. In 2-D gratings, the periodicity of the grating repeats in two perpendicular directions on the detector plane, leading to the absorption of both polarizations of incident infrared radiation. Also, experiments have been carried out for 2-D grating coupled QWIP detectors designed for wavelengths $\lambda \sim 9$ μm (Andersson and Lundqvist, 1991; Andersson et al., 1991a, 1991b; Bandara et al., 1997) and $\lambda \sim 16-17$ μm (Sarusi et al., 1994b). A factor of 2–3 responsivity enhancement relative to the standard 45° polished facet illumination was observed for large area mesas (500 × 500 μm) with a total internal reflection optical cavity which can be created with an additional AlGaAs layer (Andersson and Lundqvist, 1991; Andersson et al., 1991a, 1991b) or with a thinned substrate (Sarusi et al., 1994b). This optical cavity is responsible for an extra enhancement factor of about 2 due to the total internal reflection from the AlGaAs layer or from the thinned substrate, as shown in Fig. 24. Due to the resonance nature, the light coupling efficiency of 2-D gratings depends strongly on the wavelength and, thus, exhibits narrow-bandwidth spectral responses. The normalized responsivity spectrum for 2-D periodic grating coupled QWIP samples (with six different grating periods, D, and a fixed groove depth) and for the standard 45° sample are shown in Fig. 25. Note the normalized spectral peak shifts from 7.5 to 8.8 μm as the grating period increases from $D = 2.2$ to 3.2 μm. These measurements were repeated for three groove depths. The grating peak wavelength λ_{gp} (where the grating enhancement is maximized) and the peak enhancement (enhancement at λ_{gp}) associated with each grating period was obtained by normalizing the absolute spectral responsivity of the grating detectors relative to the 45° detector sample. As

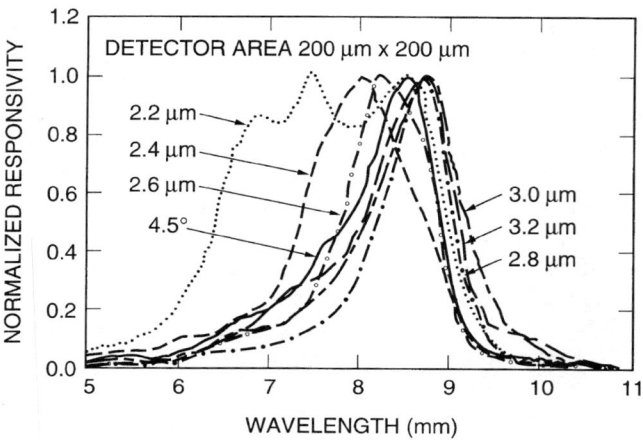

FIG. 25. Measured normalized responsivity spectra as a function of grating period D vary from 2.2 to 3.2 μm. The solid curve represents responsivity spectra of the same QWIP with 45° polished edge (from Bandara et al., 1997).

expected from the theory, measured λ_{gp} linearly depends on grating period and it is independent of the groove depth of the grating (Bandara et al., 1997). Figure 26 shows experimental responsivity enhancement due to 2-D grating at λ_{gp} for each grating period with different groove depths. One sample shows enhancement up to a factor of 3.5 (curves a and b in Fig. 26) depending on the grating period, while the other two samples show no enhancement and no dependence on the grating period. This high enhancement factor was measured in a similar (same gratings and groove depth) sample with different detector area. Scanning electron microscope (SEM) pictures of two samples, associated with curves c and d of Fig. 26, show apparent distortion in the features of the gratings (Bandara et al., 1997). This can be attributed to the partial contact between the grating mask and the wafer during the photolithography.

3. CORRUGATED STRUCTURE

Chen et al. (1997) demonstrated a new light-coupling geometry for QWIPs based on total internal reflection at the corrugated surface created within a detector pixel (Choi et al., 1998a and 1998b). In these structures, linear V-grooves are chemically etched through the active detector region down to the bottom contact layer to create a collection of angled facets within a single detector pixels as shown in Fig. 27a. These facets deflect normally

4 QUANTUM WELL INFRARED PHOTODETECTOR FOCAL PLANE ARRAYS   241

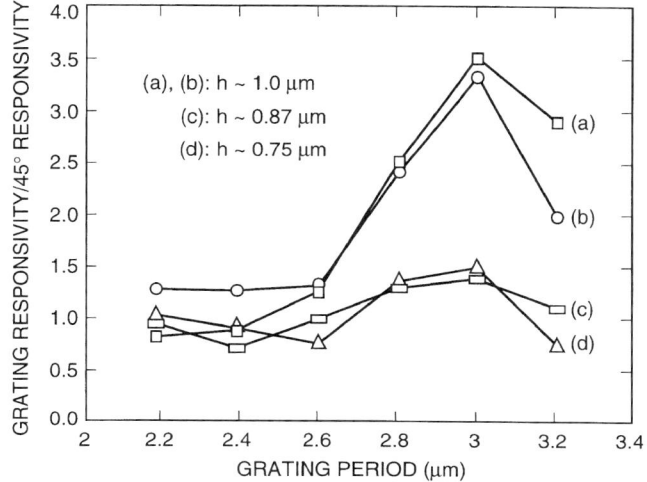

FIG. 26. The experimental responsivity enhancement at λ_{gp} for each grating period with different groove depths. Curves a and b represent gratings with same groove depth but different in detector area (a, $200 \times 200\,\mu m^2$, and b, $400 \times 400\,\mu m^2$). Curves c and d represent $200 \times 200\,\mu m^2$ area detectors with different groove depths (from Bandara et al., 1997).

incident light into the remaining QWIP active area through total internal reflection. For certain chemical solutions, such as $1H_2SO_4:8H_2O_2:10H_2O$, the etching rate is different for different crystallographic planes. As a result, triangular wires are created with sidewalls inclined around 54° with (100) surface along $(0, 1, \bar{1})$ the plane. In practice, the angle was found to be at 50° (Choi et al., 1998a, 1998b). See Fig. 27b.

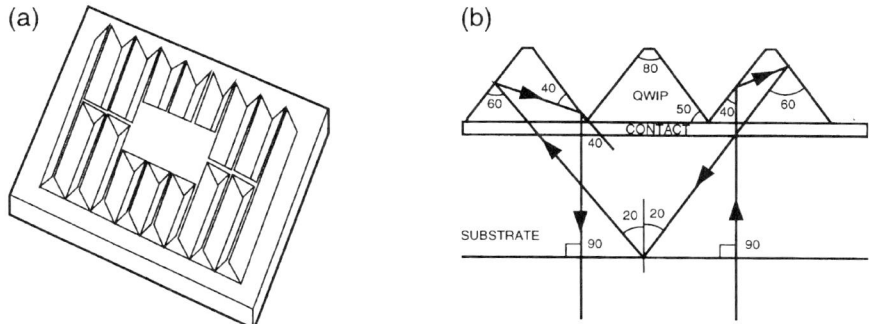

FIG. 27. (a) The 3-D perspective of a corrugated QWIP detector pixel and (b) the ray diagram in the side view (from Choi et al., 1998b).

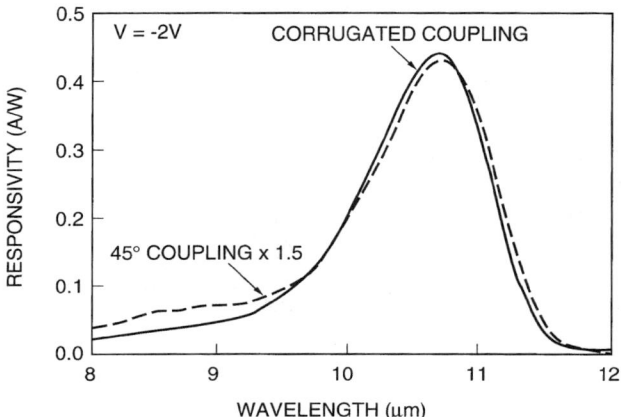

FIG. 28. Spectral responsivity of the corrugated QWIP and the same QWIP using 45° edge coupling (from Choi et al., 1998b).

Since the corrugated QWIP structure exposes the active layers of the detector, top metal contact cannot be deposited directly on the detector pixel. Two different approaches to addressing this problem were considered: (i) to leave out an unetched area for the contact, as illustrated in Fig. 27a, and (ii) isolate the active layers with a dielectric such as polymide. One of the main advantages of this light-coupling technique is reduction in pixel dark current due the smaller QWIP active area. In addition, the coupling scheme does not show significant wavelength and size dependence, and is therefore suitable for multicolor or broadband QWIPs. Figure 28 shows the spectral responsivity of the corrugated QWIP under normal incident illumination from the back of the wafer. For comparison, spectral responsivity of the same QWIP with 45° edge coupling is also illustrated in Fig. 28. It clearly shows that the corrugated structure does not change the intrinsic absorption line shape which is represented by the 45° edge-coupling response, but increases the magnitude by a factor of 1.5. Also, this structure shows about 2.6 factor lower dark current due to the less active material, resulting in about 2.4 times improved D^* (Choi et al., 1998a, 1998b).

4. MICROLENSES

The general concept of utilizing microlenses with QWIPs is to increase the signal output or decrease the dark current or both per detecting pixel. Use of a microlens to concentrate the photon flux will lead to a much smaller detector area requirement. Since the dark current is linearly related to the detector area, the dark current is decreased. Thus, the goal of increasing the D^* is achieved. The reduction of dark current can be expressed as the ratio of the area gain, G_A (Pool et al., 1998). If d is the diameter of the focal spot and D is the pixel diameter, then $G_A = (D/d)^2 = (Dn_f f/F)^2$, where n_f is the index of refraction, f is the f number of the optical system, and F is the focal length of the microlens. For example, if we let $D = 2.44$ λf, which is the diameter of an Airy disk, then at 15 μm, $D \approx 73$ μm. If we select $F = 125$ μm, then $d \approx 20$ μm. The dark current will improve by a factor of 13. If $F = 75$ μm then $d \approx 12$ μm, the dark current will improve by a factor of 37.

Monolithically integrated microlens and QWIP were demonstrated by Pool et al. (1998). Four-level binary optic microlenses were fabricated by standard contact lithography, using a two-mask photoresist pattern for plasma etching. The two-etch step process for fabrication of the four-level diffractive optics lenses created phase steps $d = \lambda/[2^2(n_r - 1)]$, where n_r is the index of refraction of GaAs, and λ is the wavelength of the incident light for which the lens is optimized. Since these lenses were designed for incident light of 15 μm the smallest feature size (outer zones) was approximately 4 μm, making lithography relatively straightforward. Fabrication errors that resulted were level misalignment, etch depth errors, and limited resolution of outer zone features. The lenses were first fabricated and the QWIPs subsequently fabricated and aligned to the lenses through use of an infrared backside aligner to within 1-μm accuracy. The theoretical efficiency of a four-level diffractive optic lens in the scalar approximation is 81%. In this study microlenses were fabricated from 150×150 to 350×350 μm^2 for infrared radiation to be focused on 75×75 μm^2 QWIP detectors, optimized for a peak wavelength of 15 μm. In addition to the four-level lenses, continuous-relief lenses were fabricated by direct-write electron beam lithography followed by transfer etching technique (Pool et al., 1998). The process for fabrication of the continuous relief lens is given by the schematic representation in Fig. 29. Figure 30 shows the surface of a 250-μm lens.

Although the effect of the microlens is evident through an increase in responsivity of approximately 2.7, the efficiency is far lower than a simple analysis would predict. The experiment was repeated with the continuous relief microlenses and the results were similar, even though the efficiency of the continuous lenses was likely higher than the four-level lenses. The poor performance of the microlens–QWIP combination can be understood by analyzing the diffraction characteristics of the grating. The lens produces a cone of converging waves that are incident on the grating. Because the grating was optimized for normal incidence, the coupling of the angled waves into the desired polarization for QWIP absorption was severely

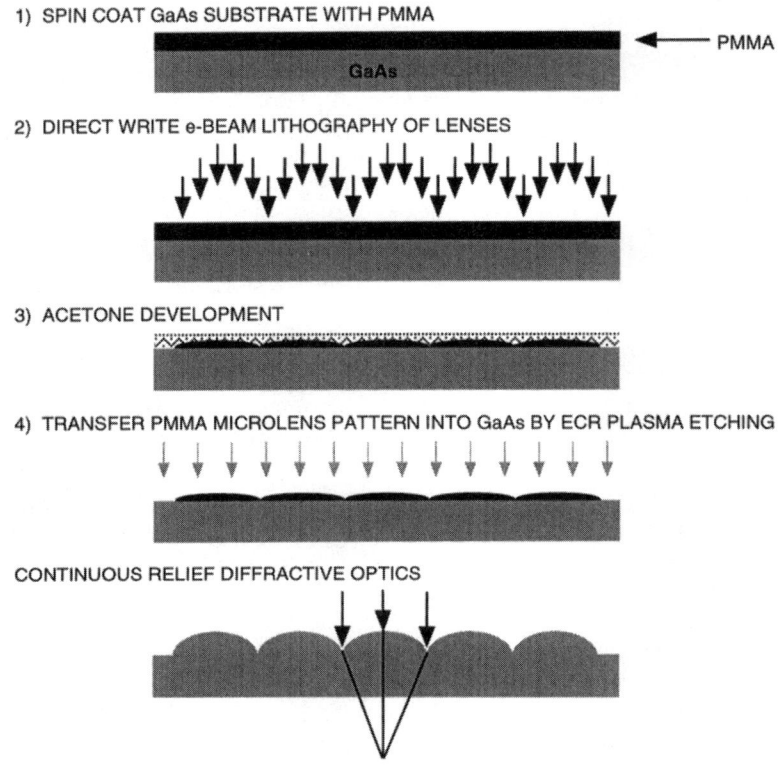

FIG. 29. Schematic representation of the fabrication process for continuous relief microlenses (from Pool et al., 1998).

degraded. There are two factors that influence the optical coupling: (i) the angles of the diffracted waves (grating orders) and (ii) the efficiencies of the diffracted orders. The angles of the diffracted orders θ_m can be found directly from the grating equation,

$$\sin(\theta_m) = m\frac{\lambda_{\text{GaAs}}}{\Lambda} - \sin(\theta_{\text{inc}}), \qquad m = 0, \pm 1, \pm 2 \qquad (24)$$

where θ_{inc} is angle of incidence, λ_{GaAs} is the wavelength inside the GaAs ($\lambda/3.1$), and $\Lambda = 4.85\,\mu\text{m}$ is the grating period.

The efficiencies of the diffracted orders must be found using a rigorous electromagnetic analysis technique. In this study, rigorous coupled-wave

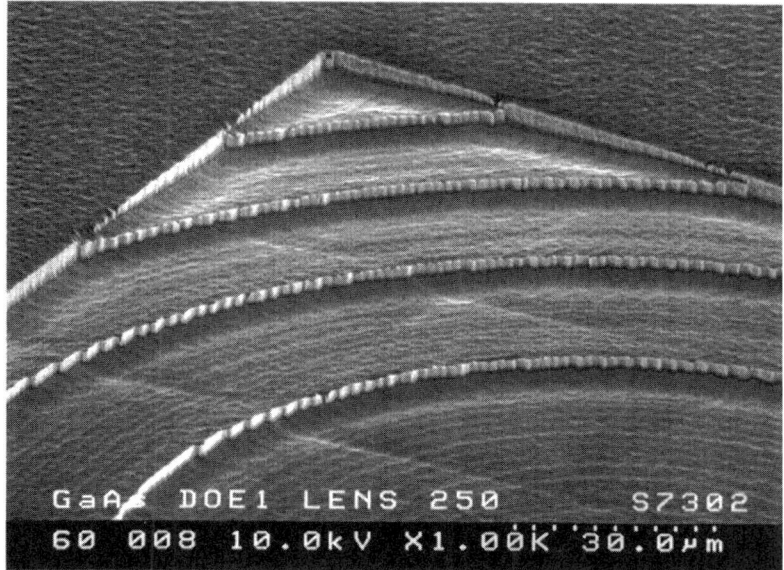

FIG. 30. Atomic force microscopy (AFM) image of a 250-μm microlens in PMMA (from Pool et al., 1998).

analysis was used to determine the efficiency of the grating as a function of incident angle and wavelength. After the angles and efficiencies of the diffracted orders were found, the fields inside the quantum well stack were calculated (Pool et al., 1998). Because the QWIP can only absorb the component of the field perpendicular to the quantum well layers $E_\perp$, the integral of $|E_\perp|^2$ inside the quantum well region was taken as a measure of QWIP coupling efficiency. At 15 μm, the coupling is strongly peaked around normal incidence even though the diffraction efficiencies are not well optimized. At a shorter wavelength of 14.5 μm, the period to wavelength ratio is larger, allowing both the +1 and −1 orders to continue to propagate up to a few degrees off-normal. This broadens the angular response, but because the angles of the diffracted orders are smaller, they do not couple as effectively into the QWIP and the peak response is reduced.

The focused field from the lens is actually not a plane wave, but it can be Fourier decomposed into an angular spectrum of plane waves. For a given lens, there is certainly an optimum grating period that maximizes the integrated product of the focused-field angular spectrum and the grating angular response. It is unlikely, however, that a lens–grating combination will achieve the same responsivity as a grating optimized for normally incident plane-wave illumination. Furthermore, for the lens–grating combination to effectively produce a higher signal-to-noise ratio, the area of the QWIP mesa must be reduced. In this situation, there may be too few grating periods for efficient diffraction. Although the microlens concept is not well suited for a QWIP detector, the concept described here can be soundly applied to normal incident absorbing infrared detectors (Pool et al., 1998).

V. Imaging Focal Plane Arrays

There are many ground-based and space-borne applications that require long-wavelength, large, uniform, reproducible, low-cost, low $1/f$ noise, low-power-dissipation, and radiation-hard infrared FPAs. For example, the absorption lines of many gas molecules, such as ozone, water, carbon monoxide, carbon dioxide, and nitrous oxide occur in the wavelength region from 3 to 18 μm. Thus, infrared imaging systems that operate in the LWIR and VLWIR regions are required in many space applications such as monitoring the global atmospheric temperature profiles, relative humidity profiles, cloud characteristics, and the distribution of minor constituents in the atmosphere, which are being planned for NASA's Earth Observing System (Chahine, 1990). In addition, 8- to 15-μm FPAs would be very useful in detecting cold objects such as ballistic missiles in midcourse (when the hot rocket engine is not burning most of the emission peaks in the 8- to 15-μm infrared region; Duston, 1995). The GaAs–Al$_x$Ga$_{1-x}$As material system enables the quantum well shape to be tweaked over a range wide enough to enable light detection at wavelengths longer than $\sim 6\,\mu$m. Thus, a GaAs-based QWIP is a potential candidate for such space-borne and ground-based applications and many research groups (Bethea et al., 1991, 1992, 1993; Kozlowski et al., 1991a; Faska et al., 1992; Beck et al., 1994; Gunapala et al., 1997a, 1997b, 1998a; Andersson et al., 1997; Choi et al., 1998a; Breiter et al., 1998) have already demonstrated large uniform FPAs of QWIPs tuned to detect light at wavelengths from 6 to 25 μm in the GaAs–Al$_x$Ga$_{1-x}$As material system.

1. Effect of Nonuniformity

The general figure of merit that describes the performance of a large imaging array is the noise equivalent temperature difference NEΔT. NEΔT is the minimum temperature difference across the target that would produce a signal-to-noise ratio of unity and it is given by (Kingston, 1978; Zussman *et al.*, 1991)

$$\text{NE}\Delta T = \frac{\sqrt{A\Delta f}}{D_B^*(dP_B/dT)} \quad (25)$$

where D_B^* is the blackbody detectivity [defined by Eq. (22)] and (dP_B/dT) is the change in the incident integrated blackbody power in the spectral range of detector with temperature. The integrated blackbody power P_B, in the spectral range from λ_1 to λ_2, can be written as

$$P_B = A \sin^2\left(\frac{\theta}{2}\right) \cos\phi \int_{\lambda_1}^{\lambda_2} W(\lambda)\, d\lambda \quad (26)$$

where θ, ϕ, and $W(\lambda)$ are the optical field of view, angle of incidence, and blackbody spectral density, respectively, and are defined by Eqs. (7) and (8) in Subsection 3 of Section III. Before discussing the array results, it is also important to understand the limitations on the FPA imaging performance due to pixel nonuniformities (Levine, 1993). This point was discussed in detail by Shepherd (1988) for the case of PtSi infrared FPAs (Mooney *et al.*, 1989), which have low response, but very high uniformity. The general figure of merit to describe the performance of a large imaging array is the noise-equivalent temperature difference NEΔT, including the spatial noise, which has been derived by Shepherd (1988), and given by

$$\text{NE}\Delta T = \frac{N_n}{dN_b/dT_b} \quad (27)$$

where T_b is the background temperature, and N_n is the total number of noise electrons per pixel, given by

$$N_n^2 = N_t^2 + N_b + u^2 N_b^2 \quad (28)$$

The photoresponse-independent temporal noise electrons are N_t, the shot noise electrons from the background radiation are N_b, and the residual nonuniformity after correction by the electronics is u. The temperature derivative of the background flux can be written to a good approximation as

$$\frac{dN_b}{dT_b} = \frac{hcN_b}{k\bar{\lambda} T_b^2} \quad (29)$$

where $\bar{\lambda} = (\lambda_1 + \lambda_2)/2$ is the average wavelength of the spectral band between λ_1 and λ_2. When temporal noise dominates, NEΔT reduces to Eq. (25). In the case where residual nonuniformity dominates, Eqs. (27) and (29) reduce to

$$\text{NE}\Delta T = \frac{u\bar{\lambda}T_b^2}{1.44} \tag{30}$$

The units of the constant is centimeters times degrees Kelvin, $\bar{\lambda}$ is in centimeters, and T_b is in degrees Kelvin. Thus, in this spatial noise limited operation NEΔT $\propto u$ and higher uniformity means higher imaging performance. Levine (1993) has shown as an example that taking $T_b = 300$ K, $\bar{\lambda} = 10\,\mu$m, and $u = 0.1\%$ leads to NEΔT $= 63$ mK, while an order of magnitude uniformity improvement (i.e., $u = 0.01\%$) gives NEΔT $= 6.3$ mK. Using the full expression of Eq. (28), Levine (1993) calculated NEΔT as a function of D^*, as shown in Fig. 31. It is important to note that when $D^* \geqslant 10^{10}$ cm $\sqrt{\text{Hz}/\text{W}}$, the performance is uniformity limited and thus essentially independent of the detectivity; that is, D^* is not the relevant figure of merit (Grave and Yariv, 1992).

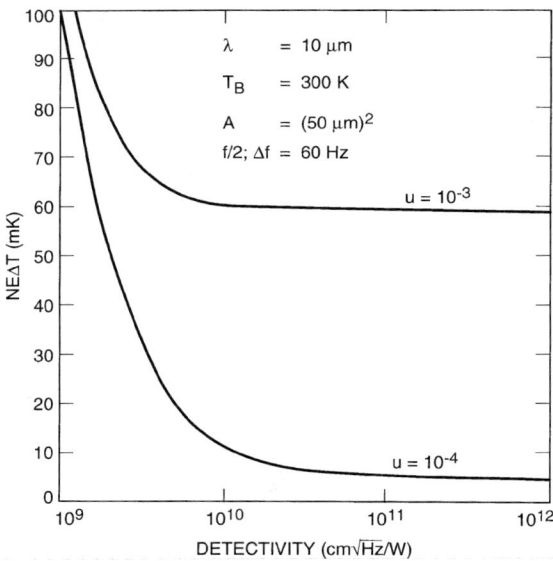

FIG. 31. Noise-equivalent temperature difference NEΔT as a function of detectivity D^*. The effects of nonuniformity are included for $u = 10^{-3}$ and 10^{-4}. Note that for $D^* > 10^{10}$ cm $\sqrt{\text{Hz}/\text{W}}$ detectivity is not the relevant figure of merit for FPAs (from Levine, 1993).

2. 128 × 128 VLWIR IMAGING CAMERA

By carefully designing the quantum well structure as well as the light coupling (as discussed in Section IV) to the detector, it is possible to optimize the material to an optical response in the desired spectral range, determine the spectral response shape, as well as reduce the parasitic dark current, and therefore increase the detector impedance. Generally, to tailor the QWIP spectral response to the VLWIR spectral region the barrier height should be lowered and the well width increased relative to the shorter-cutoff-wavelength QWIPs. For a detailed analysis of design and performance optimization of VLWIR QWIPs, see Sarusi *et al.* (1994c).

The first VLWIR FPA camera, demonstrated by Gunapala *et al.* (1997a), consisted of bound-to-quasibound QWIPs, as shown in Fig. 32. Samples were grown using MBE and their well widths L_w vary from 65 to 75 Å, while barrier widths are approximately constant at $L_b = 600$ Å. These QWIPs consisted of 50 periods of doped ($N_D = 2 \times 10^{17}$ cm^{-3}) GaAs quantum

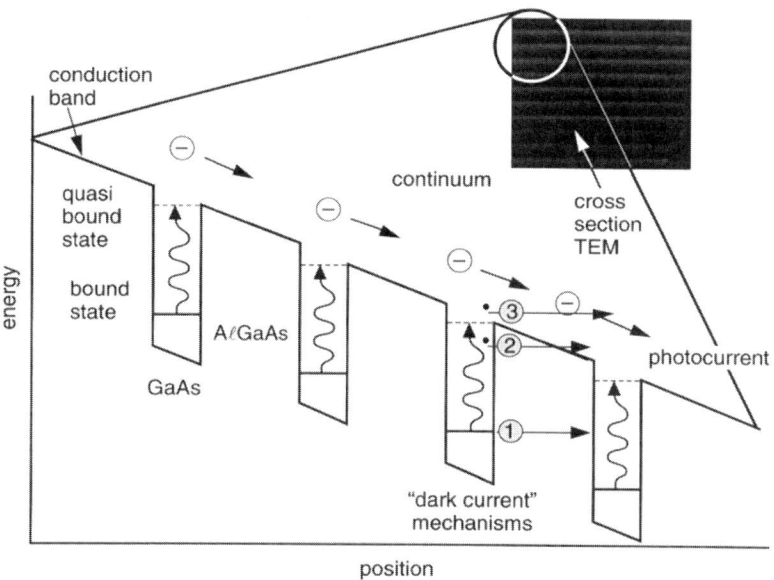

FIG. 32. Schematic diagram of the conduction band in a bound-to-quasibound QWIP in an externally applied electric field. Absorption of infrared photons can photoexcite electrons from the ground state of the quantum well into the continuum, causing a photocurrent. Three dark current mechanisms are also shown: ground state tunneling (1), thermally assisted tunneling (2), and thermionic emission (3). The inset shows a cross-section transmission electron micrograph of a QWIP sample (from Gunapala *et al.*, 1997b).

wells, and undoped $Al_xGa_{1-x}As$ barriers. Very low doping densities were used to minimize the parasitic dark current. The Al molar fraction in the $Al_xGa_{1-x}As$ barriers varies from $X = 0.15$ to 0.17 (corresponding to cutoff wavelengths of 14.9 to 15.7 μm). These QWIPs had peak wavelengths from 14 to 15.2 μm, as shown in Fig. 33. The peak quantum efficiency was 3% (lower quantum efficiency is due to the lower well doping density) for a 45° double pass.

Four device structures were grown on 3-in. GaAs wafers and each wafer processed into 35 128 × 128 FPAs. An expanded corner of a FPA is shown in Fig. 34. The pixel pitch of the FPA is 50 μm and the actual pixel size is 38 × 38 μm². Two level random reflectors used to improve the light coupling, can be seen on top of each pixel. These random reflectors, which were etched to a depth of half a peak wavelength in GaAs using reactive-ion etching, had a square profile. These reflectors are covered with Au–Ge and Au (for ohmic contact and reflection), and In bumps are evaporated on top for silicon multiplexer hybridization. A single QWIP FPA was chosen from sample number 7060 (cutoff wavelength of this sample is 14.9 μm) and bonded to a silicon readout multiplexer. The FPA was back-illuminated through the flat, thinned substrate. This initial array gave excellent images with 99.9% of the pixels working, demonstrating the high yield of GaAs technology. As mentioned earlier, this high yield is due to the excellent GaAs

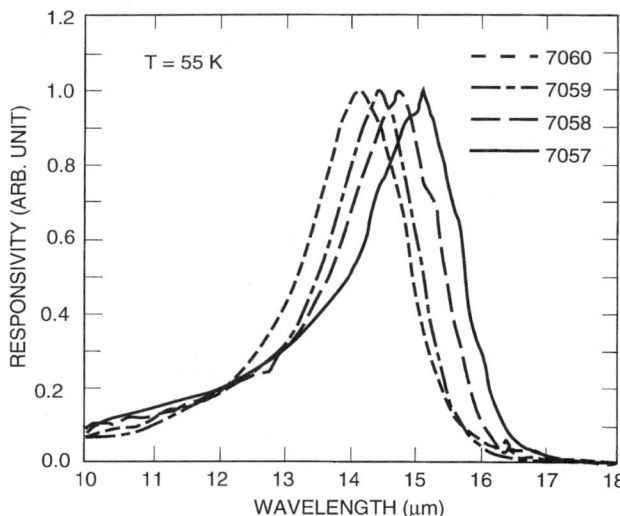

FIG. 33. Normalized responsivity spectra of four bound-to-quasibound VLWIR QWIP FPA samples at temperature $T = 55$ K (from Gunapala et al., 1997a).

FIG. 34. An expanded corner of a 128 × 128 QWIP FPA showing two level random reflectors on pixels (38 × 38 μm^2) (from Gunapala et al., 1997a).

growth uniformity and the mature GaAs processing technology. The uniformity after two-point correction was $u = 0.03\%$ (Gunapala et al., 1997a).

Video images were taken at various frame rates varying from 20 to 200 Hz with $f/2.6$ KRS-5 optics at temperatures high as $T = 45$ K, using a multiplexer having a charge capacity of 4×10^7 electrons. However, the total charge capacity was not available during the operation, since the charge storage capacitor was partly filled to provide the high operating bias voltage required by the detectors (i.e., $V_b = -3$ V). Figure 35 shows an image of a man's face with NEΔT $= 30$ mK (Gunapala et al., 1997a). Note that these initial unoptimized FPA results are far from optimum.

3. 256 × 256 LWIR IMAGING CAMERA

Infrared imaging systems that work in the 8- to 12-μm (LWIR) band have many applications, including night vision, navigation, flight control, and

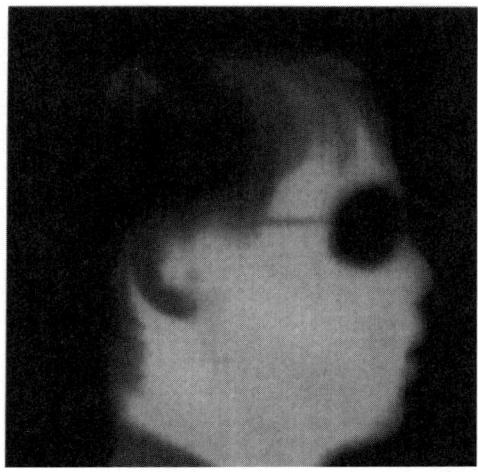

FIG. 35. One frame from a 15-μm QWIP video image of a man's face with NEΔT = 30 mK (from Gunapala et al., 1997a).

early warning systems. Several research groups have demonstrated (Bethea et al., 1991, 1992, 1993; Kozlowski et al., 1991a, 1991b; Levine et al., 1991d; Asom et al., 1991; Swaminathan et al., 1992) the excellent performance of QWIP arrays. For example, Faska et al. (1992) obtained very good images using a 256 × 256 bound-to-miniband MQW FPA. The first 256 × 256 LWIR handheld imaging camera was demonstrated by Gunapala et al. (1997b). The device structure of this FPA consisted of a bound-to-quasibound QWIP containing 50 periods of a 45-Å well of GaAs (doped $n = 4 \times 10^{17}$ cm^{-3}) and a 500-Å barrier of $Al_{0.3}Ga_{0.7}As$. Ground-state electrons are provided in the detector by doping the GaAs well layers with Si. This photosensitive MQW structure is sandwiched between 0.5-μm GaAs top and bottom contact layers doped $n = 5x10^{17}$ cm^{-3}, grown on a semi-insulating GaAs substrate by MBE. Then a 0.7-μm-thick GaAs cap layer on top of a 300-Å $Al_{0.3}Ga_{0.7}As$ stop-etch layer was grown in situ on top of the device structure to fabricate the light coupling optical cavity.

The detectors were back illuminated through a 45° polished facet, as described earlier, and a responsivity spectrum is shown in Fig. 36. The responsivity of the detector peaks at 8.5 μm and the peak responsivity (R_p) of the detector is 300 mA/W at bias $V_B = -3$ V. The spectral width and the cutoff wavelength are $\Delta\lambda/\lambda = 10\%$ and $\lambda_c = 8.9$ μm, respectively. The measured absolute peak responsivity of the detector is small, up to about $V_B = -0.5$ V. Beyond that it increases nearly linearly with bias reaching

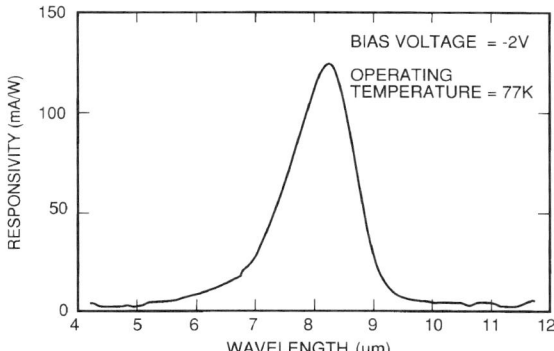

FIG. 36. Responsivity spectrum of a bound-to-quasibound LWIR QWIP test structure at temperature $T = 77$ K. The spectral response peak is at 8.5 μm and the long wavelength cutoff is at 8.9 μm (from Gunapala et al., 1997a).

$R_p = 380$ mA/W at $V_B = -5$ V. This type of behavior of responsivity versus bias is typical for a bound-to-quasibound QWIP. The peak quantum efficiency was 6.9% at bias $V_B = -1$ V for a 45° double pass. The lower quantum efficiency is due to the lower well doping density (5×10^{17} cm^{-3}) as it is necessary to suppress the dark current at the highest possible operating temperature.

After the random reflector array was defined by the lithography and dry etching, the photoconductive QWIPs of the 256 × 256 FPAs were fabricated by wet chemical etching through the photosensitive GaAs–Al$_x$Ga$_{1-x}$As MQW layers into the 0.5-μm-thick doped GaAs bottom contact layer. The pitch of the FPA is 38 μm and the actual pixel size is 28 × 28 μm^2. The random reflectors on top of the detectors were then covered with Au–Ge and Au for ohmic contact and reflection. A single QWIP FPA was chosen and hybridized (via indium bump-bonding process) to a 256 × 256 complementary metal oxide semiconductor (CMOS) readout multiplexer (Amber AE-166) and biased at $V_B = -1.0$ V. The FPA was back-illuminated through the flat thinned substrate membrane (thickness ≈ 1300 Å). This array gave excellent images with 99.98% of the pixels working (number of dead pixels ≈ 10), demonstrating the high yield of GaAs technology (Gunapala et al., 1997b). The measured NEΔT (mean value) of the FPA at an operating temperature of $T = 70$ K, bias $V_B = -1$ V, and 300 K background is 15 mK. This agrees reasonably with our estimated value of 8 mK, based on test structure data. The peak quantum efficiency of the FPA was 3.3% (lower FPA quantum efficiency is attributed to 54% fill factor and 90% charge injection efficiency) and this corresponds to an average of three

FIG. 37. Picture of the first 256 × 256 handheld long-wavelength QWIP camera (QWIP RADIANCE™) (from Gunapala et al., 1997a).

passes of infrared radiation (equivalent to a single 45° pass) through the photosensitive MQW region.

A 256 × 256 QWIP FPA hybrid was mounted onto a 250-mW integral Sterling closed-cycle cooler assembly and installed into an Amber RADIANCE 1™ camera body to demonstrate a handheld LWIR camera (shown in Fig. 37). The camera is equipped with a 32-bit floating-point digital signal processor combined with multitasking software, providing the speed and power to execute complex image processing and analysis functions inside the camera body itself. The other element of the camera is a 100-mm-focal-length germanium lens, with a 5.5° field of view. It is designed to be transparent in the 8- to 12-μm wavelength range to be compatible with the QWIP's 8.5-μm operation. The digital acquisition resolution of the camera is 12 bits, which determines the instantaneous dynamic range of the camera (i.e., 4096). However, the dynamic range of QWIP is 85 dB. Its nominal power consumption is less than 50 W.

4. 640 × 486 LWIR Imaging Camera

In this section, we discuss the demonstration of the 640 × 486 LWIR imaging camera by Gunapala et al. (1998a). Although random reflectors have achieved relatively high quantum efficiencies with large-area test device structures, it is not possible to achieve similar high quantum efficiencies with random reflectors on small-area FPA pixels due to the reduced width-to-height aspect ratios. In addition, due to fabrication difficulties of random

4 QUANTUM WELL INFRARED PHOTODETECTOR FOCAL PLANE ARRAYS 255

reflector for shorter wavelength FPAs, as described in Subsection 1 of Section IV, a 2-D periodic grating reflector was fabricated for light coupling of this 640 × 486 QWIP FPA.

After the 2-D grating array was defined by the photolithography and dry etching, the photoconductive QWIPs of the 640 × 486 FPAs were fabricated by wet chemical etching through the photosensitive GaAs–Al$_x$Ga$_{1-x}$As MQW layers into the 0.5-μm-thick doped GaAs bottom contact layer. The pitch of the FPA is 25 μm and the actual pixel size is 18 × 18 μm^2. The 2-D gratings on top of the detectors were then covered with Au–Ge and Au for ohmic contact and reflection. Figure 38 shows 12 processed QWIP FPAs on a 3-in. GaAs wafer (Gunapala et al., 1998a). Indium bumps were then evaporated on top of the detectors for silicon readout circuit (ROC) hybridization. A single QWIP FPA was chosen and hybridized (via indium bump-bonding process) to a 640 × 486 direct injection silicon readout multiplexer (Amber AE-181) and biased at $V_B = -2.0$ V. Figure 39 shows a size comparison of this large area long-wavelength QWIP FPA to a quarter. At temperatures below 70 K, the signal-to-noise ratio of the system is limited by multiplexer readout noise, and shot noise of the photocurrent. At temperatures above 70 K, temporal noise due to the QWIP's higher dark

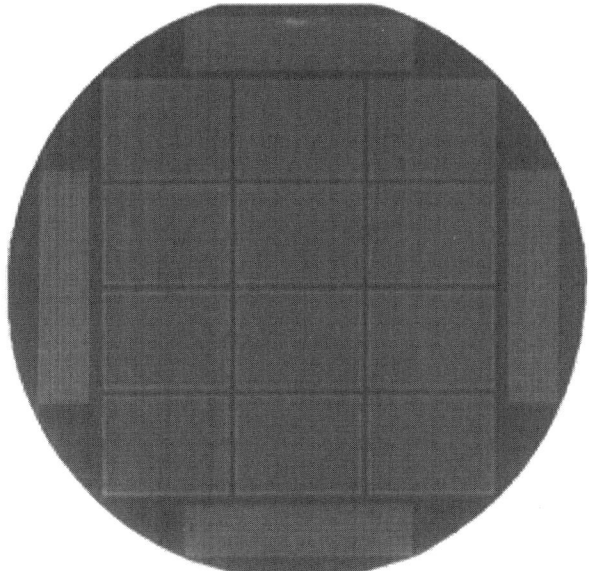

FIG. 38. Twelve 640 × 486 QWIP FPAs on a 3-in. GaAs wafer (from Gunapala et al., 1997a).

FIG. 39. A size comparison of the 640 × 486 long-wavelength QWIP FPA to a quarter (from Gunapala et al., 1997a).

current becomes the limitation. As mentioned earlier, this higher dark current is due to thermionic emission and thus causes the charge storage capacitors of the readout circuitry to saturate (Gunapala et al., 1998a). Since the QWIP is a high-impedance device, it should yield a very high charge injection coupling efficiency into the integration capacitor of the multiplexer. In fact, Bethea et al. (1993) demonstrated charge injection efficiencies approaching 90%. Charge injection efficiency can be obtained from (Bethea et al., 1993)

$$\eta_{inj} = \frac{g_m R_{det}}{1 + g_m R_{det}} \left[\frac{1}{1 + (j\omega C_{det} R_{det}/1 + g_m R_{det})} \right] \quad (31)$$

where g_m is the transconductance of the metal-oxide semiconductor field effect transistor (MOSFET) and is given by $g_m = eI_{det}/kT$. The differential resistance R_{det} of the pixels at -2 V bias is 5.4×10^{10} Ω at $T = 70$ K, and detector capacitance C_{det} is 1.4×10^{-14} F. The detector dark current $I_{det} = 24$ pA under the same operating conditions. According to Eq. (31) the charge injection efficiency $\eta_{inj} = 99.5\%$ at a frame rate of 30 Hz. The FPA was back-illuminated through the flat thinned substrate membrane (thickness ≈ 1300 Å). This thinned GaAs FPA membrane completely eliminated the thermal mismatch between the silicon CMOS readout multiplexer and the GaAs based QWIP FPA. Basically, the thinned GaAs-based QWIP

FPA membrane adapts to the thermal expansion and contraction coefficients of the silicon readout multiplexer. Thus, thinning has played an extremely important role in the fabrication of large-area FPA hybrids. In addition, this thinning has completely eliminated the pixel-to-pixel optical crosstalk of the FPA. This initial array gave very good images with 99.97% of the pixels working, demonstrating the high yield of GaAs technology. The operability was defined as the percentage of pixels having noise-equivalent differential temperature less than 100 mK at 300 K background with $f/2$ optics and, in this case, operability happened to be equal to the pixel yield.

We used the following equation to calculate NEΔT of the FPA

$$\text{NE}\Delta\text{T} = \frac{\sqrt{AB}}{D_B^*(dP_B/dT)\sin^2(\theta/2)} \tag{32}$$

where D_B^* is the blackbody detectivity, dP_B/dT is the derivative of the integrated blackbody power with respect to temperature, and θ is the field of view angle [i.e., $\sin^2(\theta/2) = (4f^2 + 1)^{-1}$, where f is the f-number of the optical system]. The background temperature $T_B = 300$ K, the area of the pixel $A = (18\ \mu\text{m})^2$, the f-number of the optical system is 2, and the frame rate is 30 Hz. Figure 40 shows the experimentally measured NEΔT histo-

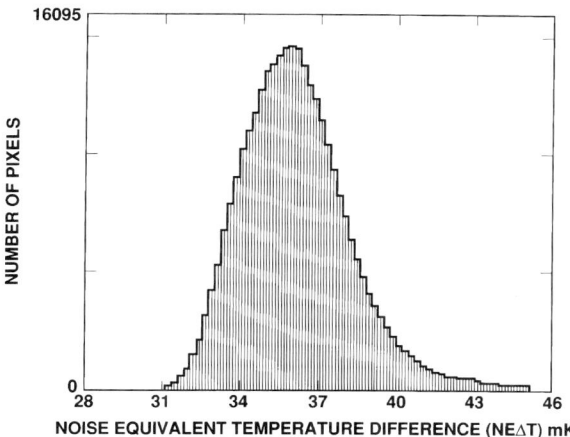

FIG. 40. Noise-equivalent temperature difference (NEΔT) histogram of the 311,040 pixels of the 640 × 486 array showing a high uniformity of the FPA. The uncorrected nonuniformity (= standard deviation/mean) of this unoptimized FPA is only 5.6% including 1% nonuniformity of ROC and 1.4% nonuniformity due to the cold stop not being able to give the same field of view to all the pixels in the FPA (from Gunapala et al., 1998a).

gram of the FPA at an operating temperature of $T = 70$ K, bias $V_B = -2$ V, and 300 K background with $f/2$ optics. The mean value of the NEΔT histogram is 36 mK (Gunapala et al., 1998a). This agrees reasonably well with our estimated value of 25 mK based on test structure data. The read noise of the multiplexer is 500 electrons. The 44% shortfall of NEΔT is mostly attributed to unoptimized detector bias (i.e., $V_B = -2$ V was used instead of optimum $V_B = -3$ V, based on detectivity data as a function of bias voltage), decrease in bias voltage across the detectors during charge accumulation (common in many direct-injection-type readout multiplexers), and read noise of the readout multiplexer. The experimentally measured peak quantum efficiency of the FPA was 2.3% (lower FPA quantum efficiency is attributed to 51% fill factor and 30% reflection loss from the GaAs back surface). Therefore, the corrected quantum efficiency of the focal plane detectors is 6.5% and this corresponds to an average of two passes of infrared radiation through the photosensitive MQW region.

A 640 × 486 QWIP FPA hybrid was mounted onto a 84-pin leadless chip carrier and installed into a laboratory dewar, which was cooled by liquid nitrogen to demonstrate a LWIR imaging camera. The FPA was cooled to 70 K by pumping on liquid nitrogen and the temperature was stabilized by regulating the pressure of gaseous nitrogen. The other element of the camera is a 100-mm-focal-length antireflection (AR) coated germanium lens, which gives a 9.2° × 6.9° field of view. It is designed to be transparent in the 8- to 12-μm wavelength range for compatibility with the QWIP's 8- to 9-μm operation. An Amber ProView™ image processing station was used to obtain clock signals for the readout multiplexer and to perform digital data acquisition and nonuniformity corrections. The digital data acquisition resolution of the camera is 12 bits, which determines the instantaneous dynamic range of the camera (i.e., 4096).

The measured mean NEΔT of the QWIP camera is 36 mK at an operating temperature of $T = 70$ K, bias $V_B = -2$ V, and 300 K background with $f/2$ optics. This is in good agreement with expected FPA sensitivity due to the practical limitations on charge handling capacity of the multiplexer, read noise, bias voltage, and operating temperature.

The uncorrected NEΔT nonuniformity (which includes a 1% nonuniformity of the ROC and a 1.4% nonuniformity due to the cold stop in front of the FPA not yielding the same field of view to all the pixels) of the 311,040 pixels of the 640 × 486 FPA is about 5.6% (= sigma/mean). The nonuniformity of the FPA after two-point (17° and 27°C) correction improves to an impressive 0.04%. As mentioned earlier, this high yield is due to the excellent GaAs growth uniformity and the mature GaAs processing technology. After two-point correction, measurements of the residual nonuniformity were made at temperatures ranging from 10°C (the cold temperature limit of the blackbody source) up to 40°C. The nonuniformity at each temperature was found by averaging 16 frames, calculating the standard deviation of the pixel-to-pixel variation of the 16-frame average and then dividing by the mean output, producing nonuniformity that may be reported as a percentage. For camera systems that have NEΔT of about 30 mK, the corrected image must have less than 0.1% nonuniformity to be standard television (TV) quality. Only at a temperature of 38°C did the camera's nonuniformity exceed the 0.1% nonuniformity threshold. Figure 41 shows residual nonuniformity plotted versus scene temperature. The 33°C window where the correction is below 0.1% is based on the measured data and one extrapolated data point at 5°C.

Video images were taken at a frame rate of 30 Hz at temperatures as high as $T = 70$ K using an ROC capacitor having a charge capacity of 9×10^6 electrons (the maximum number of photoelectrons and dark electrons that can be counted in the integration time of each detector pixel). Figure 42 show a frame of video image taken with this long-wavelength 640×486 QWIP camera. This image was taken in the night (around midnight) and it clearly shows where automobiles were parked during the day time. These high-resolution images are comparable to standard TV, and demonstrate high operability (i.e., 99.9%) and stability (i.e., lower residual uniformity and $1/f$ noise) of the 640×486 long-wavelength QWIP staring array camera (Gunapala *et al.*, 1998a).

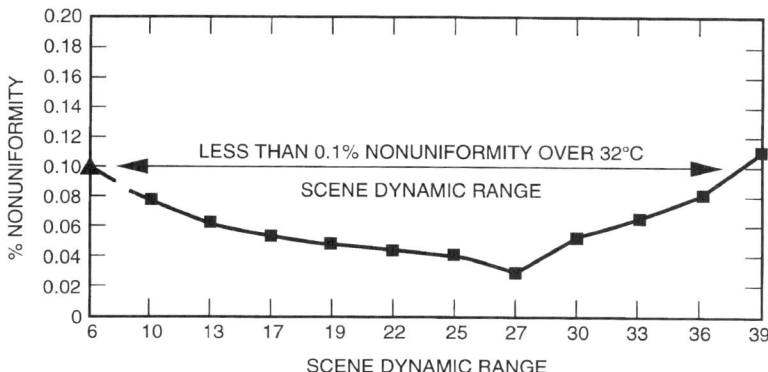

FIG. 41. Residual nonuniformity after two-point correction as a function of scene temperature. This corrected uniformity range is comparable to 3- to 5-μm IR cameras (from Gunapala *et al.*, 1998a).

FIG. 42. This picture was taken in the night (around midnight) and it clearly shows where automobiles were parked during the day time. This image demonstrates the high sensitivity of the 640 × 486 long-wavelength QWIP staring array camera (from Gunapala et al., 1998a).

5. Dualband (MWIR and LWIR) Detectors

There are several applications such as target recognition and discrimination that require monolithic mid- and long-wavelength dualband large-area, uniform, reproducible, low-cost, and low $1/f$ noise infrared FPAs. For example, a dualband FPA camera would provide the absolute temperature of the target, which is extremely important to the process of identifying temperature difference between targets, warheads, and decoys. The GaAs-based QWIP is a potential candidate for development of such a two-color FPAs. Until recently, the most developed and discussed two-color QWIP detector was the voltage-tunable two-stack QWIP. This device structure consists of two QWIP structures, one tuned for mid-wavelength detection and the other stack tuned for long-wavelength detection. This device structure utilizes the advantage of formation of electric field domains to select the response of one or the other detector (Grave et al., 1993; Mei et al., 1997). The difficulties associated with this type of two-color QWIP FPA are that these detectors need two different voltages to operate and long-wavelength sensitive segment of the device needs very high bias voltage (>8 V) to switch on the LWIR detection. A voltage tunable simplified QWIP structure consisting of three quantum wells was demonstrated by Tidrow and Bacher (1996). Due to a lower number of wells, the peak response wavelength of this detector can be tuned from 7.3 to 10 μm with smaller changes in the bias voltage. Chiang *et al.* (1996) also reported a multicolor voltage tunable triple-coupled QWIP using InGaAs–AlGaAs–GaAs for 8- to 12-μm detection. Another disadvantage of the voltage tunable scheme is that it does not provide simultaneous data from both wavelength bands.

Gunapala *et al.* (1998b) have developed a two-color, two-stack, QWIP device structure based on $In_xGa_{1-x}As$–$Al_yGa_{1-y}As$–GaAs material system for MWIR–LWIR detection (Foire *et al.*, 1994; Tidrow *et al.*, 1997). This structure can be processed in to dualband QWIP FPAs with dual or triple contacts to access the CMOS readout multiplexer. The device structure consists of a stack of 30 periods of LWIR MQW structure and another stack of 10 periods of MWIR MQW structure separated by a heavily doped 0.5-μm-thick intermediate GaAs contact layer. The first stack (LWIR) consist of 10 periods of 500-Å $Al_xGa_{1-x}As$ barrier and a GaAs well. This LWIR QWIP structure has been designed to have a bound-to-quasibound intersubband absorption peak at 8.5 μm, since the dark current of the device structure is expected to be dominated by the longer wavelength portion of the device structure. The second stack (MWIR) consist of 10 periods of 300-Å $Al_xGa_{1-x}As$ barrier and narrow $In_yGa_{1-y}As$ well sandwiched between two thin layers of GaAs. This MWIR QWIP structure was designed to have a bound-to-continuum intersubband absorption peak at 4.2 μm, since the photocurrent and dark current of the MWIR device structure are relatively small compared to the LWIR portion of the device structure. This two-color QWIP structure is then sandwiched between 0.5-μm GaAs top and bottom contact layers doped $n = 5 \times 10^{17}$ cm^{-3}. The whole device structure was grown on a semi-insulating GaAs substrate by MBE. Then a 1.0-μm-thick GaAs cap layer on top of a 300-Å $Al_{0.3}Ga_{0.7}As$ stop-etch layer is grown *in situ* on top of the device structure to fabricate the light coupling optical cavity.

The detectors were back-illuminated through a 45° polished facet, as described earlier, and a simultaneously measured responsivity spectrum of vertically integrated dualband QWIP is shown in Fig. 43. The responsivity of the MWIR detector peaks at 4.4 μm and the peak responsivity (R_p) of the detector is 140 mA/W at bias $V_B = -3$ V. The spectral width and the cutoff wavelength of the MWIR detector are $\Delta\lambda/\lambda = 20\%$ and $\lambda_c = 5$ μm, respectively. The responsivity of the LWIR detector peaks at 8.8 μm and the peak

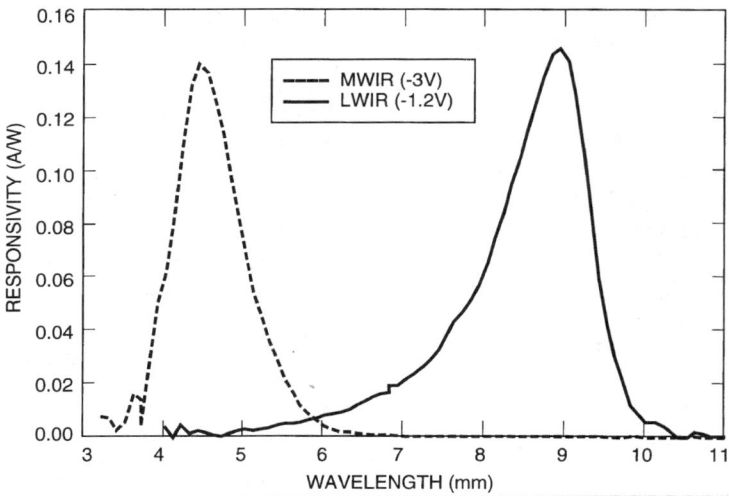

FIG. 43. Simultaneously measured responsivity spectrum of vertically integrated MWIR and LWIR dualband QWIP detector (from Gunapala et al., 1998a).

responsivity R_p of the detector is 150 mA/W at bias $V_B = -1.2$ V. The spectral width and the cutoff wavelength of the LWIR detector are $\Delta\lambda/\lambda = 14\%$ and $\lambda_c = 9.4\,\mu$m, respectively. The measured absolute peak responsivity of both MWIR and LWIR detectors are small, up to about $V_B = -0.5$ V. Beyond that it increase nearly linearly with bias in both MWIR and LWIR detectors reaching $R_P = 210$ and 440 mA/W, respectively at $V_B = -4$ V. This type of responsivity behavior versus bias is typical for a bound-to-continuum and bound-to-quasibound QWIPs in MWIR and LWIR bands, respectively. The peak quantum efficiency (45° double pass) of MWIR and LWIR detectors were 2.6 and 16.4%, respectively, at operating biases indicated in Fig. 43. The lower quantum efficiency of MWIR detector is due to the lower well doping density ($5 \times 10^{17}\,\text{cm}^{-3}$) and lower number of quantum wells in the MQW region. The peak detectivities of both MWIR and LWIR detectors were estimated at different operating temperature and bias voltages using experimentally measured noise currents (Gunapala et al., 1998b). The MWIR peak detectivity $D_\lambda^* = 3.8 \times 10^{11}\,\text{cm}\sqrt{\text{Hz}}/\text{W}$ has been achieved at $V_B = -1.8$ V and $T = 90$ K. Similarly, LWIR peak detectivity $D^* = 8 \times 10^{11}\,\text{cm}\sqrt{\text{Hz}}/\text{W}$ has achieved at $V_B = -1.1$ V and $T = 60$ K.

6. DUALBAND (LWIR AND VLWIR) DETECTORS

As discussed in the previous section of this chapter, there are many target recognition and discrimination applications that require monolithic LWIR and VLWIR dualband large-area FPAs as well. The general notion is that dualband target recognition and discrimination capability significantly improves with increasing wavelength separation between the two wavelength bands under consideration. Therefore, we are currently developing a 640 × 486 LWIR and VLWIR dualband QWIP FPA camera. Thus, we have developed the following QWIP device structure that can be processed into dualband QWIP FPAs with dual or triple contacts to access a CMOS readout multiplexer (Bois et al., 1995; Tidrow et al., 1997). Single indium bump per pixel is usable only in the case of an interlace readout scheme (i.e., odd rows for one color and the even rows for the other color) which uses an existing single color CMOS readout multiplexer. The advantages of this scheme are that it provides simultaneous data readout and enables it to use currently available single-color CMOS readout multiplexers. However, the disadvantage is that it does not provide a full fill factor for both wavelength bands. This problem can be eliminated by fabricating $(n + 1)$ terminals (e.g., three terminals for dualband) per pixel and hybridizing with a multicolor readout having n readout cells per detector pitch, where n is the number of wavelength bands.

The device structure (Gunapala et al., 1998b) consists of a stack of 25 periods of LWIR QWIP structure and another stack of 25 periods of VLWIR QWIP structure separated by a heavily doped 0.5-μm-thick intermediate GaAs contact layer. The first stack (VLWIR) consist of 25 periods of 500-Å $Al_xGa_{1-x}As$ barrier and a GaAs well. This VLWIR QWIP structure is designed to have a bound-to-quasibound intersubband absorption peak at 15 μm, since the dark current of the device structure is expected to dominated by the longer wavelength portion of the device structure. The second stack (LWIR) consist of 25 periods of 500-Å $Al_xGa_{1-x}As$ barrier and a narrow GaAs well. This LWIR QWIP structure is designed to have a bound-to-continuum intersubband absorption peak at 8.5 μm, since the photocurrent and dark current of the LWIR device structure are relatively small compared to the VLWIR portion of the device structure. This whole dualband QWIP structure is then sandwiched between 0.5-μm GaAs top and bottom contact layers doped $n = 5 \times 10^{17}$ cm^{-3}, and was grown on a semi-insulating GaAs substrate by MBE. Then a 1.0-μm-thick GaAs cap layer on top of a 300-Å $Al_{0.3}Ga_{0.7}As$ stop-etch layer must be grown *in situ* on top of the device structure for the fabrication of a light coupling optical cavity.

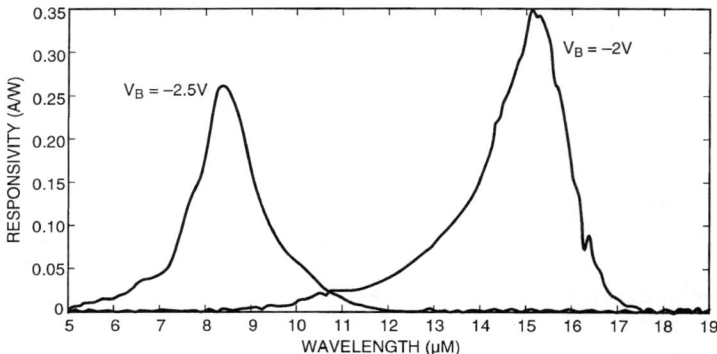

FIG. 44. Simultaneously measured responsivity spectrum of vertically integrated LWIR and VLWIR dualband QWIP detector (from Gunapala et al., 1998a).

The detectors were back illuminated through a 45° polished facet, as described earlier, and a simultaneously measured responsivity spectrum of vertically integrated dualband QWIP is shown in Fig. 44. The responsivity of the LWIR detector peaks at 8.3 μm and the peak responsivity R_p of the detector is 260 mA/W at bias $V_B = -2.5$ V. The spectral width and the cutoff wavelength of the LWIR detector are $\Delta\lambda/\lambda = 19\%$ and $\lambda_c = 9.3$ μm, respectively. The responsivity of the VLWIR detector peaks at 15.2 μm and the peak responsivity (R_p) of the detector is 340 mA/W at bias $V_B = -2.0$ V. The spectral width and the cutoff wavelength of the LWIR detector are $\Delta\lambda/\lambda = 12\%$ and $\lambda_c = 15.9$ μm, respectively. The peak quantum efficiency (45° double pass) of LWIR and VLWIR detectors were 12.6 and 9.8%, respectively, at operating biases indicated in Fig. 44. The peak detectivities of both LWIR and VLWIR detectors were estimated at different operating temperatures and bias voltages using experimentally measured noise currents (Gunapala et al., 1998b). The peak detectivities D_λ^* of LWIR and VLWIR detectors were 1.8×10^{12} cm $\sqrt{\text{Hz}}$/W (at $T = 55$ K and $V_B = -1.0$) and 1.5×10^{12} cm $\sqrt{\text{Hz}}$/W (at $T = 35$ K and $V_B = -1.0$), respectively.

7. HIGH-PERFORMANCE QWIPs FOR LOW-BACKGROUND APPLICATIONS

In this section, we discuss the demonstration of high-performance QWIPs for low-background applications by Gunapala et al. (1998b). Improving QWIP performance depends largely on minimizing the shot noise of the dark current and improving the quantum efficiency. Equations (4) and (5) in Subsection 2 of Section III were used to analyze the dark current of a QWIP, which has a intersubband absorption peak in the long-wavelength region. Figure 45 shows the estimated total dark currents (thermionic + thermionic assisted tunneling + tunneling), and experimental dark current of a QWIP sample, which has a cutoff wavelength $\lambda_c = 10\,\mu$m. According to the calculations, tunneling through the barriers dominates the dark current at temperatures below 30 K, and at temperatures above 30 K, thermionic emission into the continuum transport state dominates the dark current.

For this experiment, eight n-type doped QWIP device structures were grown using MBE. The quantum well widths L_w range from 35 to 50 Å, while the barrier widths are approximately constant at $L_b = 500$ Å. The Al molar fraction in the $Al_xGa_{1-x}As$ barriers varied from $x = 0.24$ to 0.30 (corresponding to cutoff wavelengths of $\lambda_c = 8.3$–$10.3\,\mu$m). The photosensitive doped MQW region (containing 25 to 50 periods) is sandwiched between similarly doped top (0.5-μm) and bottom (0.5-μm) ohmic contact layers. These structural parameters have been chosen to give a very wide variation in QWIP absorption and transport properties. All eight QWIP

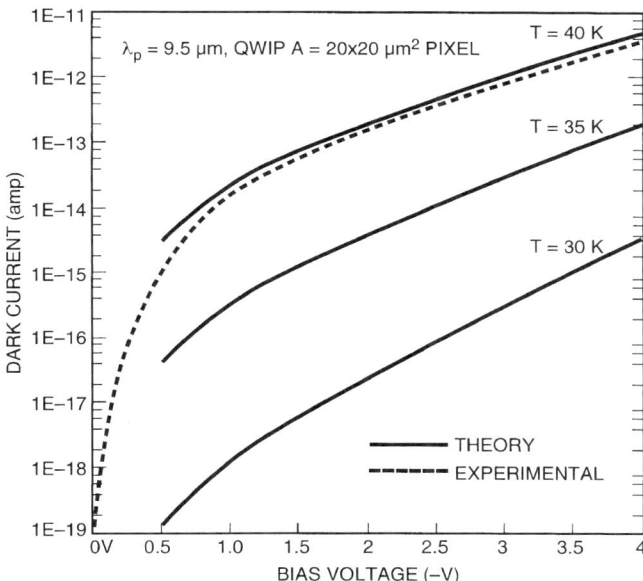

FIG. 45. Comparison of experimental (solid curves) and theoretical (dashed) dark current versus bias voltage curves at various temperatures for a 10-μm-cutoff QWIP (from Gunapala et al., 1998a).

samples are n-doped with intersubband infrared transition occurring between a single localized bound state in the well and a delocalized state in the continuum. Thus, in the presence of an electric field, the photoexcited carrier can be effectively swept out of the quantum well region.

All eight QWIP samples were processed into 200-μm-diameter mesas (area = 3.14×10^{-4} cm^2) using wet chemical etching and Au–Ge ohmic contacts were evaporated onto the top and bottom contact layers. The dark current versus voltage curves for all samples were measured as a function of temperature from $T = 40$–70 K and Fig. 45 shows the current–voltage curve of one sample. As expected, Fig. 45 clearly shows that the $T = 40$ K dark current of these QWIP devices are many orders of magnitude smaller than the dark current at $T = 70$ K. This clearly indicates that the dark current of these devices are thermionically dominant down to 40 K and the tunneling-induced dark current is insignificant.

The detectors were back-illuminated through a 45° polished facet and their normalized responsivity spectrums are shown in Fig. 46. The responsivities of all device structures peaked in the range from 7.7 to 9.7 μm. The peak responsivities (R_p), spectral widths ($\Delta\lambda$), cutoff wavelengths (λ_c), and quantum efficiency photoconductive gain products ($\eta \times g$) are listed in Table IV. It is worth noting that ηg product of sample 4 has increased to 17%. This is approximately a factor of 24 increase in ηg product compared to ηg product of QWIP devices designed for high-background and high-temperature operation. This large enhancement was achieved by (i) improv-

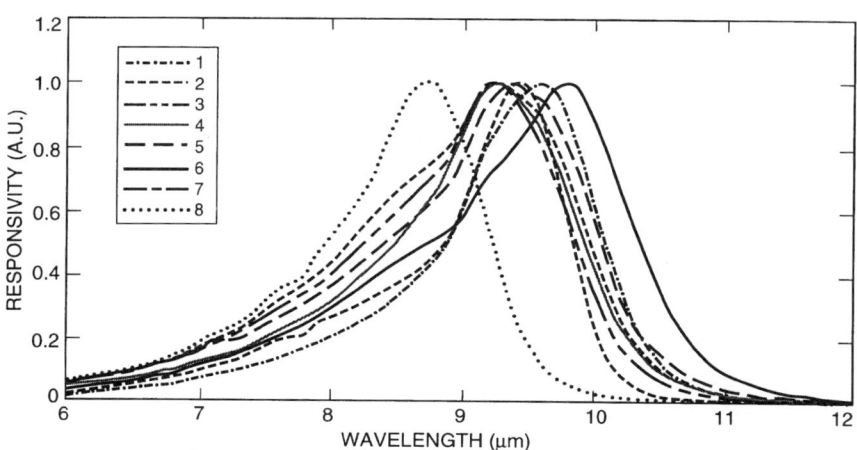

FIG. 46. Normalized responsivity spectra versus wavelength at $T = 40$ K for all samples (from Gunapala et al., 1998a).

TABLE IV
RESPONSIVITY SPECTRAL PARAMETERS OF ALL EIGHT SAMPLES (Gunapala et al., 1998b)

Sample	Spectral response (μm)			Device performance (detector area 3.1×10^{-4} cm^2)		
	λ_p	50% λ_c	$\Delta\lambda$	Peak R_p A/W at 2 V	Q.E. × gain (%)	Dark current at 50 K
1	9.6	10.1	1.1	0.65	8.4	1.1E − 11
2	9.4	10.1	1.6	0.61	8.0	8.0E − 11
3	9.3	9.8	1.6	0.38	5.2	2.0E − 11
4	9.3	9.8	1.2	1.26	17.0	9.0E − 11
5	9.4	10.0	1.8	0.46	6.1	1.0E − 10
6	9.7	10.3	1.5	0.54	5.3	3.0E − 10
7	8.5	8.9	1.0	0.15	2.2	1.0E − 12
8	7.7	8.3	1.3	0.34	5.5	2.5E − 12

ing quantum efficiency per well as a result of higher well doping density and (ii) operating at a higher electric field (i.e., higher gain). The number of periods in the structure was reduced to 30 (typically 50) to increase the applied electric field while keeping operating bias voltage at the same level. The current noise i_n was estimated using measured optical gain and dark current. The peak detectivity D^* can now be calculated from $D^* = R\sqrt{A\Delta f}/i_n$, where A is the area of the detector and $A = 27 \times 27$ μm^2. Table V shows the D^* values of both device structures at various bias voltages at $T = 40$ K. These data clearly show that detectivities of 10-μm-cutoff QWIPs reach mid 10^{13} cm$\sqrt{\text{Hz}}$/W at $T = 40$ K. As shown in Table V, these detectors are not showing background limited performance (BLIP) for a moderately low background of 2×10^9 photons/cm^2/s at $T = 40$ K operation. Since the dark current of these detectors is thermionically limited down to $T = 30$ K, these detectors should demonstrated a BLIP at $T = 35$ K for 2×10^9 photons/cm^2/s background.

8. BROADBAND QWIPs FOR THERMAL INFRARED IMAGING SPECTROMETERS

Until recently, QWIP detectors have been available only over narrow bands (typically 1.0–2.0 μm wide) in the 6- to 20-μm spectral range. Bandara et al. (1998b) developed a broadband test detector that will cover the 10- to 16-μm band and FPAs of these detectors can be mated to a 640 × 480 multiplexer (see Fig. 47). This detector was developed specifically for

TABLE V

Responsivity, Quantum Efficiency, Photoconductive Gain, and Detectivity of the First and the Fourth Samples at $T = 40$ K (Gunapala et al., 1998b)

Sample	Peak W.L. (μm)	Bias (V)	Responsivity (A/W)	Q.E. × gain (%)	ABS. Q.E. at 300 K	ABS. Q.E. at 40 K	Optical gain	Dark current at 40 K (A)	Photocurrent (A)	Detectivity (cm Hz$^{1/2}$/W)
1	9.6	1.0	0.24	3.1	14.1	18.3	0.17	1.2E − 15	7.2E − 17	5.7E + 13
		2.0	0.65	8.4	14.1	18.3	0.46	2.3E − 15	1.9E − 16	6.8E + 13
		3.0	0.66	8.5	14.1	18.3	0.47	9.3E − 15	2.0E − 16	3.4E + 13
		4.0	0.68	8.8	14.1	18.3	0.48	1.4E − 14	2.1E − 16	2.8E + 13
4	9.3	1.0	0.43	5.7	17.0	22.1	0.26	7.0E − 15	1.3E − 16	3.4E + 13
		2.0	1.26	16.8	17.0	22.1	0.76	9.3E − 15	3.9E − 16	5.1E + 13
		3.0	1.35	18.0	17.0	22.1	0.81	3.5E − 14	4.2E − 16	2.7E + 13
		4.0	1.36	18.1	17.0	22.1	0.82	2.3E − 13	4.2E − 16	1.1E + 13

Pixel area, 27×27 μm; background flux, 2×10^9 photons/cm^2/s.

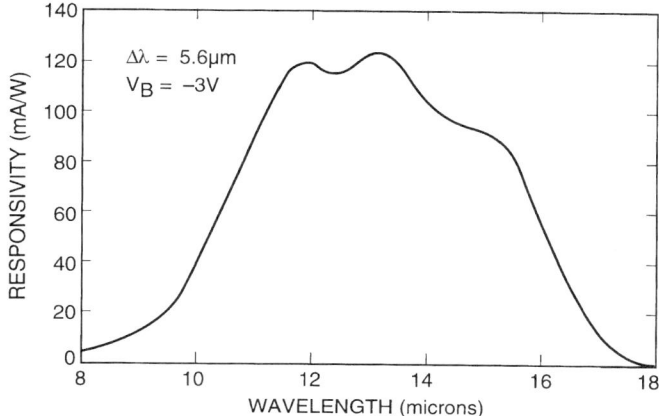

FIG. 47. Experimental measurements of the normalized responsivity spectrum of 10- to 16-μm broadband QWIP at bias voltage $V_B = -4$ V (from Gunapala et al., 1998a).

thermal infrared imaging spectrometers—anticipating a possible need for Mars exploration. The program required a thermal infrared imaging spectrometer with minimum power, mass, and volume. One attractive instrument concept involved a spatially modulated infrared spectrometer (SMIS) to cover the LWIR and VLWIR spectral ranges. This instrument does not contain any moving mirrors because it uses the spatially modulated Fourier transform spectroscopy technique. Another advantage of this concept is that it has a substantially higher signal flux because all of the photons entering the pupil are used. The high-spectral-resolution version of this instrument requires larger format FPAs (at least 640 × 486) with high pixel-to-pixel uniformity. The lack of large-format, uniform LWIR and VLWIR FPA technology has prevented the development of such highly sensitive and robust thermal infrared spectrometers. That has changed recently due to demonstration of highly uniform, large-format QWIP FPAs with lower $1/f$ noise, at a lower cost than any other LWIR detector. In addition, the use of external filters can be avoided because QWIP can be designed to have sharp spectral cutoffs at the required wavelengths. Francis Reininger, at the Jet Propulsion Laboratory (JPL), has demonstrated this concept by building a prototype laboratory instrument working with an 8- to 9-μm QWIP 640 × 486 FPA. The unique characteristic of this instrument (besides being small and efficient) is that it has one instrument line shape for all spectral colors and spatial field positions. By using broadband QWIP arrays with wavelengths out to 16 μm, the next version of this instrument could become the first compact, high-resolution thermal infrared, hyper-spectral imager

270   S. D. GUNAPALA AND S. V. BANDARA

with a single spectral line shape and zero spectral smile. Such an instrument is in strong demand by scientists studying Earth and planetary science.

VI. Applications

1. FIRE FIGHTING

Recently, 256 × 256 handheld QWIP cameras have been used to demonstrate various possible applications in science, medicine, industry, defense, etc. A 256 × 256 portable LWIR QWIP camera helped a Los Angeles TV news crew get a unique perspective on fires that raced through the Southern California seaside community of Malibu in October, 1996. The camera was used on the station's news helicopter. This portable camera features infrared detectors that cover longer wavelengths than previous portable cameras could. This enables the camera to see through smoke and pinpoint lingering hotspots that are not normally visible. This enabled the TV station to transmit live images of hotspots in areas that appeared innocuous to the naked eye. These hotspots were a source of concern for firefighters, because they could flare up even after the fire appeared to have subsided. Figure 48 shows the comparison of visible and infrared images of a just-burned area as seen by the news crew in nighttime. It works effectively in both daylight

(a) VISIBLE   (b) QWIP

FIG. 48. Comparison of visible and infrared images of a just-burned area as seen by a highly sensitive visible CCD camera and the long wavelength infrared camera in nighttime. (a) Visible image from a CCD camera. (b) Image from the 256 × 256 portable QWIP camera. Infrared imagery clearly enables firefighters to locate the hotspots in areas which appeared innocuous to the naked eye. These hotspots are a source of concern for firefighters, because fire can flare up even after it appears to have subsided (from Gunapala *et al.*, 1998a).

and nighttime conditions. The event marked the QWIP camera's debut as a fire-observing device (Gunapala *et al.*, 1998c).

2. VOLCANOLOGY

A similar 256 × 256 LWIR QWIP camera has been used to observe volcanoes, mineral formations, weather, and atmospheric conditions at the Kilauea Volcano in Hawaii. The objectives of this trip were to map geothermal features. The wide dynamic range enabled volcanologists to image volcanic features at temperatures much higher (300–1000°C) than can be imaged with conventional thermal imaging systems in the 3- to 5-μm range or in the visible (Realmuto *et al.*, 1997). Figure 49 shows the comparison of visible and infrared images of the Kilauea Volcano in Hawaii. The infrared image of the volcano clearly shows a hot lava tube running underground that is not visible to the naked eye.

3. MEDICINE

A group of researchers from the State University of New York in Buffalo and Walter Reed Army Institute of Research in Washington, D.C., used a

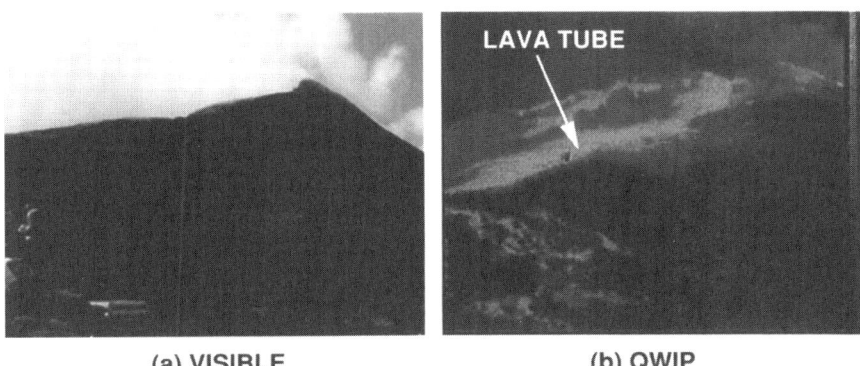

(a) VISIBLE   (b) QWIP

FIG. 49. Comparison of visible and infrared images of the Mount Kilauea Volcano in Hawaii. (a) Visible image from a highly sensitive CCD camera. (b) Image from the 256 × 256 portable QWIP camera. The wide dynamic range enabled us to image volcanic features at temperatures much higher (300–1000°C) than can be imaged with conventional thermal imaging systems in the 3- to 5-μm range or in the visible. The infrared image of the volcano clearly shows a hot lava tube running underground that is not visible to the naked eye. This demonstrates the advantages of long-wavelength infrared in geothermal mapping (from Gunapala *et al.*, 1998a).

256 × 256 LWIR QWIP camera in the dynamic area telethermometry (DAT). DAT has been used to study the physiology and pathophysiology of cutaneous perfusion, which has many clinical applications. DAT involves accumulation of hundreds of consecutive infrared images and fast Fourier transform (FFT) analysis of the biomodulation of skin temperature and of the microhomogeneity of skin temperature. The FFT analysis yields the thermoregulatory frequencies and amplitudes of temperature and the perfusion of the skin's capillaries. To obtain reliable DAT data, one needs an infrared camera in the $>8\,\mu$m range (to avoid artifacts of reflections of modulated emitters in the environment), a repetition rate of 30 Hz (enabling the accumulation of a maximal number of images during the observation period to maximize the resolution of the FFT), frame-to-frame instrumental stability (to avoid artifacts stemming from instrument modulation), and a sensitivity of less than 50 mK. According to these researchers, the longer wavelength operation, higher spatial resolution, higher sensitivity, and greater stability of the QWIP RADIANCE™ made it the best choice of all infrared cameras.

This camera was also used by a group of researchers at the Texas Heart Institute in a heart surgery experiment performed on a rabbit heart. The experiment clearly revealed that it is possible to detect arterial plaque built inside a heart by thermography. Figure 50 clearly shows arterial plaque accumulated in a rabbit heart.

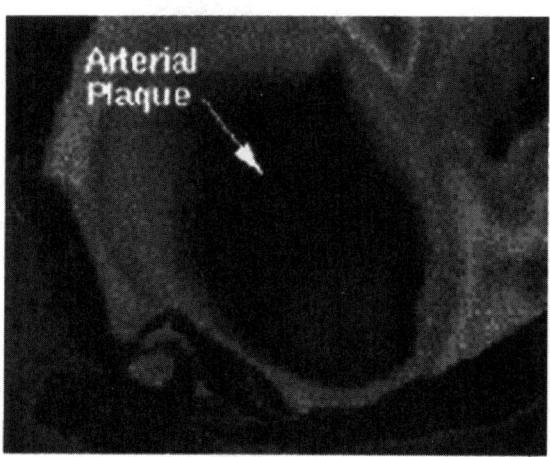

FIG. 50. This image shows arterial plaque deposited in a rabbit heart (from Gunapala et al., 1998a).

4. DEFENSE

It is not necessary to explain how real-time infrared imaging is important in surveillance, reconnaissance, and military operations. The QWIP RADIANCE™ was used by the researchers at the Ballistic Missile Defense Organization's innovative science and technology experimental facility in a unique experiment to discriminate and clearly identify the cold launch vehicle from its hot plume emanating from rocket engines.

Usually, the temperature of cold launch vehicles is about 250°C, whereas the temperatures of the hot plume emanating from launch vehicle can reach 950°C. According to the Plank blackbody emission theory, the photon flux ratio of 250 and 950°C blackbodies at 4 μm is about 25,000, whereas the same photon flux ratio at 8.5 μm is about 115 (see Fig. 51). Thus, it is very clear that one must explore longer wavelengths for better cold-body versus hot plume discrimination (Gunapala et al., 1998c), because the highest instantaneous dynamic range of infrared cameras is usually 12 bits (i.e., 4096) or less. Figure 52 shows a image of Delta-II launch taken with QWIP RADIANCE™ camera. This clearly indicates the advantage of long-wavelength QWIP cameras in the discrimination and

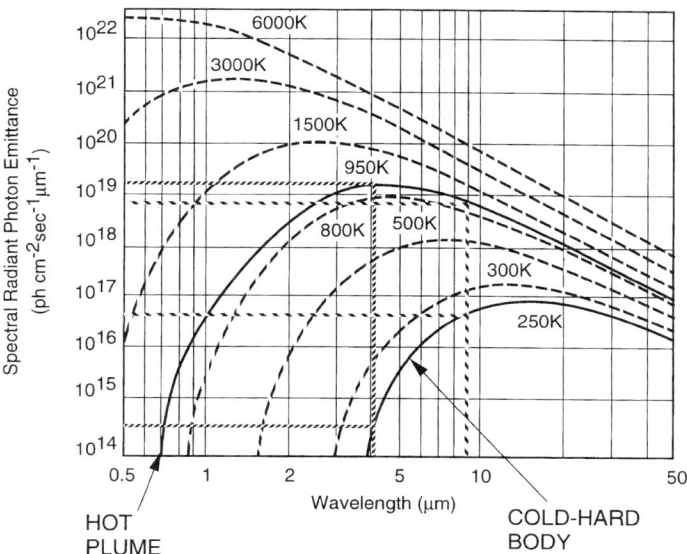

FIG. 51. Blackbody spectral radiant photon emittance at various temperatures (from Gunapala et al., 1998a).

274   S. D. GUNAPALA AND S. V. BANDARA

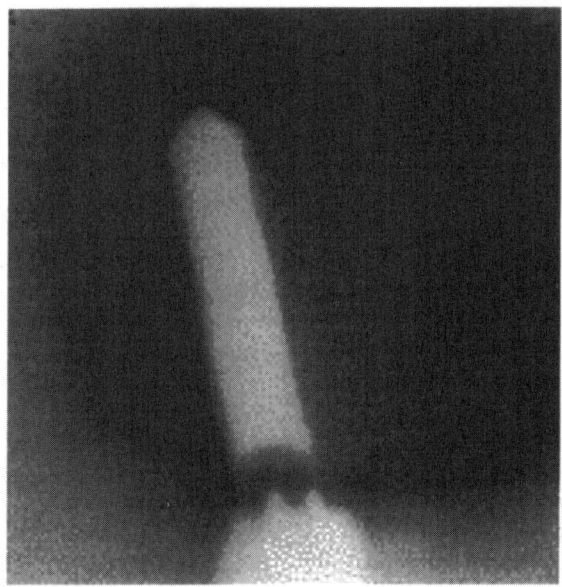

FIG. 52. Image of a Delta-II launch vehicle taken with the long-wavelength QWIP RADIANCE during the launch. This clearly indicates the advantage of long-wavelength QWIP cameras in the discrimination and identification of cold launch vehicles in the presence of hot plume during early stages of launch (from Gunapala et al., 1998a).

identification of cold launch vehicles in the presence of hot plume during early stages of launch.

5. ASTRONOMY

In this section, we discuss the first astronomical observations with a QWIP FPA. To perform this astronomical observation, we designed a QWIP wide-field imaging multicolor prime-focus infrared camera (QWICPIC). Observations were conducted at the 5-m Hale telescope at Mt. Palomar with a QWICPIC based on an 8- to 9-μm 256 × 256 QWIP FPA operating at $T = 35$ K. The ability of QWIPs to operate under high photon backgrounds without excess noise enables the instrument to observe from the prime focus with a wide 2 × 2 ft field of view, making this camera unique among the suite of infrared instruments available for astronomy. The excellent $1/f$ noise performance (see Fig. 53) of QWIP FPAs enables the

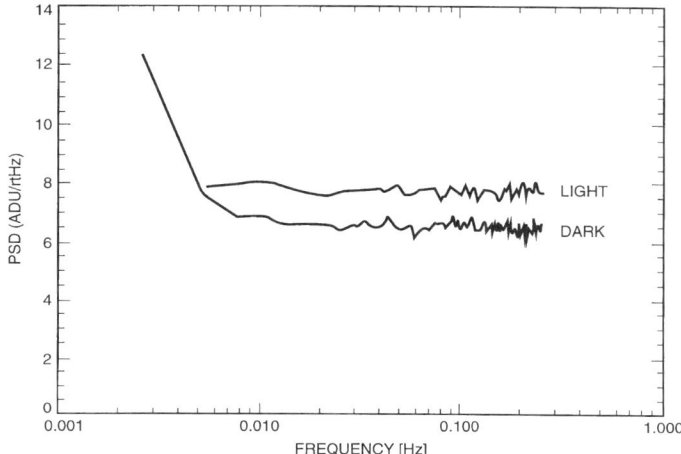

FIG. 53. The $1/f$ noise spectrum of an 8- to 9-μm 256 × 256 QWIP FPA. (1 ADU = 430 electrons.) The curves clearly show that QWIPs have no $1/f$ down to 30 mHz, which enables QWIP-based instruments to use longer integration times and frame-adding capability (from Gunapala et al., 1998a).

QWICPIC to observe in the slow scan strategy often required in infrared observations from space (Gunapala et al., 1998c).

Figure 54 compares an image of the Orion nebula obtained in a brief 30-min observation with an engineering grade QWIP FPA to a visible image of the Orion nebula taken with the wide-field planetary camera in the Hubble Space Telescope (HST). In addition to the well-known infrared bright BN-KL object in the upper right-hand corner, careful comparison of the infrared images with optical and near-infrared images obtained by the HST reveal a multitude of infrared sources that are dim or undetectable in the visible. These images demonstrate the advantage of large-format, stable (low $1/f$ noise) LWIR QWIP FPAs for surveying obscured regions in search of embedded or reddened objects such as young forming stars.

VII. Summary

Various types of GaAs–AlGaAs-based QWIPs, QWIPs with other materials systems, figures of merit, light-coupling methods, QWIP imaging arrays, and some applications of QWIP FPAs were reviewed. Our discussion of QWIPs was necessarily brief and the literature references cited

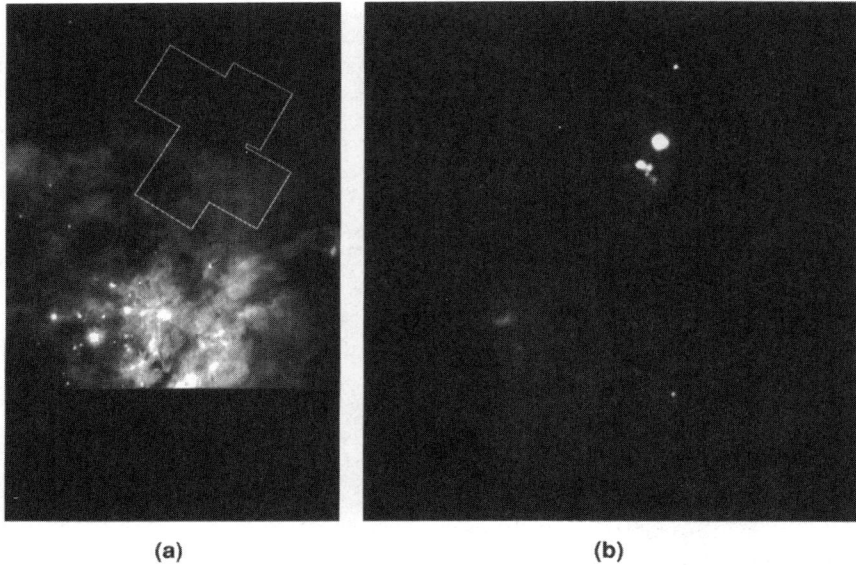

FIG. 54. (a) Visible image of the Orion nebula taken with the wide-field planetary camera in Hubble Space Telescope. (b) An 8- to 9-μm infrared image of the Orion nebula taken with an engineering grade 256 × 256 QWIP FAA at the 5-m Hale telescope at Mt. Palomar (from Gunapala et al., 1998a).

are not inclusive, but represent a selection of key articles from a historic and technical point of view. Exceptionally rapid progress has been made in the understanding of intersubband absorption and carrier transport in the QWIP device structure and practical demonstration of large, sensitive, two-dimensional QWIP imaging FPAs since they were first demonstrated only several years ago. This remarkable progress was due to the highly mature III–V growth and processing technologies. As discussed in Subsection 7 in Section V, an important advantage of QWIPs over the HgCdTe is that as the temperature is reduced, D^* increases dramatically, reaching $D^*_\lambda = 10^{14}$ cm $\sqrt{\text{Hz}}$/W at $T = 30$ K and even larger values at lower temperatures. Even though a detailed discussion of cooling methods are outside of the scope of this chapter, it should be noted that advances in cooling technologies have kept pace with the detector developments. Efficient, lightweight mechanical and sorption coolers have been developed that can achieve temperatures down to 10 K. Thus, when QWIP technology utilizes the advanced cooling technologies, it can easily meet the stringent spectroscopic and imaging requirements of ground-based and space-borne applications in the LWIR and VLWIR spectral bands. As discussed in Sections V and VI, rapid progress has been made in the performance (NEΔT) of long-wavelength QWIP FPAs, starting with bound-to-bound QWIPs, which had relatively poor sensitivity, and culminating in high-performance bound-to-quasibound QWIP FPAs with various light-coupling schemes. In conclusion, very sensitive, low-cost, large (2048 × 2048) LWIR and VLWIR QWIP FPAs can be expected in the near future.

ACKNOWLEDGMENTS

We are grateful to C. P. Bankston, M. Bothwell, G. M. Burdick, M. T. Chahine, T. Cole, D. Duston, W. Dyer, T. C. Fraschetti, S. K. Khanna, K. M. Koliwad, T. N. Krabach, C. A. Kukkonen, P. D. LeVan, R. H. Liang, T. R. Livermore, F. G. O'Callaghan, M. J. Sander, V. Sarohia, R. R. Stephenson, R. Wall, and B. A. Wilson for encouragement and support during the development and optimization of QWIP FPAs at the Jet Propulsion Laboratory for various applications. Also, we would like to give special thanks to J. K. Liu and J. M. Mumolo of the Jet Propulsion Laboratory for critical reading of the manuscript. In addition, we express our thanks to H. H. Auman, E. R. Blazejewski, C. G. Bethea, J. J. Bock, R. Carralejo, N. Chand, K. K. Choi, M. L. Eastwood, J. Frank, J. Gill, D. E. Hagan, T. R. Hamilton, R. Hamm, G. Hasnain, W. S. Hobson, T. Hoelter, M. Hong, W. Hong, J. James, D. W. Juergens, A. B. Kahle, J. T. Kenny, R. F. Kopf, J. M. Kuo, S. Laband, A. L. Lane, T. L. Lin, J. Llorens, R. A. Logan, S. S. Li, H. C. Liu, J. K. Liu, E. M. Luong, P. D. Maker, M. J. McKelvey, R. E. Muller, J. M. Mumolo, D. A. Nichols, B. Pain, M. B. Panish, J. S. Park, S. S. Pei, L. Pfeiffer, H. R. Pollock, F. S. Pool, V. J. Realmuto, F. M. Reininger, M. E. Ressler, D. Ritter, G. Sarusi, C. A. Shott, J. J. Simmonds, A. Singh, R. B. Somoano, N. Stetson, R. C. Stirbl, M. Sundaram, M. Z. Tidrow, G. Udomkesmalee, D. Vincent, M. W. Werner, and A. Zussman, who worked closely with us, for stimulating technical discussions, device processing, and crystal growth. We give our special thanks to B. F. Levine of AT&T Bell Laboratories for his support and guidance over the early years of QWIP development. We are also grateful to Dr. E. C. Stone, director of the Jet Propulsion Laboratory, for his continuing support in this endeavor. The QWIP FPA research and applications described in this chapter were performed partly by the Center for Space Microelectronics Technology, Jet Propulsion Laboratory, California Institute of Technology, and were jointly sponsored by the JPL Director's

Research and Development Fund, the Ballistic Missile Defense Organization/Innovative Science & Technology Office, the National Aeronautics and Space Administration, the Office of Space Science, and the Air Force Research Laboratory.

REFERENCES

Andersson, J. Y., Lundqvist, L., and Paska, Z. F. (1991a). *Appl. Phys. Lett.* **58**, 2264.
Andersson, J. Y., and Lundqvist, L. (1991). *Appl. Phys. Lett.* **59**, 857.
Andersson, J. Y., Lundqvist, L., and Paska, Z. F. (1991b). *J. Appl. Phys.* **71**, 3600.
Andersson, J. Y., Alverbro, J., Borglind, J., Helander, P., Martijn, H., and Ostlund, M. (1997). *Proc. SPIE* **3061**, 740.
Andersson, T. G., Chen, Z. G., Kulakovskii, V. D., Uddin, A., and Vallin, J. T. (1988). *Phys. Rev. B* **37**, 4032.
Andrews, S. R., and Miller, B. A. (1991). *J. Appl. Phys.* **70**, 993.
Asom, M. T., Bethea, C. G., Focht, M. W., Fullowan, T. R., Gault, W. A., Glogovsky, K. G., Guth, G., Leibenguth, R. E., Levine, B. F., Lievscu, G., Luther, L. C., Stayt, Jr., J. W., Swaminathan, V., Wong, Y. M., and Zussman, A. (1991). In *Proceedings of the IRIS Specialty Group on Infrared Detectors*, Vol. I, p. 13.
Bandara, K. M. S. V., Coon, D. D., O, B., Lin, Y. F., and Francombe, M. H. (1988). *Appl. Phys. Lett.* **53**, 1931.
Bandara, K. M. S. V., Choe, J.-W., Francombe, M. H., Perera, A. G. U., and Lin, Y. F. (1992). *Appl. Phys. Lett.* **60**, 3022.
Bandara, K. M. S. V., Levine, B. F., and Asom, M. T. (1993a). *J. Appl. Phys.* **74**, 346.
Bandara, K. M. S. V., Levine, B. F., Leibenguth, R. E., and Asom, M. T. (1993b). *J. Appl. Phys.* **74**, 1826.
Bandara, K. M. S. V., Levine, B. F., and Kuo, J. M. (1993c). *Phys. Rev. B* **48**, 7999.
Bandara, S., Gunapala, S., Liu, J., Hong, W., and Park, J. (1997). *Proc. SPIE* **2999**, 103.
Bandara, S. V., Gunapala, S. D., Liu, J. K., Luong, E. M., Mumolo, J. M., Hong, W., Sengupta, D. K., and McKelvey, M. J. (1998a). *Proc. SPIE* **3379**, 396.
Bandara, S. V., Gunapala, S. D., Liu, J. K., Luong, E. M., Mumolo, J. M., Hong, W., Sengupta, D. K., and McKelvey, M. J. (1998b). *Appl. Phys. Lett.* **72**, 2427.
Beck, W. A. (1993). *Appl. Phys. Lett.* **63**, 3589.
Beck, W. A., Faska, T. S., Little, J. W., Albritton, J., and Sensiper, M. (1994). In *Proceedings of the Second International Symposium on 2–20 µm Wavelength Infrared Detectors and Arrays: Physics and Applications*, Miami Beach, Florida.
Bethea, C. G., Levine, B. F., Shen, V. O., Abbott, R. R., and Hseih, S. J. (1991). *IEEE Trans. Electron Devices* **38**, 1118.
Bethea, C. G., and Levine, B. F. (1992). *Proc. SPIE.* **1735**, 198–203.
Bethea, C. G., Levine, B. F., Asom, M. T., Leibenguth, R. E., Stayt, J. W., Glogovsky, K. G., Morgan, R. A., Blackwell, J., and Parish, W. (1993). *IEEE Trans. Electron Devices* **40**, 1957.
Bois, Ph., Costard, E., Duboz, J. Y., Nagle, J., Rosencher, E., and Vinter, B. (1995). *Proc. SPIE* **2552**, 755–766.
Breiter, R., Cabanski, W., Koch, R., Rode, W., and Ziegler, J. (1998). *Proc. SPIE* **3379**, 423.
Capasso, F., Mohammed, K., and Cho, A. Y. (1986). *IEEE J. Quantum Electron.* **22**, 1853.

Chahine, M. T. (1990). In *Proceedings of Innovative Long Wavelength Infrared Detector Workshop*, Pasadena, California, April 3.
Chang, Y.-C., and James, R. B. (1989). *Phys. Rev. B* **39**, 12672.
Chen, C. J., Choi, K. K., Tidrow, M. Z., and Tsui, D. C. (1996) "Corrugated quantum well infrared photodetectors for normal incident light coupling," *Appl. Phys. Lett.* **68**, 1446.
Chiang, J. C., Li, S. S., Tidrow, M. Z., Ho, P., Tsai, C. M., and Lee, C. P. (1996). *Appl. Phys. Lett.* **69**, 2412.
Chiu, L. C., Smith, J. S., Margalit, S., Yariv, A., and Cho, A. Y. (1983). *Infrared Phys.* **23**, 93.
Choe, J. W., O, B., Bandara, K. M. S. V., and Coon, D. D. (1990). *Appl. Phys. Lett.* **56**, 1679.
Choi, K. K. (1998). *Appl. Phys. Lett.* **65**, 1266.
Choi, K. K. (1996). *J. Appl. Phys. Lett.* **80**, 1257.
Choi, K. K., Levine, B. F., Malik, R. J., Walker, J., and Bethea, C. G. (1987a). *Phys. Rev. B* **35**, 4172.
Choi, K. K., Levine, B. F., Bethea, C. G., Walker, J., and Malik, R. J. (1987b). *Phys. Rev. Lett.* **59**, 2459.
Choi, K. K., Levine, B. F., Jarosik, N., Walker, J., and Malik, R. J. (1987c). *Appl. Phys. Lett.* **50**, 1814.
Choi, K. K., Goldberg, A. C., Das, N. C., Jhabvala, M. D., Bailey, R. B., and Vural, K. (1998a). *Proc. SPIE* **3287**, 118.
Choi, K. K., Chen, C. J., Goldberg, A. C., Chang, W. H., and Tsui, D. C. (1998b). *Proc. SPIE* **3379**, 441.
Coon, D. D., and Karunasiri, R. P. G. (1984). *Appl. Phys. Lett.* **45**, 649.
Duston, D. (1995). *BMDO Monitor*, p. 180.
Elman, B., Koteles, E. S., Melman, P., Jagannath, C., Lee, J., and Dugger, D. (1989). *Appl. Phys. Lett.* **55**, 1659.
Esaki, L., and Sakaki, H. (1977). *IBM Tech. Disc. Bull.* **20**, 2456.
Faska, T. S., Little, J. W., Beck, W. A., Ritter, K. J., Goldberg, A. C., and LeBlanc, R. (1992). In *Innovative Long Wavelength Infrared Detector Workshop* (Pasadena, CA).
Foire, A., Rosencher, E., Bois, P., Nagle, J., and Laurent, N. (1994). *Appl. Phys. Lett.* **64**, 478.
Goosen, K. W., and Lyon, S. A. (1985). *Appl. Phys. Lett.* **47**, 1257.
Grave, I., and Yariv, A. (1992). *Intersubband Transitions in Quantum Wells*, eds. E. Rosencher, B. Vinter, B. Levine, (Cargese, France; Plenum, New York), p. 15.
Grave, I., Shakouri, A., Kuze, N., and Yariv, A. (1993). *Appl. Phys. Lett.* **63**, 1101.
Gunapala, S. D., and Bandara, K. M. S. V. (1995). *Phys. Thin Films* **21**, 113.
Gunapala, S. D., Levine, B. F., Logan, R. A., Tanbun-Ek, T., and Humphrey, D. A. (1990a). *Appl. Phys. Lett.* **57**, 1802.
Gunapala, S. D., Levine, B. F., Pfeiffer, L., and West, K. (1990b). *J. Appl. Phys.* **69**, 6517.
Gunapala, S. D., Levine, B. F., Ritter, D., Hamm, R. A., and Panish, M. B. (1991a). *Appl. Phys. Lett.* **58**, 2024.
Gunapala, S. D., Levine, B. F., and Chand, N. (1991b). *J. Appl. Phys.* **70**, 305.
Gunapala, S. D., Levine, B. F., Ritter, D., Hamm, R. A., and Panish, M. B. (1991c). *Proc. SPIE* **1541**, 11.
Gunapala, S. D., Levine, B. F., Ritter, D., Hamm, R. A., Panish, M. B. (1992a). *Appl. Phys. Lett.* **60**, 636.
Gunapala, S. D., Levine, B. F., Ritter, D., Hamm, R., and Panish, M. B. (1992b). *J. Appl. Phys.* **71**, 2458.
Gunapala, S. D., Bandara, K. S. M. V., Levine, B. F., Sarusi, G., Sivco, D. L., and Cho, A. Y. (1994a). *Appl. Phys. Lett.* **64**, 2288.
Gunapala, S. D., Bandara, K. S. M. V., Levine, B. F., Sarusi, G., Park, J. S., Lin, T. L., Pike, W. T., and Liu, J. K. (1994b). *Appl. Phys. Lett.* **64**, 3431.

Gunapala, S. D., Liu, J. K., Sundaram, M., Park, J. S., Shott, C. A., Hoelter, T., Lin, T. L., Massie, S. T., Maker, P. D., Muller, R. E., and Sarusi, G. (1995). *Elec. Chem. Soc.* **95-28**, 55.
Gunapala, S. D., Park, J. S., Sarusi, G., Lin, T. L., Liu, J. K., Maker, P. D., Muller, R. E., Shott, C. A., Hoelter, T., and Levine B. F. (1997a). *IEEE Trans. Electron. Devices* **44**, 45.
Gunapala, S. D., Liu, J. K., Park, J. S., Sundaram, M., Shott, C. A. Hoelter, C. A., Lin, T-L., Massie, S. T., Maker, P. D., Muller, R. E., and Sarusi, G. (1997b). *IEEE Trans. Electron. Devices* **44**, 51.
Gunapala, S. D., Bandara, S. V., Liu, J. K., Hong, W., Sundaram, M., Maker, P. D. Muller, R. E., Shott, C. A., and Carralejo R. (1998a). *IEEE Trans. Electron. Devices* **45**, 1890.
Gunapala, S. D, Bandara, S. V., Singh, A., Liu, J. K., Luong, E. M., Mumolo, J. M., and McKelvey, M. J. (1998b). *Proc. SPIE* **3379**, 225.
Gunapala, S. D., Bandara, S. V., Liu, J. K., Hong, W., Luong, E. M., Mumolo, J. M., McKelvey, M. J., Sengupta, D. K., Singh, A., Shott, C. A., Carralejo, R., Maker, P. D., Bock, J. J., Ressler, M. E., Werner, M. W., and Krabach, T. N. (1998c). *Proc. SPIE* **3379**, 382.
Harwit, A., and Harris, J. S., Jr. (1987). *Appl. Phys. Lett.* **50**, 685.
Hasnain, G., Levine, B. F., Bethea, C. G., Logan, R. A., Walker, J., and Malik, R. J. (1989). *Appl. Phys. Lett.* **54**, 2515.
Hasnain, G., Levine, B. F., Sivco, D. L., and Cho, A. Y. (1990a). *Appl. Phys. Lett.* **56**, 770.
Hasnain, G., Levine, B. F., Gunapala, S., and Chand, N. (1990b). *Appl. Phys. Lett.* **57**, 608.
Hoff, J., Kim, S., Erdtmann, M., Williams, R., Piotrowski, J., Bigan, E., and Razeghi, M. (1995a). *Appl. Phys. Lett.* **67**, 22.
Hoff, J., Jelen, C., Slivken, S., Bigan, E., Brown, G., and Razeghi, M. (1995b). *Proc. SPIE* **2397**, 445.
Janousek, B. K., Daugherty, M. J., Bloss, W. L., Rosenbluth, M. L., O'Loughlin, M. J., Kanter, H., De Luccia, F. J., and Perry, L. E. (1990). *J. Appl. Phys.* **67**, 7608.
Jelen, C., Slivken, S., Brown, G., and Razeghi, M. (1997). *Mat. Res. Soc. Symp.* **450**, 195.
Jelen, C., Slivken, S., David, T., Razeghi , M., and Brown, G. (1998a). *IEEE J. Quantum Electron.* **34**, 1124.
Jelen, C., Slivken, S., David, T., Brown, G., and Razeghi, M. (1998b). *Proc. SPIE* **3287**, 96.
Kane, M. J., Emeny, M. T., Apsley, N., and Whitehouse, C. R. (1989). *Electron. Lett.* **25**, 230.
Karunasiri, R. P. G., Park, J. S., Mii, Y. J., and Wang, K. L. (1990). *Appl. Phys. Lett.* **57**, 2585.
Kastalsky, A., Duffield, T., Allen, S. J., and Harbison, J. (1988). *Appl. Phys. Lett.* **52**, 1320.
Kiledjian, M. S., Schulman, J. N., and Wang, K. L. (1991). *Phys. Rev. B* **44**, 5616.
Kinch, M. A., and Yariv, A. (1989). *Appl. Phys. Lett.* **55**, 2093.
Kingston, R. H. (1978). *Detection of Optical and Infrared Radiation* (Springer, Berlin).
Kozlowski, L. J., Williams, G. M., Sullivan, G. J., Farley, C. W., Andersson, R. J., Chen, J., Cheung, D. T., Tennant, W. E., and DeWames, R. E. (1991a). *IEEE Trans. Electron. Devices* **ED-38**, 1124.
Kozlowski, L. J., DeWames, R. E., Williams, G. M., Cabelli, S. A., Vural, K., Cheung, D. T., Tennant, W. E., Bethea, C. G., Gault, W. A., Glogovsky, K. G., Levine, B. F., and Stayt, Jr., J. W. (1991b). In *Proceedings of the IRIS Specialty Group on Infrared Detectors*, NIST, Boulder, CO, Vol. I, p. 29.
Lacoe, R. C., O'Loughlin, M. J., Gutierrez, D. A., Bloss, W. L., Cole, R. C., Dafesh, P. A., and Isaac, M. (1992). *Proc. SPIE* **1735**, 230–240.
Levine, B. F. (1993). *J. Appl. Phys.* **74**, R1.
Levine, B. F., Malik, R. J., Walker, J., Choi, K. K., Bethea, C. G., Kleinman, D. A., and Vandenberg, J. M. (1987a). *Appl. Phys. Lett.* **50**, 273.
Levine, B. F., Choi, K. K., Bethea, C. G., Walker, J., and Malik, R. J. (1987b). *Appl. Phys. Lett.* **50**, 1092.
Levine, B. F., Choi, K. K., Bethea, C. G., Walker, J., and Malik, R. J. (1987c). *Appl. Phys. Lett.* **51**, 934.

Levine, B. F., Bethea, C. G., Choi, K. K., Walker, J., and Malik, R. J. (1987d). *J. Appl. Phys.* **64**, 1591.
Levine, B. F., Cho, A. Y., Walker, J., Malik, R. J., Kleinman, D. A., and Sivco, D. L. (1988a). *Appl. Phys. Lett.* **52**, 1481.
Levine, B. F., Bethea, C. G., Choi, K. K., Walker, J., and Malik, R. J. (1988b). *Appl. Phys. Lett.* **53**, 231.
Levine, B. F., Bethea, C. G., Hasnain, G., Walker, J., and Malik, R. J. (1988c). *Appl. Phys. Lett.* **53**, 296.
Levine, B. F., Hasnain, G., Bethea, C. G., and Chand, N. (1989). *Appl. Phys. Lett.* **54**, 2704.
Levine, B. F., Bethea, C. G., Hasnain, G., Shen, V. O., Pelve, E., Abbott, R. R., and Hsieh, S. J. (1990a). *Appl. Phys. Lett.* **56**, 851.
Levine, B. F., Bethea, C. G., Shen, V. O., and Malik, R. J. (1990b). *Appl. Phys. Lett.* **57**, 383.
Levine, B. F., Gunapala, S. D., and Kopf, R. F. (1991a). *Appl. Phys. Lett.* **58**, 1551.
Levine, B. F., Gunapala, S. D., Kuo, J. M., Pei, S. S., and Hui, S. (1991b). *Appl. Phys. Lett.* **59**, 1864.
Levine, B. F., Gunapala, S. D., and Hong, M. (1991c). *Appl. Phys. Lett.* **59**, 1969.
Levine, B. F., Bethea, C. G., Glogovsky, K. G., Stayt, J. W., and Leibenguth, R. E. (1991d). *Semicond. Sci. Technol.* **6**, C114.
Levine, B. F. (1991e). In *Proceedings of the NATO Advanced Research Workshop on Intersubband Transitions in Quantum Wells*, Cargese, France, eds. E. Rosencher, B. Vinter, and B. F. Levine (Plenum, London).
Levine, B. F., Zussman, A., Kuo, J. M., and de Jong, J. (1992a). *J. Appl. Phys.* **71**, 5130.
Levine, B. F., Zussman, A., Gunapala, S. D., Asom, M. T., Kuo, J. M., and Hobson, W. S. (1992b). *J. Appl. Phys.* **72**, 4429.
Li, S. S., Chuang, M. Y., and Yu, L. S. (1993). *Semicond. Sci. Technol.* **8**, 406.
Liu, H. C. (1992a). *Appl. Phys. Lett.* **60**, 1507.
Liu, H. C. (1992b). *Appl. Phys. Lett.* **61**, 2703.
Liu, H. C., Aers, G. C., Buchanan, M., Wasilewski, Z. R., and Landheer, D. (1991a). *Appl. Phys. Lett.* **70**, 935.
Liu, H. C., Buchanan, M., Aers, G. C., and Wasilewski, Z. R. (1991b). *Semicond. Sci. Technol.* **6**, C124.
Liu, H. C., Wasilewski, Z. R., Buchanan, M. (1993). *Appl. Phys. Lett.* **63**, 761.
Manasreh, M. O., Szmulowicz, F., Fischer, D. W., Evans, K. R., and Stutz, C. E. (1991). *Phys. Rev. B* **43**, 9996.
Martinet, E., Luc, F., Rosencher, E., Bois, Ph., Costard, E., Delaitre, S., and Bockenhoff, E. (1992). In *Intersubband Transitions in Quantum Wells*, eds. E. Rosencher, B. Vinter, and B. Levine (Plenum, New York), p. 299.
Mei, T., Karunasiri, G., and Chua, S. J. (1997). *Appl. Phys. Lett.* **71**, 2017.
Mii, Y. J., Karunasiri, R. P. G., Wang, K. L., Chen, M., and Yuh, P. F. (1989). *Appl. Phys. Lett.* **55**, 2417.
Mooney, J. M., Shepherd, F. D., Ewing, W. S., Murgia, J. E., and Silverman, J. (1989). *Opt. Eng.* **28**, 1151.
O, B., Choe, J.-W., Francombe, M. H., Bandara, K. M. S. V., Coon, D. D., Lin, Y. F., and Takei, W. J. (1990). *Appl. Phys. Lett.* **57**, 503.
Pelve, E., Beltram, F., Bethea, C. G., Levine, B. F., Shen, V. O., Hsieh, S. J., Abbott, R. R. (1989). *J. Appl. Phys.* **66**, 5656.
Pinczuk, A., Heiman, D., Sooryakumar, R., Gossard, A. C., and Wiegmann, W. (1986). *Surf. Sci.* **170**, 573.
Pool, F. S., Wilson, D. W., Maker, P. D., Muller, R. E., Gill, J. J., Sengupta, D. K., Liu, J. K.,

Bandara, S. V., and Gunapala, S. D. (1998). *Proc. SPIE* **3379**, 402.
Realmuto, V. J., Sutton, A. J., and Elias, T. (1997). *J. Geo. Phys. Res.* **102**, 15057.
Ritter, D., Hamm, R. A., Panish, M. B., Vandenberg, J. M., Gershoni, D., Gunapala, S. D., and Levine, B. F. (1991). *Appl. Phys. Lett.* **59**, 552.
Rosencher, E., Luc, F., Bois, Ph., and Delaitre, S. (1992). *Appl. Phys. Lett.* **61**, 468.
Sarusi, G., Levine, B. F., Pearton, S. J., Bandara, K. M. S. V., and Leibenguth, R. E. (1994a). *Phys. Lett.* **64**, 960.
Sarusi, G., Levine, B. F., Pearton, S. J., Bandara, K. M. S. V., and Leibenguth, R. E. (1994b). *J. Appl. Phys.* 76, 4989.
Sarusi, G., Gunapala, S. D., Park, J. S., and Levine, B. F. (1994c). *J. Appl. Phys.* **76**, 6001.
Sengupta, D. K., Jackson, S. L., Ahmari, D., Kuo, H. C., Malin, J. I., Thomas, S., Feng, M., Stillman, G. E., Chang, Y. C., Li, L., and Liu, H. C. (1996). *Appl. Phys. Lett.* **69**, 3209.
Shepherd, F. D. (1988). *Proc. SPIE* **930**, 2.
Smith, J. S., Chiu, L. C., Margalit, S., Yariv, A., and Cho, A. Y. (1983). *J. Vac. Sci. Technol. B* **1**, 376.
Steele, A. G., Liu, H. C., Buchanan, M., and Wasilewski, Z. R. (1991). *Appl. Phys. Lett.* **59**, 3625.
Swaminathan, V., Stayt, Jr., J. W., Zilko, J. L., Trapp, K. D. C., Smith, L. E., Nakahara, S., Luther, L. C., Livescu, G., Levine, B. F., Leibenguth, R. E., Glogovsky, K. G., Gault, W. A., Focht, M. W., Buiocchi, C., and Asom, M. T. (1992). In *Proceedings of the IRIS Specialty Group on Infrared Detectors*, Moffet Field, CA.
Tidrow, M. Z., and Bacher, K. (1996). *Appl. Phys. Lett.* **69**, 3396.
Tidrow, M. Z., Chiang, J. C., Li, S. S., and Bacher, K. (1997). *Appl. Phys. Lett.* **70**, 859.
Vodjdani, N., Vinter, B., Berger, V., Bockenhoff, E., and Costard, E. (1991). *Appl. Phys. Lett.* **59**, 555.
Watanabe, M. O., and Ohba, Y. (1987). *Appl. Phys. Lett.* **50**, 906.
Weisbuch, C. (1987). *Semicond. Semimet.* **24**, 1.
West, L. C., and Eglash, S. J. (1985). *Appl. Phys. Lett.* **46**, 1156.
Wheeler, R. G., and Goldberg, H. S. (1975). *IEEE Trans. Electron. Devices* **ED-22**, 1001.
Wieck, A. D., Batke, E., Heitman, D., and Kotthaus, J. P. (1984). *Phys. Rev. B* **30**, 4653.
Wu, C. S., Wen, C. P., Sato, R. N., Hu, M., Tu, C. W., Zhang, J., Flesner, L. D., Pham, L., and Nayer, P. S. (1992). *IEEE Trans. Electron. Devices* **39**, 234.
Xing, B., and Liu, H. C. (1996). *J. Appl. Phys.* **80**, 1214.
Yao, J. Y., Andersson, T. G., and Dunlop, G. L. (1991). *Appl. Phys.* **69**, 2224.
Yu, L. S., and Li, S. S. (1991). *Appl. Phys. Lett.* **59**, 1332.
Yu, L. S., Li, S. S., and Ho, P. (1991). *Appl. Phys. Lett.* **59**, 2712.
Yu, L. S., Wang, Y. H., Li, S. S., and Ho, P. (1992). *Appl. Phys. Lett.* **60**, 992.
Zhou, X., Bhattacharya, P. K., Hugo, G., Hong, S. C., and Gulari, E. (1989). *Appl. Phys. Lett.* **54**, 855.
Zussman, A., Levine, B. F., Kuo, J. M., and de Jong, J. (1991). *J. Appl. Phys.* **70**, 5101.

Index

Note: Page numbers in italics refer to a figure on that page.

A

Absorption frequency, Coulomb effects, 39
Absorption spectra
   bound-continuum for multiquantum well, *19*
   intersubband for single quantum well, *18*
   multi-quantum well system with and without Bragg reflectors, *107*
   QWIP comparison, 221–222
   two-pass waveguide, *111*
Anticrossing
   and coupled-well system, 110
   and depolarization shift, 83
Astronomy and QWIPs, 274–275
Asymmetric coupled quantum well (ACQW), 28
   absorption spectra, *30*
Asymmetric quantum well, 26–31

B

Band structure effects, 39
Band structure engineering, 2
Bandedge profile
   integrated QWIP with LED, *183*
   stepped quantum well infrared photodetector, *174*
   three color QWIP, *172*
Beck's quantum efficiency, 155
Bound state from Bragg reflection, 102–103, 105–108
Bound-to-bound QWIPs, 201–202
   absorption spectra, 221–222
   dark current, 222–224
   detectivity, 234–236
   noise gain, 229–232
   photoconductive gain, 229–232
   quantum efficiency, 232–234
   responsivity, 225–227
Bound-to-continuum QWIPs, 202–203
   absorption spectra, 221–222
   dark current, 222–224
   detectivity, 234–236
   noise gain, 229–232
   photoconductive gain, 229–232
   quantum efficiency, 232–234
   responsivity, 225–227
Bound-to-quasibound QWIPs, 203–205
   absorption spectra, 221–222
   dark current, 222–224
   detectivity, 234–236
   noise gain, 229–232
   photoconductive gain, 229–232
   quantum efficiency, 232–234
   responsivity, 225–227
Bragg confinement, 18
Bragg reflectors
   and bound state, 102–103, 105–108
   and multi-quantum well system, 106–107
   multi-quantum well system, *107*
Brewster's angle, 20
Brillouin zone, 59

C

Coherence between paths
   applications, 102
   external manipulation, 102
Conduction band
   bound-to-quasibound QWIP, *249*
   broadband QWIPs, *206*

284 INDEX

Conduction band (*Continued*)
 description, 5
 in-plane polarization, 54–56
 nonparabolicity, 60
 offset values, 136
Conduction band diagram
 bound-to-bound QWIPs, 202
 bound-to-continuum QWIPs, *203*
 bound-to-miniband QWIP, *209*
 bound-to-miniband QWIPs, *209*
 indirect bandgap, *213*
Conductivity, preference in 2D case, 12
Constructive interference
 absorption experiments, 118–122
 emission experiments, 122–126
Corrugated structure, and QWIP light coupling, 240–242
Coulomb effects, 39, 42–45
Coupled quantum wells, 102–103, 108–113
Critical incidence coupling, 25

D

Dark current, 137–150
 bound-continuum vs. bound-to-quasibound QWIPs, *204*
 bound-to-bound state QWIPs, 201–202
 characteristics for n-type QWIPs, *142*
 characteristics for p-type QWIPs, *143*
 control of, *139*
 decreasing with microlenses, 242
 derivation, 144–145
 minimizing, 203
 origin, 224
 QWIP comparison, 222–224
Defense and QWIPs, 273–274
Density of states, 11
Depolarization shift
 and anticrossing, 83
 calculation, 48
 definition, 47
 and nonparabolicity cancellation, 51–52
 physical content, 48
Destructive interference
 absorption experiments, 118–122
 emission experiments, 122–126
Detectivity, QWIP comparison, 234–236
DFM. *see* Difference frequency mixing
Dielectric function
 preference in 2D case, 12
 and waveguide transmission, 23
Diffraction gratings for QWIPs, 133
Drude–Lorentz oscillator, 13

E

Effective mass
 energy dependence, 40
 nonparabolic effects, 41
 and polarization selection rule, 132
 spatial variation, 56–57
Effective mass approximation, 9
Eigenvalues
 definition, 199
 and intersubband absorption coefficient, 6
 symmetric quantum well, 15
Electromagnetic wave, description, 7
Emitters, 3
Energy level
 Coulomb effects, 39
 and quantum well depth and thickness, 2
 in quantum wells, 199–200
Envelope function approach to intersubband coefficient, 5–6
Envelope wave function, symmetric quantum well, 15
Exciton correction
 calculation, 49–50
 definition, 47
 example, 50–51

F

Fabry–Pérot effect, 17
Fano effect, 102–103, 112–116
 practical applications, 103
Fermi wavelength, 77–78
Fermi's golden rule, 4, 5, 6
Final-state interaction. see Exciton correction
Fire fighting and QWIPs, 270–271
Focal plane arrays. *see* FPA (focal plane arrays)
FPA (focal plane arrays)
 128×128 VLWIR imaging camera, 249–251

256 × 256 LWIR imaging camera, 251–254
640 × 486 LWIR imaging camera, 254–259
applications, 246
dualband (LWIR and VLWIR), 263–264
dualband (MWIR and LWIR) detectors, 260–262
effect of nonuniformity, 246–248
limitations on performance, 247
noise equivalent temperature difference (NEΔT), 246
pixel-to-pixel uniformity, 238–239
two-color, 260–262

G

GaAs–Al$_x$Ga$_{1-x}$, and infrared detection, 200
GaAs structures, wavelengths, 135–137
Geometries and intersubband transitions, 19–26
  Brewster's angle, 20
  Si/Ge prism, 24–25
  standard, 20
  transmission line arrangement, 24
  two-pass arrangement, 20–21
  use of grating coupler, 25
  for weak absorption, 20
Grating coupling for intersubband absorption, 25

H

Hartree–Fock method, 45
Hartree potential, 42–45
Heavy holes (HH), 59
Homogeneous broadening, 77–78

I

Impurities, 86–88
In-plane absorption mechanisms, 54–59
  spatial variation of effective mass, 56–57
  valence band coupling, 57–59
Incidence of radiation
  oblique, *14*

oblique and absorption coefficient, 13
Indirect bandgap QWIPs, 212–213
Indirect-gap semiconductors, in-plane polarization, 54–56
Infinite quantum well, intersubband transitions, 16
Infrared detectors, 54. *see also* QWIP
  band diagram, *199*
  p-type doped quantum wells, 59
  and quantum wells, 129–130
Infrared imaging systems. *see* FPA (focal plane arrays); QWIP
Inhomogeneous broadening, 77–78
Injection mechanism, *154*
Interband transitions, 2
Interference. *see* Quantum interference
Intersubband absorption
  electric field-induced suppression, 108
  and Fano interference, 112–116
  and interference, 118–122
  in metal-coated multi-quantum well, 22–23
  and quantum well position, 22
  red shift with high quantum numbers, 41
  relationship to emission, 126
  and Stark effect, 27
  in strongly coupled superlattices, 37–38
  valence band, 59–73
Intersubband absorption coefficient, 5–14
  definition, 10
  finite, symmetric quantum well, 6
  for indirect-gap semiconductors, 55
  multi-quantum well system, 13–14
  one-band model, 7–8
  superlattices, 35, *36*
  transfer matrix method, 6
  for two-dimension layer, 11
Intersubband absorption spectra. *see* Absorption spectra
Intersubband emission, relationship to absorption, 126
Intersubband transitions, 2
  absorption example, 135
  absorption spectra for single quantum well, *18*
  definition, 1
  density of states, 11
  description, 132–135
  energy as function of applied field, *124*
  finite, symmetric quantum well, 16–19

Intersubband transitions (*Continued*)
  geometries, 19–26
  history of study, 2–3
  infinite quantum well, *16*
  location, *199*
  materials, 3
  optical studies, 129
  peak absorption strength, 11
  photoinduced absorption, 3
  physics, 134
  and quantum well depth, 15
  spectral range, 4
  technological uses, 3
  valence band optical matrix elements, 67–69

K

Kane–Bastard two-band model, 40
Kane's theory, 40
  model of nonparabolicity, 60
Kohn-Sham density functional theory, 45
Kohn's theorem, 85
k · p theory
  and nonparabolicity, 41
  and valence band coupling, 57
Kronig–Penney model of superlattices, 9, 33

L

Landau levels, 80
Laser without population inversion (LWI)
  and Fano effect, 103
  and interference by tunneling into a continuum, 122
LEDs, integration with QWIPs, 183–188
Light coupling
  geometry, *133*
  to QWIPs, 237–246
Light holes (LH), 59
Line broadening, 73–81
  due to nonparabolicity, 42
  homogeneous and inhomogeneous, 77–78
  Lorenzian model, 135
Line width, applications, 73

Local-density approximation (LDA), 45
Luttinger–Kohn (LK) model, 60–61
  for bulk semiconductors, 61–64
  Hamiltonian, 61
  Hamiltonian for quantum wells, 65
  Hamiltonian for QW from strained structures, 67
  Hamiltonian matrix, 63
  for quantum wells, 64–67
LWI. *see* Laser without population inversion (LWI)

M

Magnetic field effects, 80–84
Many-body effects, 24, 45–46
  nonparabolicity cancellation, 51–52, 78
  and polarizability, 47
  studies of, 53
MBE. *see* Molecular beam epitaxy
Medicine and QWIPS, 271–272
Microlenses
  fabrication, *244*
  and QWIP light coupling, 242–246
Minibands
  and constructed bound states, 104
  definition, 32
  infrared detection, 206
  interminiband absorption, 38
Mixing of states, two QW, 102–103, 116–118
Molecular beam epitaxy, 2
MOSFET, 31
  and intersubband transition study, 3
Multi-quantum well (MQW), 32–38
  energy spectrum, *33*
  infrared detection, 198–199
  intersubband absorption coefficient calculation, 13–14
  metal coating and electric field distribution, 22–23
  nonlocal effective-medium approach, 24
Multicolor quantum well infrared photodetectors, 169–176
  designs, *170*
Multipass waveguide structure
  absorption calculation, 21
  transmission through, 22–23

N

Negative photoconductivity, 151
Noise and QWIPs, 160–161
Nonlinear optical elements, 3
Nonparabolicity, 40–42
  cancellation by many-body effects, 51–52, 78
  and subband energy shift in quantum wells, 41
Novel infrared detectors, 3

O

Optical coupling in QWIPs, 244
Optical matrix elements, for magnetic fields, 82
Optical phonon scattering, 76
Optical transitions, 2
  relaxation process, 73–74
Oscillator strength
  equation, 134
  and intersubband absorption coefficient calculation, 9
  and magnetic limit, 82
  vs miniband width, 36

P

p-type doped quantum wells
  intersubband studies, 71–73
  valence band intersubband transitions, 59
Parabolic quantum wells, 84–85
Pauli exclusion principle, 123
Peak absorption strength, 11
Phase-breaking processes, 101–102
Photoconductive gain, 151–158
  comparison with experimental results, 156–158
  derivation, 152–156
  mechanism, *152*
  vs number of wells, *157*
  physical mechanisms, 151–152
  QWIP comparison, 229–232
  time scale estimate, 158
Photocurrent, 150–159
  detector responsivity, 158–159
  and number of quantum wells, *153*

photoconductive gain, 151–158
Photoinduced intersubband absorption, 3
Photon drag effect, 88–89
Photon energy, 4
Pixelless large-area imaging device, 184–185
Polarization selection rule, 13, 19, 54, 104, 132, 237
  and QWIPs, 132
Positive photoconductivity, 151

Q

Quantized states in quantum well, 2–3
Quantum efficiency
  Beck's definition, 155
  QWIP comparison, 232–234
Quantum interference
  control by tunneling to a continuum, 116–118
  in coupled-well system, 110
  description, 101
  and intersubband optical absorption, 102–103
Quantum well
  asymmetric, 26–31
  description, 2, 198–199
  diagram, *199*
  doping and Coulomb potential, 42–45
  Luttinger-Kohn (LK) model of nonparabolicity, 64–67
  vs multi-quantum well system, 31–32
  shallow donor states, 87
  symmetric, 15–19
Quantum well infrared photodetectors. *see* QWIP
QWIP. *see also* FPA (focal plane arrays)
  3D carrier drift model, 139–143
  advantages, 130–131, 200
  applications, 270–275
  background limited spectral peak, 164
  background limited spectral peak derivation, 166–167
  blip temperature derivation, 166
  blip temperature equation, 164
  broadband for thermal infrared imaging, 267–270
  conduction band-edge profile, *137*
  detectivity derivation, 165–166

QWIP (*Continued*)
  detectivity expressions, 162
  detector noise, 160–161
  emission-capture model, 143–148
  figures of merit, 221–222
  high-frequency detectors, 176–182
  high performance for low-background applications, 264–267
  indirect bandgap, 212–213
  infrared heterodyne/microwave mixing experiment, *180*
  InGaAsP quaternary, 215–216
  integration with LEDs, 183–188
  light coupling, 237–246
  microwave rectification experiment, *177*
  models in literature, 143–144
  multicolor detectors, 169–176
  multispectral detectors, 169–176
  n-doped GaAs–Al$_{0.5}$In$_{0.5}$P, 217
  n-doped GaAsG–a$_{0.5}$In$_{0.5}$P, 216–217
  n-doped In$_{0.15}$Ga$_{0.85}$As–GaAs, 218–219
  n-doped In$_{0.53}$Ga$_{0.47}$As–In$_{0.52}$Al$_{0.48}$AS, 213–214
  n-doped In$_{0.53}$Ga$_{0.47}$As–InP, 214–215
  numerical Monte Carlo model, 150
  optical heterodyne experiment, *179*
  optimum design, 135–136, 167–169
  p-doped In$_{0.53}$Ga$_{0.47}$As–InP, 218–219
  physical models, 138
  rectified current vs. frequency, *178*
  self-consistent emission capture model, 150
  single quantum well, 211–212
  stepped quantum well, 174
  three color, *173*
  tuning with material selection, 198
  two-color, 172, 173
QWIP light coupling
  corrugated structure, 240–242
  microlenses, 242–246
  random reflectors, 238–239
  two-dimensional periodic gratings, 239–240
QWIP types
  n-doped asymmetrical GaAs–Al$_x$Ga$_{1-x}$AS, 208–209
  n-doped asymmetrical GaAs–Al$_x$Ga$_{1-x}$As, 209–210
  n-doped bound-to-bound, 201–202
  n-doped bound-to-bound miniband QWIPs, 206–207
  n-doped bound-to-continuum miniband QWIPs, 207–208
  n-doped bound-to-continuum QWIPs, 202–203
  n-doped bound-to-miniband, 208–209
  n-doped bound-to-quasibound QWIPs, 203–205
  n-doped broadband, 205–206
  p-doped, 210–211

R

Random reflectors, and QWIP light coupling, 238–239
Relaxation, 73–81, *75*
Relaxation time, 78–80
Responsivity, QWIP comparison, 225–227

S

Scattering processes in semiconductors, 74–76
Schottky gate, 26
Schrodinger equation
  for complete bloch wave function, 61
  for envelope function, 5
Selection rule. *see* Polarization selection rule
Semiconductors
  and band-structure engineering, 102
  indirect-gap and in-plane polarization, 54–56
  Luttinger–Kohn (LK) model of nonparabolicity, 61–64
  scattering processes, 74–76
Spatially modulated infrared spectrometer (SMIS), 269
Spectral range for intersubband transitions, 4
Standing electron wave effects, 18
Stark effect
  for square quantum wells, 27
  step quantum wells, *29*
Step quantum wells, 28
  as quantum well infrared photodetectors, 174
  and spatial variation of effective mass, 57

Stark effect, 29
Subband, definition, 134
Superlattices
  definition, 32
  depolarization effect, 51
  exciton correction, 51
  interminiband absorption, 38
  intersubband absorption coefficient calculation, 35, 36
  Kronig–Penney, 33
  Kronig–Penney model, 9
  lattice matched, 213
  miniband detector, 206
  oscillator strength, 37
  oscillator strength vs. miniband width, 36
  strongly coupled and intersubband absorption, 37–38
Symmetric quantum well, 15–19
Symmetry breaking of quantum wells, 26

T

Thermal imaging
  broadband QWIPs, 267–270
  QWIP application, 183
  wavelengths, 136
Transfer matrix method, 6
Two-band models, 40
Two-dimensional layers, intersubband absorption coefficient calculation, 11–13
Two-dimensional periodic gratings, and QWIP light coupling, 239–240

U

Up-conversion approach to QWIP–LED integration, 187

V

Valence band
  intersubband absorption, 59–73
  nonparabolicity, 60–61
  selection rules, 69–71
  structure, 60
  transition matrix elements, 69–71
  warping, 64
Valence band coupling, 57–59
Voigt geometry, 85
Volcanology and QWIPs, 271

W

Waveguide geometries, 21. see also Geometries and intersubband transitions
  multipass structure and absorption, 21
  self-consistent drift-diffusion model, 148–150
Waveguide transmission, and dielectric function, 23
Wavelength energy, 4
Wavelengths of GaAs-based structures, 135–137
Wide quantum wells
  broadening mechanism, 78–79
  Coulomb effects, 40

Contents of Volumes in This Series

Volume 1 Physics of III–V Compounds

C. *Hilsum*, Some Key Features of III–V Compounds
F. *Bassani*, Methods of Band Calculations Applicable to III–V Compounds
E. O. *Kane*, The k-p Method
V. L. *Bonch-Bruevich*, Effect of Heavy Doping on the Semiconductor Band Structure
D. *Long*, Energy Band Structures of Mixed Crystals of III–V Compounds
L. M. *Roth and P. N. Argyres*, Magnetic Quantum Effects
S. M. *Puri and T. H. Geballe*, Thermomagnetic Effects in the Quantum Region
W. M. *Becker*, Band Characteristics near Principal Minima from Magnetoresistance
E. H. *Putley*, Freeze-Out Effects, Hot Electron Effects, and Submillimeter Photoconductivity in InSb
H. *Weiss*, Magnetoresistance
B. *Ancker-Johnson*, Plasma in Semiconductors and Semimetals

Volume 2 Physics of III–V Compounds

M. G. *Holland*, Thermal Conductivity
S. I. *Novkova*, Thermal Expansion
U. *Piesbergen*, Heat Capacity and Debye Temperatures
G. *Giesecke*, Lattice Constants
J. R. *Drabble*, Elastic Properties
A. U. *Mac Rae and G. W. Gobeli*, Low Energy Electron Diffraction Studies
R. *Lee Mieher*, Nuclear Magnetic Resonance
B. *Goldstein*, Electron Paramagnetic Resonance
T. S. *Moss*, Photoconduction in III–V Compounds
E. *Antoncik and J. Tauc*, Quantum Efficiency of the Internal Photoelectric Effect in InSb
G. W. *Gobeli and I. G. Allen*, Photoelectric Threshold and Work Function
P. S. *Pershan*, Nonlinear Optics in III–V Compounds
M. *Gershenzon*, Radiative Recombination in the III–V Compounds
F. *Stern*, Stimulated Emission in Semiconductors

Volume 3 Optical of Properties III–V Compounds

M. Hass, Lattice Reflection
W. G. Spitzer, Multiphonon Lattice Absorption
D. L. Stierwalt and R. F. Potter, Emittance Studies
H. R. Philipp and H. Ehrenveich, Ultraviolet Optical Properties
M. Cardona, Optical Absorption above the Fundamental Edge
E. J. Johnson, Absorption near the Fundamental Edge
J. O. Dimmock, Introduction to the Theory of Exciton States in Semiconductors
B. Lax and J. G. Mavroides, Interband Magnetooptical Effects
H. Y. Fan, Effects of Free Carries on Optical Properties
E. D. Palik and G. B. Wright, Free-Carrier Magnetooptical Effects
R. H. Bube, Photoelectronic Analysis
B. O. Seraphin and H. E. Bennett, Optical Constants

Volume 4 Physics of III–V Compounds

N. A. Goryunova, A. S. Borschevskii, and D. N. Tretiakov, Hardness
N. N. Sirota, Heats of Formation and Temperatures and Heats of Fusion of Compounds $A^{III}B^{V}$
D. L. Kendall, Diffusion
A. G. Chynoweth, Charge Multiplication Phenomena
R. W. Keyes, The Effects of Hydrostatic Pressure on the Properties of III–V Semiconductors
L. W. Aukerman, Radiation Effects
N. A. Goryunova, F. P. Kesamanly, and D. N. Nasledov, Phenomena in Solid Solutions
R. T. Bate, Electrical Properties of Nonuniform Crystals

Volume 5 Infrared Detectors

H. Levinstein, Characterization of Infrared Detectors
P. W. Kruse, Indium Antimonide Photoconductive and Photoelectromagnetic Detectors
M. B. Prince, Narrowband Self-Filtering Detectors
I. Melngalis and T. C. Harman, Single-Crystal Lead-Tin Chalcogenides
D. Long and J. L. Schmidt, Mercury-Cadmium Telluride and Closely Related Alloys
E. H. Putley, The Pyroelectric Detector
N. B. Stevens, Radiation Thermopiles
R. J. Keyes and T. M. Quist, Low Level Coherent and Incoherent Detection in the Infrared
M. C. Teich, Coherent Detection in the Infrared
F. R. Arams, E. W. Sard, B. J. Peyton, and F. P. Pace, Infrared Heterodyne Detection with Gigahertz IF Response
H. S. Sommers, Jr., Macrowave-Based Photoconductive Detector
R. Sehr and R. Zuleeg, Imaging and Display

Volume 6 Injection Phenomena

M. A. Lampert and R. B. Schilling, Current Injection in Solids: The Regional Approximation Method
R. Williams, Injection by Internal Photoemission
A. M. Barnett, Current Filament Formation

R. Baron and J. W. Mayer, Double Injection in Semiconductors
W. Ruppel, The Photoconductor-Metal Contact

Volume 7 Application and Devices
Part A

J. A. Copeland and S. Knight, Applications Utilizing Bulk Negative Resistance
F. A. Padovani, The Voltage-Current Characteristics of Metal-Semiconductor Contacts
P. L. Hower, W. W. Hooper, B. R. Cairns, R. D. Fairman, and D. A. Tremere, The GaAs Field-Effect Transistor
M. H. White, MOS Transistors
G. R. Antell, Gallium Arsenide Transistors
T. L. Tansley, Heterojunction Properties

Part B

T. Misawa, IMPATT Diodes
H. C. Okean, Tunnel Diodes
R. B. Campbell and Hung-Chi Chang, Silicon Junction Carbide Devices
R. E. Enstrom, H. Kressel, and L. Krassner, High-Temperature Power Rectifiers of $GaAs_{1-x}P_x$

Volume 8 Transport and Optical Phenomena

R. J. Stirn, Band Structure and Galvanomagnetic Effects in III–V Compounds with Indirect Band Gaps
R. W. Ure, Jr., Thermoelectric Effects in III–V Compounds
H. Piller, Faraday Rotation
H. Barry Bebb and E. W. Williams, Photoluminescence I: Theory
E. W. Williams and H. Barry Bebb, Photoluminescence II: Gallium Arsenide

Volume 9 Modulation Techniques

B. O. Seraphin, Electroreflectance
R. L. Aggarwal, Modulated Interband Magnetooptics
D. F. Blossey and Paul Handler, Electroabsorption
B. Batz, Thermal and Wavelength Modulation Spectroscopy
I. Balslev, Piezopptical Effects
D. E. Aspnes and N. Bottka, Electric-Field Effects on the Dielectric Function of Semiconductors and Insulators

Volume 10 Transport Phenomena

R. L. Rhode, Low-Field Electron Transport
J. D. Wiley, Mobility of Holes in III–V Compounds
C. M. Wolfe and G. E. Stillman, Apparent Mobility Enhancement in Inhomogeneous Crystals
R. L. Petersen, The Magnetophonon Effect

Volume 11 Solar Cells

H. J. Hovel, Introduction; Carrier Collection, Spectral Response, and Photocurrent; Solar Cell Electrical Characteristics; Efficiency; Thickness; Other Solar Cell Devices; Radiation Effects; Temperature and Intensity; Solar Cell Technology

Volume 12 Infrared Detectors (II)

W. L. Eiseman, J. D. Merriam, and R. F. Potter, Operational Characteristics of Infrared Photodetectors
P. R. Bratt, Impurity Germanium and Silicon Infrared Detectors
E. H. Putley, InSb Submillimeter Photoconductive Detectors
G. E. Stillman, C. M. Wolfe, and J. O. Dimmock, Far-Infrared Photoconductivity in High Purity GaAs
G. E. Stillman and C. M. Wolfe, Avalanche Photodiodes
P. L. Richards, The Josephson Junction as a Detector of Microwave and Far-Infrared Radiation
E. H. Putley, The Pyroelectric Detector — An Update

Volume 13 Cadmium Telluride

K. Zanio, Materials Preparations; Physics; Defects; Applications

Volume 14 Lasers, Junctions, Transport

N. Holonyak, Jr. and M. H. Lee, Photopumped III–V Semiconductor Lasers
H. Kressel and J. K. Butler, Heterojunction Laser Diodes
A Van der Ziel, Space-Charge-Limited Solid-State Diodes
P. J. Price, Monte Carlo Calculation of Electron Transport in Solids

Volume 15 Contacts, Junctions, Emitters

B. L. Sharma, Ohmic Contacts to III–V Compounds Semiconductors
A. Nussbaum, The Theory of Semiconducting Junctions
J. S. Escher, NEA Semiconductor Photoemitters

Volume 16 Defects, (HgCd)Se, (HgCd)Te

H. Kressel, The Effect of Crystal Defects on Optoelectronic Devices
C. R. Whitsett, J. G. Broerman, and C. J. Summers, Crystal Growth and Properties of $Hg_{1-x}Cd_xSe$ alloys
M. H. Weiler, Magnetooptical Properties of $Hg_{1-x}Cd_x$Te Alloys
P. W. Kruse and J. G. Ready, Nonlinear Optical Effects in $Hg_{1-x}Cd_x$Te

Volume 17 CW Processing of Silicon and Other Semiconductors

J. F. Gibbons, Beam Processing of Silicon
A. Lietoila, R. B. Gold, J. F. Gibbons, and L. A. Christel, Temperature Distributions and Solid Phase Reaction Rates Produced by Scanning CW Beams

A. Leitoila and J. F. Gibbons, Applications of CW Beam Processing to Ion Implanted Crystalline Silicon
N. M. Johnson, Electronic Defects in CW Transient Thermal Processed Silicon
K. F. Lee, T. J. Stultz, and J. F. Gibbons, Beam Recrystallized Polycrystalline Silicon: Properties, Applications, and Techniques
T. Shibata, A. Wakita, T. W. Sigmon, and J. F. Gibbons, Metal-Silicon Reactions and Silicide
Y. I. Nissim and J. F. Gibbons, CW Beam Processing of Gallium Arsenide

Volume 18 Mercury Cadmium Telluride

P. W. Kruse, The Emergence of $(Hg_{1-x}Cd_x)Te$ as a Modern Infrared Sensitive Material
H. E. Hirsch, S. C. Liang, and A. G. White, Preparation of High-Purity Cadmium, Mercury, and Tellurium
W. F. H. Micklethwaite, The Crystal Growth of Cadmium Mercury Telluride
P. E. Petersen, Auger Recombination in Mercury Cadmium Telluride
R. M. Broudy and V. J. Mazurczyck, (HgCd)Te Photoconductive Detectors
M. B. Reine, A. K. Soad, and T. J. Tredwell, Photovoltaic Infrared Detectors
M. A. Kinch, Metal-Insulator-Semiconductor Infrared Detectors

Volume 19 Deep Levels, GaAs, Alloys, Photochemistry

G. F. Neumark and K. Kosai, Deep Levels in Wide Band-Gap III–V Semiconductors
D. C. Look, The Electrical and Photoelectronic Properties of Semi-Insulating GaAs
R. F. Brebrick, Ching-Hua Su, and Pok-Kai Liao, Associated Solution Model for Ga-In-Sb and Hg-Cd-Te
Y. Ya. Gurevich and Y. V. Pleskon, Photoelectrochemistry of Semiconductors

Volume 20 Semi-Insulating GaAs

R. N. Thomas, H. M. Hobgood, G. W. Eldridge, D. L. Barrett, T. T. Braggins, L. B. Ta, and S. K. Wang, High-Purity LEC Growth and Direct Implantation of GaAs for Monolithic Microwave Circuits
C. A. Stolte, Ion Implantation and Materials for GaAs Integrated Circuits
C. G. Kirkpatrick, R. T. Chen, D. E. Holmes, P. M. Asbeck, K. R. Elliott, R. D. Fairman, and J. R. Oliver, LEC GaAs for Integrated Circuit Applications
J. S. Blakemore and S. Rahimi, Models for Mid-Gap Centers in Gallium Arsenide

Volume 21 Hydrogenated Amorphous Silicon
Part A

J. I. Pankove, Introduction
M. Hirose, Glow Discharge; Chemical Vapor Deposition
Y. Uchida, di Glow Discharge
T. D. Moustakas, Sputtering
I. Yamada, Ionized-Cluster Beam Deposition
B. A. Scott, Homogeneous Chemical Vapor Deposition

F. J. Kampas, Chemical Reactions in Plasma Deposition
P. A. Longeway, Plasma Kinetics
H. A. Weakliem, Diagnostics of Silane Glow Discharges Using Probes and Mass Spectroscopy
L. Gluttman, Relation between the Atomic and the Electronic Structures
A. Chenevas-Paule, Experiment Determination of Structure
S. Minomura, Pressure Effects on the Local Atomic Structure
D. Adler, Defects and Density of Localized States

Part B

J. I. Pankove, Introduction
G. D. Cody, The Optical Absorption Edge of a-Si:H
N. M. Amer and W. B. Jackson, Optical Properties of Defect States in a-Si:H
P. J. Zanzucchi, The Vibrational Spectra of a-Si:H
Y. Hamakawa, Electroreflectance and Electroabsorption
J. S. Lannin, Raman Scattering of Amorphous Si, Ge, and Their Alloys
R. A. Street, Luminescence in a-Si:H
R. S. Crandall, Photoconductivity
J. Tauc, Time-Resolved Spectroscopy of Electronic Relaxation Processes
P. E. Vanier, IR-Induced Quenching and Enhancement of Photoconductivity and Photo luminescence
H. Schade, Irradiation-Induced Metastable Effects
L. Ley, Photoelectron Emission Studies

Part C

J. I. Pankove, Introduction
J. D. Cohen, Density of States from Junction Measurements in Hydrogenated Amorphous Silicon
P. C. Taylor, Magnetic Resonance Measurements in a-Si:H
K. Morigaki, Optically Detected Magnetic Resonance
J. Dresner, Carrier Mobility in a-Si:H
T. Tiedje, Information about band-Tail States from Time-of-Flight Experiments
A. R. Moore, Diffusion Length in Undoped a-Si:H
W. Beyer and J. Overhof, Doping Effects in a-Si:H
H. Fritzche, Electronic Properties of Surfaces in a-Si:H
C. R. Wronski, The Staebler-Wronski Effect
R. J. Nemanich, Schottky Barriers on a-Si:H
B. Abeles and T. Tiedje, Amorphous Semiconductor Superlattices

Part D

J. I. Pankove, Introduction
D. E. Carlson, Solar Cells
G. A. Swartz, Closed-Form Solution of I–V Characteristic for a a-Si:H Solar Cells
I. Shimizu, Electrophotography
S. Ishioka, Image Pickup Tubes

P. G. LeComber and W. E. Spear, The Development of the a-Si:H Field-Effect Transistor and Its Possible Applications
D. G. Ast, a-Si:H FET-Addressed LCD Panel
S. Kaneko, Solid-State Image Sensor
M. Matsumura, Charge-Coupled Devices
M. A. Bosch, Optical Recording
A. D'Amico and G. Fortunato, Ambient Sensors
H. Kukimoto, Amorphous Light-Emitting Devices
R. J. Phelan, Jr., Fast Detectors and Modulators
J. I. Pankove, Hybrid Structures
P. G. LeComber, A. E. Owen, W. E. Spear, J. Hajto, and W. K. Choi, Electronic Switching in Amorphous Silicon Junction Devices

Volume 22 Lightwave Communications Technology
Part A

K. Nakajima, The Liquid-Phase Epitaxial Growth of InGaAsP
W. T. Tsang, Molecular Beam Epitaxy for III–V Compound Semiconductors
G. B. Stringfellow, Organometallic Vapor-Phase Epitaxial Growth of III–V Semiconductors
G. Beuchet, Halide and Chloride Transport Vapor-Phase Deposition of InGaAsP and GaAs
M. Razeghi, Low-Pressure Metallo-Organic Chemical Vapor Deposition of $Ga_xIn_{1-x}AsP_{1-y}$ Alloys
P. M. Petroff, Defects in III–V Compound Semiconductors

Part B

J. P. van der Ziel, Mode Locking of Semiconductor Lasers
K. Y. Lau and A. Yariv, High-Frequency Current Modulation of Semiconductor Injection Lasers
C. H. Henry, Special Properties of Semiconductor Lasers
Y. Suematsu, K. Kishino, S. Arai, and F. Koyama, Dynamic Single-Mode Semiconductor Lasers with a Distributed Reflector
W. T. Tsang, The Cleaved-Coupled-Cavity (C^3) Laser

Part C

R. J. Nelson and N. K. Dutta, Review of InGaAsP InP Laser Structures and Comparison of Their Performance
N. Chinone and M. Nakamura, Mode-Stabilized Semiconductor Lasers for 0.7–0.8- and 1.1–1.6-μm Regions
Y. Horikoshi, Semiconductor Lasers with Wavelengths Exceeding 2 μm
B. A. Dean and M. Dixon, The Functional Reliability of Semiconductor Lasers as Optical Transmitters
R. H. Saul, T. P. Lee, and C. A. Burus, Light-Emitting Device Design
C. L. Zipfel, Light-Emitting Diode-Reliability
T. P. Lee and T. Li, LED-Based Multimode Lightwave Systems
K. Ogawa, Semiconductor Noise-Mode Partition Noise

Part D

F. *Capasso*, The Physics of Avalanche Photodiodes
T. P. *Pearsall and M. A. Pollack*, Compound Semiconductor Photodiodes
T. *Kaneda*, Silicon and Germanium Avalanche Photodiodes
S. R. *Forrest*, Sensitivity of Avalanche Photodetector Receivers for High-Bit-Rate Long-Wavelength Optical Communication Systems
J. C. *Campbell*, Phototransistors for Lightwave Communications

Part E

S. *Wang*, Principles and Characteristics of Integrable Active and Passive Optical Devices
S. *Margalit and A. Yariv*, Integrated Electronic and Photonic Devices
T. *Mukai, Y. Yamamoto, and T. Kimura*, Optical Amplification by Semiconductor Lasers

Volume 23 Pulsed Laser Processing of Semiconductors

R. F. *Wood, C. W. White, and R. T. Young*, Laser Processing of Semiconductors: An Overview
C. W. *White*, Segregation, Solute Trapping, and Supersaturated Alloys
G. E. *Jellison, Jr.*, Optical and Electrical Properties of Pulsed Laser-Annealed Silicon
R. F. *Wood and G. E. Jellison, Jr.*, Melting Model of Pulsed Laser Processing
R. F. *Wood and F. W. Young, Jr.*, Nonequilibrium Solidification Following Pulsed Laser Melting
D. H. *Lowndes and G. E. Jellison, Jr.*, Time-Resolved Measurement During Pulsed Laser Irradiation of Silicon
D. M. *Zebner*, Surface Studies of Pulsed Laser Irradiated Semiconductors
D. H. *Lowndes*, Pulsed Beam Processing of Gallium Arsenide
R. B. *James*, Pulsed CO_2 Laser Annealing of Semiconductors
R. T. *Young and R. F. Wood*, Applications of Pulsed Laser Processing

Volume 24 Applications of Multiquantum Wells, Selective Doping, and Superlattices

C. *Weisbuch*, Fundamental Properties of III–V Semiconductor Two-Dimensional Quantized Structures: The Basis for Optical and Electronic Device Applications
H. *Morkoc and H. Unlu*, Factors Affecting the Performance of (Al,Ga)As/GaAs and (Al,Ga)As/InGaAs Modulation-Doped Field-Effect Transistors: Microwave and Digital Applications
N. T. *Linh*, Two-Dimensional Electron Gas FETs: Microwave Applications
M. *Abe et al.*, Ultra-High-Speed HEMT Integrated Circuits
D. S. *Chemla, D. A. B. Miller, and P. W. Smith*, Nonlinear Optical Properties of Multiple Quantum Well Structures for Optical Signal Processing
F. *Capasso*, Graded-Gap and Superlattice Devices by Band-Gap Engineering
W. T. *Tsang*, Quantum Confinement Heterostructure Semiconductor Lasers
G. C. *Osbourn et al.*, Principles and Applications of Semiconductor Strained-Layer Superlattices

Volume 25 Diluted Magnetic Semiconductors

W. Giriat and J. K. Furdyna, Crystal Structure, Composition, and Materials Preparation of Diluted Magnetic Semiconductors

W. M. Becker, Band Structure and Optical Properties of Wide-Gap $A_{1-x}^{II} Mn_x B_{IV}$ Alloys at Zero Magnetic Field

S. Oseroff and P. H. Keesom, Magnetic Properties: Macroscopic Studies

T. Giebultowicz and T. M. Holden, Neutron Scattering Studies of the Magnetic Structure and Dynamics of Diluted Magnetic Semiconductors

J. Kossut, Band Structure and Quantum Transport Phenomena in Narrow-Gap Diluted Magnetic Semiconductors

C. Riquaux, Magnetooptical Properties of Large-Gap Diluted Magnetic Semiconductors

J. A. Gaj, Magnetooptical Properties of Large-Gap Diluted Magnetic Semiconductors

J. Mycielski, Shallow Acceptors in Diluted Magnetic Semiconductors: Splitting, Boil-off, Giant Negative Magnetoresistance

A. K. Ramadas and R. Rodriquez, Raman Scattering in Diluted Magnetic Semiconductors

P. A. Wolff, Theory of Bound Magnetic Polarons in Semimagnetic Semiconductors

Volume 26 III–V Compound Semiconductors and Semiconductor Properties of Superionic Materials

Z. Yuanxi, III–V Compounds

H. V. Winston, A. T. Hunter, H. Kimura, and R. E. Lee, InAs-Alloyed GaAs Substrates for Direct Implantation

P. K. Bhattacharya and S. Dhar, Deep Levels in III–V Compound Semiconductors Grown by MBE

Y. Ya. Gurevich and A. K. Ivanov-Shits, Semiconductor Properties of Supersonic Materials

Volume 27 High Conducting Quasi-One-Dimensional Organic Crystals

E. M. Conwell, Introduction to Highly Conducting Quasi-One-Dimensional Organic Crystals

I. A. Howard, A Reference Guide to the Conducting Quasi-One-Dimensional Organic Molecular Crystals

J. P. Pouquet, Structural Instabilities

E. M. Conwell, Transport Properties

C. S. Jacobsen, Optical Properties

J. C. Scott, Magnetic Properties

L. Zuppiroli, Irradiation Effects: Perfect Crystals and Real Crystals

Volume 28 Measurement of High-Speed Signals in Solid State Devices

J. Frey and D. Ioannou, Materials and Devices for High-Speed and Optoelectronic Applications

H. Schumacher and E. Strid, Electronic Wafer Probing Techniques

D. H. Auston, Picosecond Photoconductivity: High-Speed Measurements of Devices and Materials

J. A. Valdmanis, Electro-Optic Measurement Techniques for Picosecond Materials, Devices, and Integrated Circuits.

J. M. Wiesenfeld and R. K. Jain, Direct Optical Probing of Integrated Circuits and High-Speed Devices

G. Plows, Electron-Beam Probing

A. M. Weiner and R. B. Marcus, Photoemissive Probing

Volume 29 Very High Speed Integrated Circuits: Gallium Arsenide LSI

M. Kuzuhara and T. Nazaki, Active Layer Formation by Ion Implantation
H. Hasimoto, Focused Ion Beam Implantation Technology
T. Nozaki and A. Higashisaka, Device Fabrication Process Technology
M. Ino and T. Takada, GaAs LSI Circuit Design
M. Hirayama, M. Ohmori, and K. Yamasaki, GaAs LSI Fabrication and Performance

Volume 30 Very High Speed Integrated Circuits: Heterostructure

H. Watanabe, T. Mizutani, and A. Usui, Fundamentals of Epitaxial Growth and Atomic Layer Epitaxy
S. Hiyamizu, Characteristics of Two-Dimensional Electron Gas in III–V Compound Heterostructures Grown by MBE
T. Nakanisi, Metalorganic Vapor Phase Epitaxy for High-Quality Active Layers
T. Nimura, High Electron Mobility Transistor and LSI Applications
T. Sugeta and T. Ishibashi, Hetero-Bipolar Transistor and LSI Application
H. Matsueda, T. Tanaka, and M. Nakamura, Optoelectronic Integrated Circuits

Volume 31 Indium Phosphide: Crystal Growth and Characterization

J. P. Farges, Growth of Discoloration-free InP
M. J. McCollum and G. E. Stillman, High Purity InP Grown by Hydride Vapor Phase Epitaxy
T. Inada and T. Fukuda, Direct Synthesis and Growth of Indium Phosphide by the Liquid Phosphorous Encapsulated Czochralski Method
O. Oda, K. Katagiri, K. Shinohara, S. Katsura, Y. Takahashi, K. Kainosho, K. Kohiro, and R. Hirano, InP Crystal Growth, Substrate Preparation and Evaluation
K. Tada, M. Tatsumi, M. Morioka, T. Araki, and T. Kawase, InP Substrates: Production and Quality Control
M. Razeghi, LP-MOCVD Growth, Characterization, and Application of InP Material
T. A. Kennedy and P. J. Lin-Chung, Stoichiometric Defects in InP

Volme 32 Strained-Layer Superlattices: Physics

T. P. Pearsall, Strained-Layer Superlattices
F. H. Pollack, Effects of Homogeneous Strain on the Electronic and Vibrational Levels in Semiconductors
J. Y. Marzin, J. M. Gerárd, P. Voisin, and J. A. Brum, Optical Studies of Strained III–V Heterolayers
R. People and S. A. Jackson, Structurally Induced States from Strain and Confinement
M. Jaros, Microscopic Phenomena in Ordered Superlattices

Volume 33 Strained-Layer Superlattices: Materials Science and Technology

R. Hull and J. C. Bean, Principles and Concepts of Strained-Layer Epitaxy
W. J. Schaff, P. J. Tasker, M. C. Foisy, and L. F. Eastman, Device Applications of Strained-Layer Epitaxy

S. T. Picraux, B. L. Doyle, and J. Y. Tsao, Structure and Characterization of Strained-Layer Superlattices
E. Kasper and F. Schaffer, Group IV Compounds
D. L. Martin, Molecular Beam Epitaxy of IV–VI Compounds Heterojunction
R. L. Gunshor, L. A. Kolodziejski, A. V. Nurmikko, and N. Otsuka, Molecular Beam Epitaxy of II–VI Semiconductor Microstructures

Volume 34   Hydrogen in Semiconductors

J. I. Pankove and N. M. Johnson, Introduction to Hydrogen in Semiconductors
C. H. Seager, Hydrogenation Methods
J. I. Pankove, Hydrogenation of Defects in Crystalline Silicon
J. W. Corbett, P. Deák, U. V. Desnica, and S. J. Pearton, Hydrogen Passivation of Damage Centers in Semiconductors
S. J. Pearton, Neutralization of Deep Levels in Silicon
J. I. Pankove, Neutralization of Shallow Acceptors in Silicon
N. M. Johnson, Neutralization of Donor Dopants and Formation of Hydrogen-Induced Defects in n-Type Silicon
M. Stavola and S. J. Pearton, Vibrational Spectroscopy of Hydrogen-Related Defects in Silicon
A. D. Marwick, Hydrogen in Semiconductors: Ion Beam Techniques
C. Herring and N. M. Johnson, Hydrogen Migration and Solubility in Silicon
E. E. Haller, Hydrogen-Related Phenomena in Crystalline Germanium
J. Kakalios, Hydrogen Diffusion in Amorphous Silicon
J. Chevalier, B. Clerjaud, and B. Pajot, Neutralization of Defects and Dopants in III–V Semiconductors
G. G. DeLeo and W. B. Fowler, Computational Studies of Hydrogen-Containing Complexes in Semiconductors
R. F. Kiefl and T. L. Estle, Muonium in Semiconductors
C. G. Van de Walle, Theory of Isolated Interstitial Hydrogen and Muonium in Crystalline Semiconductors

Volume 35   Nanostructured Systems

M. Reed, Introduction
H. van Houten, C. W. J. Beenakker, and B. J. van Wees, Quantum Point Contacts
G. Timp, When Does a Wire Become an Electron Waveguide?
M. Büttiker, The Quantum Hall Effects in Open Conductors
W. Hansen, J. P. Kotthaus, and U. Merkt, Electrons in Laterally Periodic Nanostructures

Volume 36   The Spectroscopy of Semiconductors

D. Heiman, Spectroscopy of Semiconductors at Low Temperatures and High Magnetic Fields
A. V. Nurmikko, Transient Spectroscopy by Ultrashort Laser Pulse Techniques
A. K. Ramdas and S. Rodriguez, Piezospectroscopy of Semiconductors
O. J. Glembocki and B. V. Shanabrook, Photoreflectance Spectroscopy of Microstructures
D. G. Seiler, C. L. Littler, and M. H. Wiler, One- and Two-Photon Magneto-Optical Spectroscopy of InSb and $Hg_{1-x}Cd_xTe$

Volume 37 The Mechanical Properties of Semiconductors

A.-B. Chen, A. Sher and W. T. Yost, Elastic Constants and Related Properties of Semiconductor Compounds and Their Alloys
D. R. Clarke, Fracture of Silicon and Other Semiconductors
H. Siethoff, The Plasticity of Elemental and Compound Semiconductors
S. Guruswamy, K. T. Faber and J. P. Hirth, Mechanical Behavior of Compound Semiconductors
S. Mahajan, Deformation Behavior of Compound Semiconductors
J. P. Hirth, Injection of Dislocations into Strained Multilayer Structures
D. Kendall, C. B. Fleddermann, and K. J. Malloy, Critical Technologies for the Micromachining of Silicon
I. Matsuba and K. Mokuya, Processing and Semiconductor Thermoelastic Behavior

Volume 38 Imperfections in III/V Materials

U. Scherz and M. Scheffler, Density-Functional Theory of sp-Bonded Defects in III/V Semiconductors
M. Kaminska and E. R. Weber, El2 Defect in GaAs
D. C. Look, Defects Relevant for Compensation in Semi-Insulating GaAs
R. C. Newman, Local Vibrational Mode Spectroscopy of Defects in III/V Compounds
A. M. Hennel, Transition Metals in III/V Compounds
K. J. Malloy and K. Khachaturyan, DX and Related Defects in Semiconductors
V. Swaminathan and A. S. Jordan, Dislocations in III/V Compounds
K. W. Nauka, Deep Level Defects in the Epitaxial III/V Materials

Volume 39 Minority Carriers in III–V Semiconductors: Physics and Applications

N. K. Dutta, Radiative Transitions in GaAs and Other III–V Compounds
R. K. Ahrenkiel, Minority-Carrier Lifetime in III–V Semiconductors
T. Furuta, High Field Minority Electron Transport in p-GaAs
M. S. Lundstrom, Minority-Carrier Transport in III–V Semiconductors
R. A. Abram, Effects of Heavy Doping and High Excitation on the Band Structure of GaAs
D. Yevick and W. Bardyszewski, An Introduction to Non-Equilibrium Many-Body Analyses of Optical Processes in III–V Semiconductors

Volume 40 Epitaxial Microstructures

E. F. Schubert, Delta-Doping of Semiconductors: Electronic, Optical, and Structural Properties of Materials and Devices
A. Gossard, M. Sundaram, and P. Hopkins, Wide Graded Potential Wells
P. Petroff, Direct Growth of Nanometer-Size Quantum Wire Superlattices
E. Kapon, Lateral Patterning of Quantum Well Heterostructures by Growth of Nonplanar Substrates
H. Temkin, D. Gershoni, and M. Panish, Optical Properties of $Ga_{1-x}In_xAs/InP$ Quantum Wells

Volume 41 High Speed Heterostructure Devices

F. Capasso, F. Beltram, S. Sen, A. Pahlevi, and A. Y. Cho, Quantum Electron Devices: Physics and Applications
P. Solomon, D. J. Frank, S. L. Wright, and F. Canora, GaAs-Gate Semiconductor–Insulator–Semiconductor FET
M. H. Hashemi and U. K. Mishra, Unipolar InP-Based Transistors
R. Kiehl, Complementary Heterostructure FET Integrated Circuits
T. Ishibashi, GaAs-Based and InP-Based Heterostructure Bipolar Transistors
H. C. Liu and T. C. L. G. Sollner, High-Frequency-Tunneling Devices
H. Ohnishi, T. More, M. Takatsu, K. Imamura, and N. Yokoyama, Resonant-Tunneling Hot-Electron Transistors and Circuits

Volume 42 Oxygen in Silicon

F. Shimura, Introduction to Oxygen in Silicon
W. Lin, The Incorporation of Oxygen into Silicon Crystals
T. J. Schaffner and D. K. Schroder, Characterization Techniques for Oxygen in Silicon
W. M. Bullis, Oxygen Concentration Measurement
S. M. Hu, Intrinsic Point Defects in Silicon
B. Pajot, Some Atomic Configurations of Oxygen
J. Michel and L. C. Kimerling, Electical Properties of Oxygen in Silicon
R. C. Newman and R. Jones, Diffusion of Oxygen in Silicon
T. Y. Tan and W. J. Taylor, Mechanisms of Oxygen Precipitation: Some Quantitative Aspects
M. Schrems, Simulation of Oxygen Precipitation
K. Simino and I. Yonenaga, Oxygen Effect on Mechanical Properties
W. Bergholz, Grown-in and Process-Induced Effects
F. Shimura, Intrinsic/Internal Gettering
H. Tsuya, Oxygen Effect on Electronic Device Performance

Volume 43 Semiconductors for Room Temperature Nuclear Detector Applications

R. B. James and T. E. Schlesinger, Introduction and Overview
L. S. Darken and C. E. Cox, High-Purity Germanium Detectors
A. Burger, D. Nason, L. Van den Berg, and M. Schieber, Growth of Mercuric Iodide
X. J. Bao, T. E. Schlesinger, and R. B. James, Electrical Properties of Mercuric Iodide
X. J. Bao, R. B. James, and T. E. Schlesinger, Optical Properties of Red Mercuric Iodide
M. Hage-Ali and P. Siffert, Growth Methods of CdTe Nuclear Detector Materials
M. Hage-Ali and P Siffert, Characterization of CdTe Nuclear Detector Materials
M. Hage-Ali and P. Siffert, CdTe Nuclear Detectors and Applications
R. B. James, T. E. Schlesinger, J. Lund, and M. Schieber, $Cd_{1-x}Zn_xTe$ Spectrometers for Gamma and X-Ray Applications
D. S. McGregor, J. E. Kammeraad, Gallium Arsenide Radiation Detectors and Spectrometers
J. C. Lund, F. Olschner, and A. Burger, Lead Iodide
M. R. Squillante, and K. S. Shah, Other Materials: Status and Prospects
V. M. Gerrish, Characterization and Quantification of Detector Performance
J. S. Iwanczyk and B. E. Patt, Electronics for X-ray and Gamma Ray Spectrometers
M. Schieber, R. B. James, and T. E. Schlesinger, Summary and Remaining Issues for Room Temperature Radiation Spectrometers

Volume 44   II–IV Blue/Green Light Emitters: Device Physics and Epitaxial Growth

J. Han and R. L. Gunshor, MBE Growth and Electrical Properties of Wide Bandgap ZnSe-based II–VI Semiconductors
S. Fujita and S. Fujita, Growth and Characterization of ZnSe-based II–VI Semiconductors by MOVPE
E. Ho and L. A. Kolodziejski, Gaseous Source UHV Epitaxy Technologies for Wide Bandgap II–VI Semiconductors
C. G. Van de Walle, Doping of Wide-Band-Gap II–VI Compounds — Theory
R. Cingolani, Optical Properties of Excitons in ZnSe-Based Quantum Well Heterostructures
A. Ishibashi and A. V. Nurmikko, II–VI Diode Lasers: A Current View of Device Performance and Issues
S. Guha and J. Petruzzello, Defects and Degradation in Wide-Gap II–VI-based Structures and Light Emitting Devices

Volume 45   Effect of Disorder and Defects in Ion-Implanted Semiconductors: Electrical and Physiochemical Characterization

H. Ryssel, Ion Implantation into Semiconductors: Historical Perspectives
You-Nian Wang and Teng-Cai Ma, Electronic Stopping Power for Energetic Ions in Solids
S. T. Nakagawa, Solid Effect on the Electronic Stopping of Crystalline Target and Application to Range Estimation
G. Müller, S. Kalbitzer and G. N. Greaves, Ion Beams in Amorphous Semiconductor Research
J. Boussey-Said, Sheet and Spreading Resistance Analysis of Ion Implanted and Annealed Semiconductors
M. L. Polignano and G. Queirolo, Studies of the Stripping Hall Effect in Ion-Implanted Silicon
J. Stoemenos, Transmission Electron Microscopy Analyses
R. Nipoti and M. Servidori, Rutherford Backscattering Studies of Ion Implanted Semiconductors
P. Zaumseil, X-ray Diffraction Techniques

Volume 46   Effect of Disorder and Defects in Ion-Implanted Semiconductors: Optical and Photothermal Characterization

M. Fried, T. Lohner and J. Gyulai, Ellipsometric Analysis
A. Seas and C. Christofides, Transmission and Reflection Spectroscopy on Ion Implanted Semiconductors
A. Othonos and C. Christofides, Photoluminescence and Raman Scattering of Ion Implanted Semiconductors. Influence of Annealing
C. Christofides, Photomodulated Thermoreflectance Investigation of Implanted Wafers. Annealing Kinetics of Defects
U. Zammit, Photothermal Deflection Spectroscopy Characterization of Ion-Implanted and Annealed Silicon Films
A. Mandelis, A. Budiman and M. Vargas, Photothermal Deep-Level Transient Spectroscopy of Impurities and Defects in Semiconductors
R. Kalish and S. Charbonneau, Ion Implantation into Quantum-Well Structures
A. M. Myasnikov and N. N. Gerasimenko, Ion Implantation and Thermal Annealing of III-V Compound Semiconducting Systems: Some Problems of III-V Narrow Gap Semiconductors

Volume 47 Uncooled Infrared Imaging Arrays and Systems

R. G. Buser and M. P. Tompsett, Historical Overview
P. W. Kruse, Principles of Uncooled Infrared Focal Plane Arrays
R. A. Wood, Monolithic Silicon Microbolometer Arrays
C. M. Hanson, Hybrid Pyroelectric-Ferroelectric Bolometer Arrays
D. L. Polla and J. R. Choi, Monolithic Pyroelectric Bolometer Arrays
N. Teranishi, Thermoelectric Uncooled Infrared Focal Plane Arrays
M. F. Tompsett, Pyroelectric Vidicon
T. W. Kenny, Tunneling Infrared Sensors
J. R. Vig, R. L. Filler and Y. Kim, Application of Quartz Microresonators to Uncooled Infrared Imaging Arrays
P. W. Kruse, Application of Uncooled Monolithic Thermoelectric Linear Arrays to Imaging Radiometers

Volume 48 High Brightness Light Emitting Diodes

G. B. Stringfellow, Materials Issues in High-Brightness Light-Emitting Diodes
M. G. Craford, Overview of Device issues in High-Brightness Light-Emitting Diodes
F. M. Steranka, AlGaAs Red Light Emitting Diodes
C. H. Chen, S. A. Stockman, M. J. Peanasky, and C. P. Kuo, OMVPE Growth of AlGaInP for High Efficiency Visible Light-Emitting Diodes
F. A. Kish and R. M. Fletcher, AlGaInP Light-Emitting Diodes
M. W. Hodapp, Applications for High Brightness Light-Emitting Diodes
I. Akasaki and H. Amano, Organometallic Vapor Epitaxy of GaN for High Brightness Blue Light Emitting Diodes
S. Nakamura, Group III-V Nitride Based Ultraviolet-Blue-Green-Yellow Light-Emitting Diodes and Laser Diodes

Volume 49 Light Emission in Silicon: from Physics to Devices

D. J. Lockwood, Light Emission in Silicon
G. Abstreiter, Band Gaps and Light Emission in Si/SiGe Atomic Layer Structures
T. G. Brown and D. G. Hall, Radiative Isoelectronic Impurities in Silicon and Silicon-Germanium Alloys and Superlattices
J. Michel, L. V. C. Assali, M. T. Morse, and L. C. Kimerling, Erbium in Silicon
Y. Kanemitsu, Silicon and Germanium Nanoparticles
P. M. Fauchet, Porous Silicon: Photoluminescence and Electroluminescent Devices
C. Delerue, G. Allan, and M. Lannoo, Theory of Radiative and Nonradiative Processes in Silicon Nanocrystallites
L. Brus, Silicon Polymers and Nanocrystals

Volume 50 Gallium Nitride (GaN)

J. I. Pankove and T. D. Moustakas, Introduction
S. P. DenBaars and S. Keller, Metalorganic Chemical Vapor Deposition (MOCVD) of Group III Nitrides
W. A. Bryden and T. J. Kistenmacher, Growth of Group III-A Nitrides by Reactive Sputtering
N. Newman, Thermochemistry of III-N Semiconductors
S. J. Pearton and R. J. Shul, Etching of III Nitrides

S. M. Bedair, Indium-based Nitride Compounds
A. Trampert, O. Brandt, and K. H. Ploog, Crystal Structure of Group III Nitrides
H. Morkoc, F. Hamdani, and A. Salvador, Electronic and Optical Properties of III–V Nitride based Quantum Wells and Superlattices
K. Doverspike and J. I. Pankove, Doping in the III-Nitrides
T. Suski and P. Perlin, High Pressure Studies of Defects and Impurities in Gallium Nitride
B. Monemar, Optical Properties of GaN
W. R. L. Lambrecht, Band Structure of the Group III Nitrides
N. E. Christensen and P. Perlin, Phonons and Phase Transitions in GaN
S. Nakamura, Applications of LEDs and LDs
I. Akasaki and H. Amano, Lasers
J. A. Cooper, Jr., Nonvolatile Random Access Memories in Wide Bandgap Semiconductors

Volume 51A Identification of Defects in Semiconductors

G. D. Watkins, EPR and ENDOR Studies of Defects in Semiconductors
J.-M. Spaeth, Magneto-Optical and Electrical Detection of Paramagnetic Resonance in Semiconductors
T. A. Kennedy and E. R. Glaser, Magnetic Resonance of Epitaxial Layers Detected by Photoluminescence
K. H. Chow, B. Hitti, and R. F. Kiefl, μSR on Muonium in Semiconductors and Its Relation to Hydrogen
K. Saarinen, P. Hautojärvi, and C. Corbel, Positron Annihilation Spectroscopy of Defects in Semiconductors
R. Jones and P. R. Briddon, The *Ab Initio* Cluster Method and the Dynamics of Defects in Semiconductors

Volume 51B Identification of Defects in Semiconductors

G. Davies, Optical Measurements of Point Defects
P. M. Mooney, Defect Identification Using Capacitance Spectroscopy
M. Stavola, Vibrational Spectroscopy of Light Element Impurities in Semiconductors
P. Schwander, W. D. Rau, C. Kisielowski, M. Gribelyuk, and A. Ourmazd, Defect Processes in Semiconductors Studied at the Atomic Level by Transmission Electron Microscopy
N. D. Jager and E. R. Weber, Scanning Tunneling Microscopy of Defects in Semiconductors

Volume 52 SiC Materials and Devices

K. Järrendahl and R. F. Davis, Materials Properties and Characterization of SiC
V. A. Dmitriev and M. G. Spencer, SiC Fabrication Technology: Growth and Doping
V. Saxena and A. J. Steckl, Building Blocks for SiC Devices: Ohmic Contacts, Schottky Contacts, and p-n Junctions
M. S. Shur, SiC Transistors
C. D. Brandt, R. C. Clarke, R. R. Siergiej, J. B. Casady, A. W. Morse, S. Sriram, and A. K. Agarwal, SiC for Applications in High-Power Electronics
R. J. Trew, SiC Microwave Devices

J. Edmond, H. Kong, G. Negley, M. Leonard, K. Doverspike, W. Weeks, A. Suvorov, D. Waltz, and C. Carter, Jr., SiC-Based UV Photodiodes and Light-Emitting Diodes
H. Morkoç, Beyond Silicon Carbide! III–V Nitride-Based Heterostructures and Devices

Volume 53 Cumulative Subject and Author Index Including Tables of Contents for Volume 1–50

Volume 54 High Pressure in Semiconductor Physics I

W. Paul, High Pressure in Semiconductor Physics: A Historical Overview
N. E. Christensen, Electronic Structure Calculations for Semiconductors under Pressure
R. J. Neimes and M. I. McMahon, Structural Transitions in the Group IV, III-V and II-VI Semiconductors Under Pressure
A. R. Goni and K. Syassen, Optical Properties of Semiconductors Under Pressure
P. Trautman, M. Baj, and J. M. Baranowski, Hydrostatic Pressure and Uniaxial Stress in Investigations of the EL2 Defect in GaAs
M. Li and P. Y. Yu, High-Pressure Study of DX Centers Using Capacitance Techniques
T. Suski, Spatial Correlations of Impurity Charges in Doped Semiconductors
N. Kuroda, Pressure Effects on the Electronic Properties of Diluted Magnetic Semiconductors

Volume 55 High Pressure in Semiconductor Physics II

D. K. Maude and J. C. Portal, Parallel Transport in Low-Dimensional Semiconductor Structures
P. C. Klipstein, Tunneling Under Pressure: High-Pressure Studies of Vertical Transport in Semiconductor Heterostructures
E. Anastassakis and M. Cardona, Phonons, Strains, and Pressure in Semiconductors
F. H. Pollak, Effects of External Uniaxial Stress on the Optical Properties of Semiconductors and Semiconductor Microstructures
A. R. Adams, M. Silver, and J. Allam, Semiconductor Optoelectronic Devices
S. Porowski and I. Grzegory, The Application of High Nitrogen Pressure in the Physics and Technology of III-N Compounds
M. Yousuf, Diamond Anvil Cells in High Pressure Studies of Semiconductors

Volume 56 Germanium Silicon: Physics and Materials

J. C. Bean, Growth Techniques and Procedures
D. E. Savage, F. Liu, V. Zielasek, and M. G. Lagally, Fundamental Crystal Growth Mechanisms
R. Hull, Misfit Strain Accommodation in SiGe Heterostructures
M. J. Shaw and M. Jaros, Fundamental Physics of Strained Layer GeSi: Quo Vadis?
F. Cerdeira, Optical Properties
S. A. Ringel and P. N. Grillot, Electronic Properties and Deep Levels in Germanium-Silicon
J. C. Campbell, Optoelectronics in Silicon and Germanium Silicon
K. Eberl, K. Brunner, and O. G. Schmidt, $Si_{1-y}C_y$ and $Si_{1-x-y}Ge_xC_y$ Alloy Layers

Volume 57 Gallium Nitride (GaN) II

R. J. Molnar, Hydride Vapor Phase Epitaxial Growth of III-V Nitrides
T. D. Moustakas, Growth of III-V Nitrides by Molecular Beam Epitaxy
Z. Liliental-Weber, Defects in Bulk GaN and Homoepitaxial Layers
C. G. Van de Walle and N. M. Johnson, Hydrogen in III-V Nitrides
W. Götz and N. M. Johnson, Characterization of Dopants and Deep Level Defects in Gallium Nitride
B. Gil, Stress Effects on Optical Properties
C. Kisielowski, Strain in GaN Thin Films and Heterostructures
J. A. Miragliotta and D. K. Wickenden, Nonlinear Optical Properties of Gallium Nitride
B. K. Meyer, Magnetic Resonance Investigations on Group III-Nitrides
M. S. Shur and M. Asif Khan, GaN and AlGaN Ultraviolet Detectors
C. H. Qiu, J. I. Pankove, and C. Rossington, III-V Nitride-Based X-ray Detectors

Volume 58 Nonlinear Optics in Semiconductors I

A. Kost, Resonant Optical Nonlinearities in Semiconductors
E. Garmire, Optical Nonlinearities in Semiconductors Enhanced by Carrier Transport
D. S. Chemla, Ultrafast Transient Nonlinear Optical Processes in Semiconductors
M. Sheik-Bahae and E. W. Van Stryland, Optical Nonlinearities in the Transparency Region of Bulk Semiconductors
J. E. Millerd, M. Ziari, and A. Partovi, Photorefractivity in Semiconductors

Volume 59 Nonlinear Optics in Semiconductors II

J. B. Khurgin, Second Order Nonlinearities and Optical Rectification
K. L. Hall, E. R. Thoen, and E. P. Ippen, Nonlinearities in Active Media
E. Hanamura, Optical Responses of Quantum Wires/Dots and Microcavities
U. Keller, Semiconductor Nonlinearities for Solid-State Laser Modelocking and Q-Switching
A. Miller, Transient Grating Studies of Carrier Diffusion and Mobility in Semiconductors

Volume 60 Self-Assembled InGaAs/GaAs Quantum Dots

Mitsuru Sugawara, Theoretical Bases of the Optical Properties of Semiconductor Quantum Nano-Structures
Yoshiaki Nakata, Yoshihiro Sugiyama, and Mitsuru Sugawara, Molecular Beam Epitaxial Growth of Self-Assembled InAs/GaAs Quantum Dots
Kohki Mukai, Mitsuru Sugawara, Mitsuru Egawa, and Nobuyuki Ohtsuka, Metalorganic Vapor Phase Epitaxial Growth of Self-Assembled InGaAs/GaAs Quantum Dots Emitting at 1.3 μm
Kohki Mukai and Mitsuru Sugawara, Optical Characterization of Quantum Dots
Kohki Mukai and Mitsuru Sugawara, The Photon Bottleneck Effect in Quantum Dots
Hajime Shoji, Self-Assembled Quantum Dot Lasers
Hiroshi Ishikawa, Applications of Quantum Dot to Optical Devices
Mitsuru Sugawara, Kohki Mukai, Hiroshi Ishikawa, Koji Otsubo, and Yoshiaki Nakata, The Latest News

Volume 61 **Hydrogen in Semiconductors II**

Norbert H. Nickel, Introduction to Hydrogen in Semiconductors II
Noble M. Johnson and Chris G. Van de Walle, Isolated Monatomic Hydrogen in Silicon
Yurij V. Gorelkinskii, Electron Paramagnetic Resonance Studies of Hydrogen and Hydrogen-Related Defects in Crystalline Silicon
Norbert H. Nickel, Hydrogen in Polycrystalline Silicon
Wolfhard Beyer, Hydrogen Phenomena in Hydrogenated Amorphous Silicon
Chris G. Van de Walle, Hydrogen Interactions with Polycrystalline and Amorphous Silicon—Theory
Karen M. McNamara Rutledge, Hydrogen in Polycrystalline CVD Diamond
Roger L. Lichti, Dynamics of Muonium Diffusion, Site Changes and Charge-State Transitions
Matthew D. McCluskey and Eugene E. Haller, Hydrogen in III-V and II-VI Semiconductors
S. J. Pearton and J. W. Lee, The Properties of Hydrogen in GaN and Related Alloys
Jörg Neugebauer and Chris G. Van de Walle, Theory of Hydrogen in GaN

ISBN 0-12-752171-2